The Changing Status of Arable Habitats in Europe

Clive Hurford • Phil Wilson • Jonathan Storkey
Editors

The Changing Status of Arable Habitats in Europe

A Nature Conservation Review

"One of the penalties of an ecological education is that one lives alone in a world of wounds. Much of the damage inflicted on land is quite invisible to laymen. An ecologist must either harden his shell and make believe that the consequences of science are none of his business, or he must be the doctor who sees the marks of death in a community that believes itself well and does not want to be told otherwise."
— Aldo Leopold, A Sand County Almanac

Springer

Editors
Clive Hurford
Serapias (Ecological Monitoring Consultancy), Ty Llaeth, Y Fraich
Fishguard, UK

Phil Wilson
Pennyhayes Farm
Shute, Axminster, Devon, UK

Jonathan Storkey
Sustainable Agriculture Sciences
Rothamsted Research
Harpenden, Hertfordshire, UK

ISBN 978-3-030-59874-7 ISBN 978-3-030-59875-4 (eBook)
https://doi.org/10.1007/978-3-030-59875-4

This Springer imprint is published by the registered company Springer Nature Switzerland AG
The registered company address is: Gewerbestrasse 11, 6330 Cham, Switzerland

Preface

For a large proportion of the European population, our immediate encounter with the natural world is in the agricultural landscapes that surround our towns and cities. The plants and animals that are adapted to these habitats, poppies and cornflowers, skylarks and lapwings, take on a familiarity that gives us a sense of place and belonging to our local environments. For many years now, this rich cultural fabric that we associate with farmland has been systematically eroded with catastrophic declines in the diversity and abundance of plant and animal life documented across Europe. Seeking an escape from our urban environments in the surrounding countryside, many of us can now expect to encounter just a fraction of the wildlife our parents and grandparents enjoyed just five decades ago. The headlong rush for productivity in the post–World War II years has led to fewer crop species being grown, larger fields with less semi-natural habitat and the increasing use of inorganic fertilisers and chemical pesticides. Where once there was a fine scale patchwork of habitats supporting a rich diversity of plants and animals, the dominant characteristic of much of Europe's agricultural landscapes is now large homogenous areas of major crops in which only generalist fauna and flora can persist.

This volume celebrates the diversity of life adapted to arable habitats, the plants and animals that have been our travelling companions since we first learned to farm. For some of these species, they have been pushed to the fringes of our fields and arable landscapes and are hanging on in isolated populations and communities. Here, we show where they can still be found and the efforts being made to retain the rich cultural heritage they represent. But as well as the inherent value of biodiversity, we are increasingly realising that without the 'ecosystem services' it provides, crop production itself would not be possible. If we do not take the appropriate steps to conserve and enhance communities of natural predators of crop pests, pollinators and soil organisms, it is likely the current stagnation of crop yields will continue. It is, therefore, in all our interests to work towards diverse, functioning arable ecosystems that reconcile crop production with a healthy environment.

Harpenden, UK Jon Storkey

Acknowledgements

Initially our thanks are due to the authors who made the book possible, and especially those who took up our invitations to contribute to the book, particularly Alan Brown, Chris Stoate, Christer Olsson, Emily Swan, Felicity Crotty, Gabrielle Rosquist, Guillaume Fried, Gyula Pinke, Jordi Recasens, Julie Ewald, Marian Reed, Mark Brown, Miroslav Šálek, Nick Sotherton, Nicholas Aebischer Sabrina Gaba, Santi Mañosa and colleagues, Stefan Meyer and Terho Hyvonen.

We are also grateful to the publishing editors and technicians at Springer NL, particularly to Malini Arumugam, our Project Coordinator, for her help, patience and support during all phases of the publication process.

Thanks also to our friends and colleagues who have shared our enthusiasm and concern for the fauna and flora of arable habitats over the years, especially Dom Price of the Species Recovery Trust, Cath Shellswell and Kate Still of Plantlife, Peter Thompson of the Game and Wildlife Conservation Trust, Stanislav Březina of Krkonoše NPA, Julian Woodman and Dan Guest of Natural Resources Wales, the free spirit that is Jan Albert Šturma and the late Dick Potts.

Last but not least, we would like to thank our families, not only for their endless patience and just being there of course but, in the cases of Marian Reed and Chris Millican, for also playing an active role in helping to proofread and edit the manuscript.

Contents

Part I Introduction

1 **Historical and Ecological Background to the Arable Habitats of Europe** 3
Chris Stoate and Phil Wilson

Part II National and International Nature Conservation Reviews

2 **A Weed's Eye View of Arable Habitats** 17
Jonathan Storkey

3 **The Status of Arable Plant Habitats in Northwestern Europe** 31
Guillaume Fried

4 **The Status of Arable Plant Habitats in Scandinavian Countries** . . . 47
Terho Hyvönen, Lars Andersson, and Christian Andreasen

5 **The Status of Arable Plant Habitats in Central Europe** 55
Stefan Meyer

6 **The Status of Arable Plant Habitats in Eastern Europe** 75
Gyula Pinke

7 **The Arable Flora of Mediterranean Agricultural Systems in the Iberian Peninsula: Current Status, Threats and Perspectives** 89
Jordi Recasens, Alejandro Juárez-Escario, Bàrbara Baraibar, and Xavier O. Solé-Senan

8 **The Status of Arable Plant Habitats in Greece – The Cradle of Arable Farming in Europe** 111
Stefan Meyer and Erwin Bergmeier

Part III Research and Surveillance Projects

9 Soil Organisms Within Arable Habitats 123
Felicity V. Crotty

10 Agricultural Intensification, Sustainable Farming and the Fate of Arable Bryophytes in Switzerland 139
Irene Bisang and Ariel Bergamini

11 Invertebrate Trends in an Arable Environment: Long-Term Changes from the Sussex Study in Southern England 157
J. A. Ewald, N. J. Aebischer, and N. W. Sotherton

12 Ex Situ Conservation and Reintroduction of Vulnerable Arable Plants in Skåne, Sweden 173
Gabrielle Rosquist and Jakob Sandberg

13 Wild Pollinators in Arable Habitats: Trends, Threats and Opportunities 187
Jonathan Storkey, Mark J. F. Brown, Claire Carvell, Lynn V. Dicks, and Deepa Senapathi

14 Designing Multifunctional and Resilient Agricultural Landscapes: Lessons from Long-Term Monitoring of Biodiversity and Land Use 203
Sabrina Gaba and Vincent Bretagnolle

Part IV Farmland Bird Case Studies

15 Historic Overview and Conservation Perspectives of the Czech Grey Partridge (*Perdix perdix*) Population 227
Miroslav Šálek and Václav Zámečník

16 Perspectives on the Declining Ortolan Bunting (*Emberiza hortulana*) Population in Northern Sweden 245
Christer Olsson

17 The Impact of Land-Use Change on Arable Plant Habitats and Wintering Farmland Birds on a Farm in South-West Wales, UK 255
Clive Hurford

18 The Conservation of the Grey Partridge (*Perdix perdix*) on Farmland in Southern England 279
N. W. Sotherton, N. J. Aebischer, and J. A. Ewald

19 The Changing Status of Steppe-Land Birds in the Lleida Plain of Catalonia 291
Santi Mañosa, Gerard Bota, David Giralt, and Joan Estrada

Part V Applications, Considerations and Recommendations for the Future

20 Monitoring Arable Landscapes Using Free Satellite Imagery 319
Alan Brown and Iain Cameron

21 The Art of Agri-environment in the UK, an Advisor's Perspective 343
Emily Swan

22 The Future of Europe's Arable Wildlife 353
Phil Wilson, Clive Hurford, Marian Reed, Chris Stoate, and Jonathan Storkey

Editors and Contributors

Editors

Clive Hurford Serapias (Ecological Monitoring Consultancy), Ty Llaeth, Y Fraich, Fishguard, UK

Phil Wilson Pennyhayes Farm, Shute, Axminster, Devon, UK

Jonathan Storkey Sustainable Agriculture Sciences, Rothamsted Research, Harpenden, Hertfordshire, UK

Contributors

Nicholas J. Aebischer Game & Wildlife Conservation Trust, Hampshire, UK

Lars Andersson Swedish University of Agricultural Sciences, Uppsala, Sweden

Christian Andreasen University of Copenhagen, Taastrup, Denmark

Bàrbara Baraibar Research Group of Weed Science and Plant Ecology, Department of Horticulture, Botany and Landscaping, University of Lleida, Lleida, Catalonia, Spain

Ariel Bergamini WSL Swiss Federal Research Institute, Birmensdorf, Switzerland

Erwin Bergmeier Albrecht-von-Haller-Institute for Plant Sciences, Vegetation and Phytodiversity Research, University of Göttingen, Göttingen, Germany

Irene Bisang Department of Botany, Swedish Museum of Natural History, Stockholm, Sweden

Gerard Bota Centre de Ciència i Tecnologia Forestal de Catalunya (CTFC), Solsona, Catalonia, Spain

Vincent Bretagnolle Centre d'Etudes Biologiques de Chizé, UMR7372, CNRS & Université de La Rochelle, Villiers-en-Bois, France
LTSER Zone Atelier Plaine & Val de Sèvre, Villiers-en-Bois, France

Alan Brown Upper Thorn, Evenjobb, Presteigne, Powys, UK

Mark J. F. Brown School of Biological Sciences, Royal Holloway, University of London, Egham, Surrey, UK

Iain Cameron Environment Systems, Aberystwyth, Ceredigion, UK

Claire Carvell UK Centre for Ecology & Hydrology, Wallingford, UK

Felicity V. Crotty Royal Agricultural University, Cirencester, UK

Lynn V. Dicks Department of Zoology, University of Cambridge, Cambridge, UK

Joan Estrada Institució per a l'Estudi, Gestió i Recuperació dels Ecosistemes Lleidatans (EGRELL), Centre Cívic del Centre Històric, Lleida, Catalonia, Spain

Julie A. Ewald Game & Wildlife Conservation Trust, Hampshire, UK

Guillaume Fried Anses, Laboratoire de la Santé des Végétaux, Unité Entomologie et Plantes Invasives, Montferrier-sur-Lez, France

Sabrina Gaba Centre d'Etudes Biologiques de Chizé, UMR7372, CNRS & Université de La Rochelle, Villiers-en-Bois, France
USC 1339 Centre d'Etudes Biologiques de Chizé, INRAE, Villiers-en-Bois, France
LTSER Zone Atelier Plaine & Val de Sèvre, Villiers-en-Bois, France

David Giralt Centre de Ciència i Tecnologia Forestal de Catalunya (CTFC), Solsona, Catalonia, Spain

Clive Hurford Serapias (Ecological Monitoring Consultancy), Ty Llaeth, Y Fraich, Fishguard, UK

Terho Hyvönen Natural Resources Institute Finland (Luke), Jokioinen, Finland

Alejandro Juárez-Escario Research Group of Weed Science and Plant Ecology, Department of Horticulture, Botany and Landscaping, University of Lleida, Lleida, Catalonia, Spain

Santi Mañosa Departament de Biologia Evolutiva, Ecologia i Ciències Ambientals, Facultat de Biologia, Institut de Recerca de la Biodiversitat (IRBio), Universitat de Barcelona, Barcelona, Catalonia, Spain

Stefan Meyer Albrecht-von-Haller-Institute for Plant Sciences, Plant Ecology and Ecosystems Research, University of Göttingen, Göttingen, Germany

Christer Olsson Fågelperspektiv, Umeå, Sweden

Gyula Pinke Faculty of Agricultural and Food Sciences, Széchenyi István University, Mosonmagyaróvár, Hungary

Jordi Recasens Research Group of Weed Science and Plant Ecology, Department of Horticulture, Botany and Landscaping, University of Lleida, Lleida, Catalonia, Spain

Marian Reed Pennyhayes Farm, Shute, Axminster, Devon, UK

Gabrielle Rosquist County Administrative Board Skåne, Malmö, Sweden

Miroslav Šálek Department of Ecology, Faculty of Environmental Sciences, Czech University of Life Sciences Prague, Praha Suchdol, Czechia

Jakob Sandberg Fredriksdal Museums and Gardens, Helsingborg, Sweden

Deepa Senapathi School of Agriculture, Policy and Development, University of Reading, Reading, UK

Xavier O. Solé-Senan Research Group of Weed Science and Plant Ecology, Department of Horticulture, Botany and Landscaping, University of Lleida, Lleida, Catalonia, Spain

Nick W. Sotherton Game & Wildlife Conservation Trust, Hampshire, UK

Chris Stoate Game and Wildlife Conservation Trust Allerton Project, Loddington, Leicestershire, UK

Jonathan Storkey Sustainable Agriculture Sciences, Rothamsted Research, Harpenden, Hertfordshire, UK

Emily Swan Pennyhayes Farm, Shute, Axminster, Devon, UK

Phil Wilson Pennyhayes Farm, Shute, Axminster, Devon, UK

Václav Zámečník Czech Society for Ornithology, Prague, Czechia

Part I
Introduction

Chapter 1
Historical and Ecological Background to the Arable Habitats of Europe

Chris Stoate and Phil Wilson

Abstract Arable agriculture is the principal form of land use in Europe, occupying nearly 30% of the land area. While much of this is farmed very intensively, there are arable systems of high value to wildlife where cropping is diverse and interstitial habitats such as hedges, lines of trees, grassland patches and ponds are present.

Arable farming practices have changed greatly since the end of the nineteenth century, and particularly rapidly since the 1940s. These changes have included improvements to seed purity, the introduction of herbicides for control of non-crop plants, large increases in the application rates of fertilisers together with the development of high-yielding crop varieties and the ever-increasing size of farm machinery causing the removal of field boundary habitats. In recent years, irrigation has become more widespread, ploughing has been replaced by minimum cultivation and there has been an overall homogenisation of arable land. While these processes have been slower in the south and east of Europe, all countries are now affected.

These changes have had profound impacts on farmland wildlife through the direct and indirect effects and the loss of heterogeneity. These have included the catastrophic population declines of farmland birds such as corn bunting, little bustard and grey partridge, the loss of many formerly widespread plants characteristic of arable land, and declines of invertebrate numbers and diversity.

The rising demand for food to supply an increasing global population is the main driver of agricultural production. A slowing population growth and increasing agricultural efficiency may reduce pressures on arable land, while better understanding of the environmental costs of agricultural intensification coupled with payments to farmers will provide incentives to reduce inputs and mitigate their impacts.

C. Stoate (✉)
Game and Wildlife Conservation Trust Allerton Project, Loddington, Leicestershire, UK
e-mail: cstoate@gwct.org.uk

P. Wilson
Pennyhayes Farm, Shute, Axminster, Devon, UK

C. Hurford et al. (eds.), *The Changing Status of Arable Habitats in Europe*,
https://doi.org/10.1007/978-3-030-59875-4_1

Keywords High nature-value farming · Farmland ecology · Agricultural intensification · Landscape homogenisation · Agricultural economics · Conservation management

1 Arable Farming in Europe

Agricultural land covers almost half of the European land area. Within this is an enormous number of different farming systems, each associated with different species assemblages which have changed through time as the farming systems themselves have experienced changes. Although declines in biodiversity on farmland are well documented, the fact remains that many systems still support numerous species that have adapted to the farmed landscape as it has evolved over more than 10,000 years, and some species are primarily associated with this habitat (Stoate 2011).

Within the total European farmland area, the proportion classed as arable varies between countries (Table 1.1), although the distinction between arable and other forms of land use is not always clear as arable land is integrated into livestock systems and the ratio between cropping and grassland changes through time. This is particularly the case where short-term grass leys form part of the arable rotation.

The proportion of arable land has fluctuated through history. In medieval Europe, much of the land was cultivated in rotation with fallow periods and common grazing land delivering yields that were very low by modern standards, both for direct human consumption and as winter feed for livestock. Taking the UK as an example, the combination of plague, which may have killed 50% of Europe's population in the mid fourteenth century alone, and subsequent enclosure of land by private land owners, caused a move from arable rotations to grazing ruminant systems, especially sheep for wool production, but later also cattle for beef and dairy production. The most dramatic changes in the UK were the result of parliamentary enclosure acts between 1760 and 1870 under which nearly 3 million ha of land previously farmed in common were brought under private management (Fairlie 2009). These in combination with high prices for arable produce which resulted from the Napoleonic Wars and subsequent agricultural subsidies under the 1815–1846 Corn Laws increased the arable area of the country to 3.8 million ha (Olson Jr and Harris 1959). The importation of cheap wheat from the USA from 1873 onwards caused a massive decline to around two million ha by 1900, which persisted with only a temporary increase during the First World War (Borchert 1948).

Although 'High Nature Value' farming systems are often associated with extensive grazing, some arable cropping systems are also categorised as such. Traditional low-input arable farming persists in many countries. Much of the arable area in Spain, Portugal and Hungary is managed extensively and supports high biodiversity. In Hungary it includes numerous small diverse Tanya farms on the Great Hungarian Plain. In Portugal and Spain, the agroforestry systems of montado and dehesa, comprising arable crops and livestock grazing under cork and holm oaks remains

Table 1.1 Areas of arable land in Europe

	Utilised agricultural area (X1000ha)	Arable land (X1000ha)	Percentage arable land
Estonia	892.7	597.1	76
Latvia	1838.2	1182.7	67
Lithuania	2711.9	1904.5	64
Poland	15815.2	12036.2	76
Finland	2288.7	2256.9	99
Belgium	1369.9	840.2	61
Czechia	3556.6	2595.9	73
Denmark	2682.2	2446.7	91
Germany	16930.2	11905.2	70
Ireland	4231.6	1100.8	26
France	31129.5	20019.9	65
Luxembourg	130.3	61.1	47
Netherlands	1902.3	1041.8	55
Austria	3204.5	1371.7	43
Sweden	3097.5	2627.6	85
United Kingdom	17573.9	6070.1	35
Norway	1030.2	852.5	83
Switzerland	1521.0	405.3	27
Greece	3963.9	2312.5	58
Spain	24663.8	12745.6	52
Italy	13914.0	7347.6	53
Cyprus	140.8	103.1	73
Malta	10.3	8.0	78
Portugal	3703.7	1161.6	31
FYR of Macedonia	1120.5	441.3	39
Albania	975.9	579.0	59
Montenegro	518.0	44.8	9
Bulgaria	4087.3	2467.9	60
Hungary	5797.3	4492.3	77
Romania	13758.1	8755.0	64
Slovakia	1931.6	1348.5	70
Slovenia	487.5	175.5	36
Croatia	1254.1	862.1	69
Bosnia-Herzegovina	2194.4	534.1	24
Serbia	5112.0	3329.0	65
Kosovo	369.0	136.1	37

From Hucorne (2012)

widespread although decreasing in area. Localised centres of "High Nature Value" arable farming are present in many other places throughout Europe. Even in the mountains of Transylvania, small arable fields close to villages produce winter food for livestock, alongside hay meadows further away, and summer-grazed pasture on the hill sides. In Scandinavia and the Baltic countries, a short growing season restricts arable land to spring sown crops, but these represent an important part of the farming systems.

Across most of lowland Europe, crops such as wheat, oilseed rape, barley, rye and field beans are mainly autumn-sown, but spring sown crops such as sugar beet, maize, peas, linseed, sunflower and some cereals are also grown in large-scale intensive arable systems. In small scale systems the diversity of crops grown can be much greater with minor crops such as buckwheat and borage also present in some areas, and small plots of horticulture and vegetables.

Within arable systems there are frequently considerable areas taken out of cultivation for one or more years, traditionally to help maintain fertility and to facilitate weed-control. In cool temperate regions where there is sufficient rainfall short-term grass leys are established, while in regions where rainfall is a limiting factor, land is frequently fallowed with vegetation allowed to regenerate naturally. Orchards and vineyards are also a feature of some arable landscapes. This diversity of cropping is extremely important in the maintenance of the biodiversity of cropped land.

Interstitial components of the agricultural landscape also influence its ecology. In the UK and parts of France, field boundary hedges are an obvious feature of landscapes, originally created to contain livestock. In today's arable systems these are recognised and maintained not only for their wildlife and aesthetic value but for many other agronomically important functions (Dover 2019). In the Netherlands and Baltic countries, land drainage channels form important components of the landscape which influence the wildlife present in otherwise arable landscapes. In many parts of Europe, field boundaries are marked by tree lines. In the remaining dry plains of Catalunya for example, old almond and olive trees in field margins are valuable nest sites for endangered birds (Mañosa et al., this volume – Chap. 19). In other countries such as Poland and Hungary however, both large and small fields have no clear boundary features creating an open landscape in which only the diversity of crops and size of fields contribute to the landscape scale heterogeneity.

2 Twentieth Century Changes in Arable Farming – Intensification and Homogenisation

Changes in European agricultural systems around the turn of the century have been reviewed from an environmental perspective by Stoate et al. (2001, 2009).

During and after the Second World War there was a substantial government-led change in the UK from livestock to arable systems to improve food security so that arable and livestock areas became more polarised into the east and west respectively. The Agriculture Act (1947) guaranteed prices and stability for British farmers. Similar arable expansion had occurred in other countries under fascist regimes in the 1930s. In Portugal, the 'Wheat Campaign' (Campanha do Trigo) and in Italy the

"Battle of Wheat" (Battaglia del Grano) (Saraiva 2010) resulted in the expansion of the arable area, even into large areas of hilly land with thin soils where production could not be maintained for more than a few years and the land was subsequently abandoned. Mechanisation increased rapidly in the post-war period in northwest Europe, and at a slower pace in the south and east, but the size of agricultural machinery used in most areas has continued to increase into the current century, initially with the consequent removal of non-cropped features such as hedges and ponds.

Most twenty-first century arable systems have high external inputs, based on crop cultivars developed to respond to these with high yields. Such inputs include herbicides, insecticides and fungicides, of which some are applied prophylactically, and others when weed, pest or disease thresholds are reached. The first modern selective herbicides, 2-4-D, 2-4-5-T and MCPA, were introduced in the mid-1940s (Makepeace 1980), and by the 1970s most arable land in lowland north-western Europe had received a herbicide treatment (Fryer and Chancellor 1970). The first modern insecticide, the organochlorine DDT, was introduced for agricultural use in 1945, followed by other organochlorines, carbamates, organophosphates, synthetic pyrethroids and other compounds (Oberemok et al. 2015).

Similarly, fertilisers (mainly nitrogen and phosphate) are applied according to nutrient availability in the soil and the estimated requirements of specific crops. Although nitrogen- and phosphorus-containing artificial fertilisers had been available since the late-nineteenth century, until the 1940s the bulk of nutrients were supplied to arable crops as animal manure. The availability and application of large quantities of cheap and concentrated manufactured fertilisers from the 1940s onwards (Fig. 1.1), was accompanied by the development of high-yielding and highly competitive crop varieties (Fischbeck 1990).

Irrigation has become more widely adopted for high value crops in the north, and for a wider range of crops in the south. In southern Europe, the area of irrigated crops has increased considerably since the 1960s. In Italy for example 25% of the arable land area was irrigated in 2000 and 55% of arable produce was from irrigated land (Bazzani et al. 2004). Pivot irrigation of crops such as maize is associated with increased fertiliser and pesticide applications and the environmental impacts of irrigation are at least partially those of these inputs, but also include the loss of fallows from the rotation. Mañosa et al. (this volume – Chap. 19) describe the loss of some of the remaining areas of dry arable land to irrigation in Catalunya.

Cultivation systems have also changed in recent decades, with adoption of reduced tillage using discs and tines, and direct drilling using specialised drills in many parts of Europe instead of the traditional plough and associated cultivations to produce a seedbed. The development of these systems throughout the world has been documented by Derpsch (1998). In 2015 it was estimated that approximately 55% of arable crops in the UK were established by minimum or no-tillage (Alskaf et al. 2019). These changes have had substantial impacts on the above and below ground ecology of the farmed environment, and the effects of these on soil organisms are described by Crotty (this volume – Chap. 10).

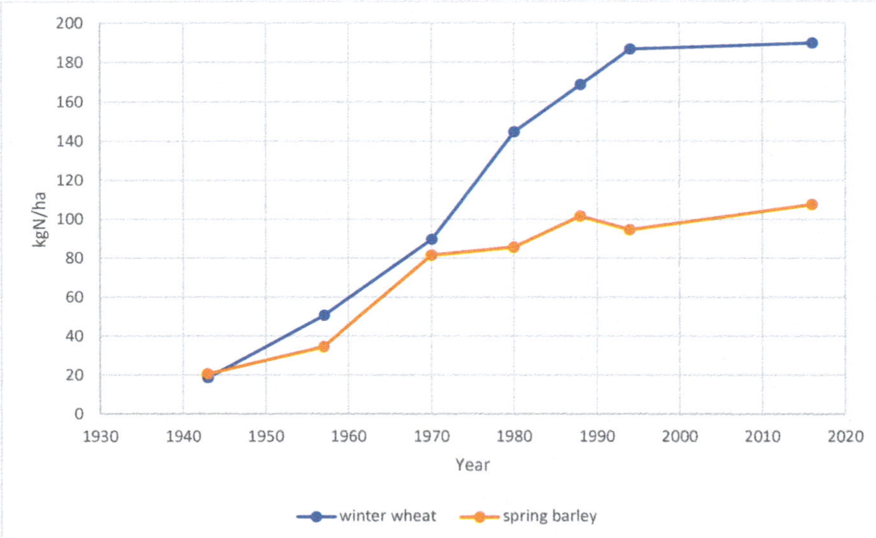

Fig. 1.1 Application rates of synthetic nitrogen fertilisers to cereal crops (Church 1981; Anon 2017)

Homogenisation of farmland landscapes has been characteristic of Eastern Europe, especially in the period of collectivisation in the 1950s and '60s. Socialist agriculture supported through state ownership and planning in the 1970 – 1990s was as intensive as in western countries in terms of use of inputs but received less financial support than in the EU. In many eastern European countries however the economic drivers, poor communications and mountainous terrain permitted the localised survival of a rich arable biodiversity (Pinke, this volume – Chap. 2). Most collectivised land was privatised in the 1990s, input costs increased, and livestock numbers declined.

Enlargement of the EU in 2004 and 2007 incorporated 12 eastern and southern European countries including much high nature value extensively-managed farmland and EU agricultural policy was therefore extended. This has resulted in convergence of agricultural intensification across east and western Europe, and this has been compounded by the general drift of the rural population to the cities, resulting in abandonment of farmland and intensification of management on the remainder. Agricultural commodity prices showed an unprecedented increase in 2007/8, partly in response to demand for biofuels, and partly because of global food insecurity associated with climate change and increasing human population and per capita consumption. Global commodity prices subsequently declined, reducing the profitability of farm businesses, but continue to be the main driving force for the management of crops in European countries.

EU financial support for farmers is another major influence on the way arable land is managed. Successive reforms of the CAP from the 1980s onwards, but particularly in 2005 and 2008 included decoupling of payments from production, introduction of cross-compliance, modulation, and extension of the Rural Development Programme. Cross-compliance comprised a range of basic environmental criteria for land management, while agri-environment schemes under the Rural Development Regulation provided funding for more targeted management of the farmed environment to meet specific environmental objectives, including those associated with biodiversity. The first agri-environment schemes were introduced in 1985 (in the UK, Environmentally Sensitive Areas, but subsequently a series of stewardship schemes). Adoption of these was at the discretion of member states and consequently differed across the continent, but numerous studies have demonstrated their importance in the conservation of arable biodiversity.

3 Impacts of Changes in Arable Management on Farmland Ecology

These changes in structure and management of the arable environment have influenced farmland ecology in several ways. These include loss of crop and non-crop habitat heterogeneity, loss of intrinsic habitat value at a range of scales, and intensification in the form of increased external inputs such as nutrients which change the composition of vegetation through competition, herbicides, fungicides and insecticides all of which have direct toxicity towards target organisms and some also to non-target organisms. Adoption of highly nitrogen-responsive crop varieties, abandonment of arable land, including the conversion to forestry and intensively managed grassland, irrigation, land drainage and changes to tillage systems have all had profound effects on the arable ecosystem.

Brown hare (Tapper and Barnes 1986), lapwing (Henderson et al. 2012), yellowhammer (Stoate et al. 1998; McHugh et al. 2016) and little bustard (Mañosa, this volume – Chap. 19) are all examples of vertebrate species that use different features of the arable landscape for different purposes or at different times of the year and consequently rely on landscape-scale heterogeneity. Santana et al. (2017) demonstrate that both habitat diversity and complexity can influence landscape scale avian biodiversity across Portuguese farming systems. Headage payments for livestock provide an example of how economic support can influence the composition of the landscape, in this case by reducing the arable area in favour of grazing livestock, and increasing livestock density, to the detriment of regenerating trees in traditional agroforestry systems such as Montado and Dehesa. Payments under agri-environment schemes for the enhancement of habitats for wildlife can be important as drivers in agricultural systems where profit margins are small.

Intensification of management has contributed to biodiversity loss in many ways. Increased fertiliser use, and the vigorous crop cultivars developed for high input systems, radically change the plant assemblage associated with crops, while misplaced fertiliser in field edges has also resulted in eutrophication of field margin habitats and a decline in plant species richness. These competitive effects, together with direct toxicity from herbicides, have combined to transform and impoverish the species-rich and diverse plant communities formerly associated with arable cultivation thoughout Europe.

The plant species which have suffered most are those segetal species with life-cycles most closely synchronised with traditional cereal growing. They are often poorly competitive with a fully fertilised modern cereal variety and very susceptible to a wide range of herbicides. Some of these such as *Agrostemma githago* (Firbank 1988), *Ranunculus arvensis* and *Scandix pecten-veneris* (Wilson 1990) produce relatively few seeds with limited dormancy in the soil and are consequently particularly vulnerable to change. Many of these species have experienced massive declines throughout Europe. In Britain for instance, of the 100 species of plant exhibiting the greatest declines in the latter half of the twentieth century, 47 were typical species of arable land (Preston et al. 2002) (Storkey, this volume – Chap. 2). These declines were noted as early as the 1900s as improved methods of seed-cleaning removed contaminants such as *Agrostemma githago, Centaurea cyanus, Lolium temulentum* and *Adonis spp* (Meyer, this volume – Chap. 6) and the suite of species associated with flax cultivation (Kornas 1961). Declines continued through the end of the twentieth century and into the 21st with the near-disappearance by the 1970s from many northern European countries and the collectivised parts of eastern Europe of segetal communities associated with nutrient-poor soils such as the *Teesdalio-Arnoseridetum* and *Caucalido-Adonidetum*. Intensification of arable management in much of southern Europe occurred later (Meyer and Bergmeier, this volume – Chap. 8; Recasens, this volume – Chap. 5) and parts of Eastern Europe (Pinke, this volume – Chap. 3), but even in these areas, there have been great losses of species diversity and of individual species in recent years.

Direct effects of insecticides on vertebrates have been greatly reduced since the phasing out of organochlorines in the 1980s (Newton 2015). However, long-running research in southern England demonstrates that direct effects on non-target invertebrates persist, and direct effects of fungicides on invertebrates have also been documented (Ewald et al., this volume – Chap. 11). Indirect effects of herbicide use on farmland invertebrates have been demonstrated in the same study. Other work has implicated the most widely used modern insecticides, the neo-nicotinoids, in the continuing global declines in farmland invertebrates and vertebrates (Goulson 2013).

Sotherton et al. (this volume – Chap. 19) describe the strong link between grey partridge chick survival and subsequent breeding abundance and the abundance of arable invertebrates and their plant hosts. Throughout Europe, nestling survival and breeding abundance of other characteristic birds of arable farming systems such as corn bunting (Brickle et al. 2000) and little bustard (Mañosa et al., this volume – Chap. 19) have been shown to be higher in extensively managed areas supporting weeds and invertebrates.

4 Current Threats and Opportunities

The key influences on arable biodiversity are therefore heterogeneity at a range of scales and the intensity of management in terms of use of external inputs. Loss of heterogeneity and intensification are major threats, while abandonment to other land uses is an increasing threat in marginal areas.

Demand for food is increasing globally as the human population increases, alongside increasing consumption, especially of meat. Intensive production of the latter requires a large arable area for animal feed, compounding pressure on land from other markets, including biofuels such as maize for anaerobic digestion. High input costs and low global commodity prices contribute to the loss of small farms, amalgamation into larger units, increased use of contractors, and adoption of methods that take little account of site-specific circumstances or environmental considerations including biodiversity. Increases in farm, field and equipment size have combined to simplify arable landscapes, while the increasing prevalence of contractors rather than owner occupiers on enlarged holdings can contribute to reduced attention to detail and increased prophylactic use of pesticides.

On the other hand, economic drivers for nutrient use efficiency can also result in reduced nutrient emissions to the environment and therefore a decline in negative impacts on biodiversity as well as other externalities such as aquatic eutrophication and climate change. A combination of regulation and high input costs can also encourage adoption of holistic approaches to crop management such as integrated pest control and organic systems. Encouraging invertebrate predators of crop pests can potentially reduce insecticide use with benefits to farmland biodiversity resulting both from the habitats created and from less frequent use of insecticides. Crop cultivars are also being developed that are resistant to some insect pests (e.g. orange wheat blossom midge in wheat), as well as the well-established breeding of resistance to or tolerance of fungal diseases. Although challenging on some soil types, the widespread move towards reduced tillage and direct drilling can also have benefits to the ecology of aquatic and terrestrial habitats, as well as to the soil itself. In addition, reduced tillage can reduce costs and soil erosion, while also reducing fossil fuel use and carbon dioxide flux, and allowing the accumulation of carbon stores in the soil. At the same time however it can lead to impeded drainage, accumulation of perennial weeds and grasses, the loss of uncommon segetal species and an increase in the use of broad-spectrum herbicides.

There remains a considerable need to support farmers to achieve both specific and systemic environmental objectives through agri-environment schemes such as those demonstrated in this volume (Meyer – Chap. 6; Mañosa et al. – Chap. 19; Swan – Chap. 21) to benefit wildlife associated with arable habitats. There is increasing interest amongst policy makers in the potential of 'payments by results' in which the level of payments received by farmers is determined by the quantified delivery of wildlife, rather than, or in addition to, management-based payments made for creating and managing habitats. Such an approach encourages a genuine understanding by farmers of the life cycle requirements of species and the management required to meet these, and can stimulate innovation to achieve targets. However, this approach may be more readily applied to segetal plant communities than to mobile and less easily measurable animals associated with arable landscapes.

Increasingly, food processors are providing incentives to their farmer suppliers to adopt management practices that benefit wildlife or have other environmental benefits, and some products from farming systems that support wildlife continue to attain a premium that provides farmers with an economic incentive. There is also a move towards collaboration between farmers to manage their land to benefit certain iconic species such as wading birds in the Netherlands and a range of species that are valued by Farmer Clusters in the UK. Such ‘ownership’ of species at the farm level is fundamental to their conservation. In the end, it is farmers working individually and collaboratively within a robust, scientifically-guided regulatory and funding system who will determine the future of arable ecosystems.

References

Alskaf K, Sparkes DL, Mooney SJ, Sjögersten S, Wilson P (2019) The uptake of different tillage practices in England. Soil Use Manag. https://doi.org/10.1111/sum.12542

Anon (2017) The British survey of fertiliser practice: fertiliser use on farm crops for crop year 2017. Defra, York

Bazzani GM, Di Pasquale S, Gallerani V, Viaggi D (2004) Irrigated agriculture in Italy and water regulation under the European Union water framework directive. Water Resour Res 40: W07S04. https://doi.org/10.1029/2003WR002201

Borchert JR (1948) The agriculture of England and Wales, 1939–1946. Agric Hist 22(1):56–62

Brickle NW, Harper DGC, Aebischer NJ, Cockayne SH (2000) Effects of agricultural intensification on the breeding success of corn buntings *Miliaria calandra*. J Appl Ecol 37:742–755

Church BM (1981) Use of fertiliser in England and Wales 1980. Report of the Rothamsted Experimental Station for 1980 2: 115–122

Derpsch R (1998) Historical review of No-tillage cultivation of crops. The 1st JIRCAS seminar on soybean research. No-tillage cultivation and future research needs. JIRCAS Work Rep 13:1–18

Dover J (2019) Introduction to hedgerows and field margins. In: Dover J (ed) The ecology of hedgerows and field margins. Routledge, Abingdon

Fairlie S (2009) A short history of enclosure in Britain. Land 7:16–31

Firbank LG (1988) Biological Flora of the British Isles, *Agrostemma githago L.* J Ecol 76:1232–1246

Fischbeck G (1990) The evolution of cereal crops. In: Firbank LG, Carter N, Darbyshire JF, Potts GR (eds) The ecology of temperate cereal fields

Fryer JD, Chancellor RJ (1970) Herbicides and our changing arable weeds. In: Perring F (ed) The Flora of a changing Britain, BSBI Conference Reports 11, pp 105–119

Goulson D (2013) An overview of the environmental risks posed by neo-nicotinoid insecticides. J Appl Ecol 50:977–987

Henderson IG, Holland JM, Storkey J, Lutman PJW, Orson J, Simper JN (2012) Effects of the proportion and spatial arrangement of un-cropped land on breeding bird abundance in arable rotations. J Appl Ecol 49:883–891

Hucorne P (2012) The actual distribution of crops in Europe. https://www.eppo.int/media/uploaded_images/ACTIVITIES/plant_protect_products/zonal_assessment/Hucorne_2012_crop_distribution.pdf

Kornas J (1961) The extinction of the association Sperguleto-Lolietum remoti in flax cultures in the Gorce (Polish Western Carpathian Mountains). Bulletin De L'Academie Polonaise des Sciences, Ser Sci Biol 9:37–40
Makepeace RJ (1980) Herbicides: past, present and future. In: Hurd RG, Biscoe PV, Dennis C (eds) Opportunities for increasing crop yields. Pitman, London
McHugh NM, Goodwin CED, Hughes S, Leather SR, Holland JM (2016) Agri-environment scheme habitat preferences of Yellowhammer *Emberiza citrinella* on English farmland. Acta Ornithologica 51:199–209
Newton I (2015) Pesticides and birds of prey – the breakthrough. In: Nature's conscience, the life and legacy of Derek Ratcliffe. Langford Press, Narborough
Oberemok VV, Laikova KV, Gninenko YI, Zaitsev AS, Nyadar PM, Adeyemi TA (2015) A short history of insecticides. J Plant Prot Res 55(3):221–226
Olson M Jr, Harris CC (1959) Free trade in "Corn": a statistical study of the prices and production of wheat in Great Britain from 1873 to 1914. Q J Econ 73(1):145–168
Preston CD, Pearman DA, Dines TD (2002) New atlas of the British and Irish flora. Oxford University Press, Oxford
Santana J, Reino L, Stoate C, Moreira F, Ribeiro PF, Santos JL, Rotenberry JT, Beja P (2017) Combined effects of landscape composition and heterogeneity on farmland avian diversity. Ecol Evol 7:1212–1223
Saraiva T (2010) Fascist labscapes: geneticists, wheat and the landscapes of Fascism in Italy and Portugal. Hist Stud Nat Sci 40(4):457–498
Stoate C (2011) Biogeography of agricultural environments. In: The Sage handbook of biogeography. Sage, London, pp 338–356
Stoate C, Moreby SJ, Szczur J (1998) Breeding ecology of farmland Yellowhammers Emberiza citrinella. Bird Study 45:109–121
Stoate C, Boatman ND, Borralho R, Rio Carvalho C, de Snoo G, Eden P (2001) Ecological impacts of arable intensification in Europe. J Environ Manag 63:337–365
Stoate C, Baldi A, Beja P, Boatman ND, Herzon I, van Doorn A, de Snoo GR, Rakosy L, Ramwell C (2009) Ecological impacts of early 21st century agriculture in Europe – a review. J Environ Manag 91:22–46
Tapper SC, Barnes RFW (1986) Influence of farming practice on the ecology of the Brown Hare (*Lepus europaeus*). J Appl Ecol 23:39–52
The Agriculture Act (1947). http://www.legislation.gov.uk/ukpga/Geo6/10-11/48/contents/enacted
Wilson PJ (1990) The ecology and conservation of rare arable weed species and communities. PhD Thesis, Southampton A short history of insecticides

Part II
National and International Nature Conservation Reviews

Chapter 2
A Weed's Eye View of Arable Habitats

Jonathan Storkey

Abstract The diversity of plant communities adapted to arable habitats has declined across Europe with many species becoming nationally rare or extinct since the 1960s. Common trends in the response of these communities can be understood in the context of plant ecological principles. The assembly of plant communities is driven by the interaction of the frequency of disturbance with the availability of resources; determining the prevalence of contrasting plant strategies. Arable plants are characterised by traits that make them well adapted to the highly disturbed environment of cultivated fields (being dominated by annuals) and can be negatively impacted by changes in the disturbance regime either through shifts in cropping patterns or land abandonment. The intensification of agricultural production has also seen a dramatic increase in the amount of fertiliser applied to crops to realise the greater yield potential of modern cultivars. This has selected against more 'stress tolerant' arable plants that are less able to compete with the crop for above ground resources and for faster growing, more competitive species that are adapted to soils with high fertility. When combined with the increased use of herbicides, these changes in crop management have effectively narrowed the available habitat niche for arable plants with a tendency for more specialized species to decline and generalists to increase. In seeking to conserve the ecological value and cultural heritage that arable plant communities represent, space needs to be found in intensively farmed landscapes to support the range of species that are currently excluded from our fields.

Keywords Rare arable weeds · Plant ecological strategies · Fertilisers · Cultivation · Herbicides · Broadbalk long-term experiment

---

J. Storkey (✉)
Sustainable Agriculture Sciences, Rothamsted Research, Harpenden, Hertfordshire, UK
e-mail: jonathan.storkey@rothamsted.ac.uk

C. Hurford et al. (eds.), *The Changing Status of Arable Habitats in Europe*,
https://doi.org/10.1007/978-3-030-59875-4_2

1 Introduction

Arable fields are a novel habitat for floral communities that represent a combination of environmental conditions not previously encountered by plants and animals in evolutionary history. However, the inherent diversity of life and its capacity to adapt has meant that these new environments have become a new opportunity for colonisation for a group of plants we now call 'weeds' such that they have become our constant travelling companions since the advent of agriculture in Europe (Jones et al. 2005). In some cases, this has even meant that weed species have co-evolved with agriculture (Vigueira et al. 2013) to the extent that they include some examples of anecophytes (plants that are only found in human-created habitats). Previous generations had no choice but to learn to accommodate and live with these hitchhikers in their crops (Shakespeare's King Lear was "*Crowned with rank fumiter and furrow-weeds, with burdocks, hemlock, nettles, cuckoo-flowers, Darnel, and all the idle weeds that grow in our sustaining corn*"). However, in more recent history, in the face of the growing world population and need to increase productivity, they have been perceived almost exclusively as a problem that needs to be removed (Oerke 2006). As a result, the diversity and abundance of weeds in arable fields has declined dramatically across Europe, especially in the post World War II era after the advent of herbicides, and specialist weed floras of national cultural and ecological importance are now under threat across the continent (Storkey et al. 2012). These declines have refocussed attention on the inherent cultural value of weeds and their potential contribution to maintaining ecosystem function by providing resources for higher trophic groups (Marshall et al. 2003).

This collection of chapters has been written by agricultural botanists and weed scientists representing different regions in Europe all of whom share a concern for the conservation of arable floras. Because of the negative connotations associated with the term 'weed' (implying it is unwanted), these species tend to be referred to henceforth as 'arable plants'. However, whatever we decide to call them, these plants are defined by the fact that their fortunes are closely wedded to human activity. The historical trends in arable plant communities can, therefore, only be understood in the context of changes in the management of arable fields and associated impacts on their habitat. In this introductory chapter, this concept is explored in the context of general ecological principles that underlie the region-specific patterns discussed in subsequent chapters. Additional examples and case studies are introduced from a UK perspective to illustrate these principles.

2 Monitoring the Status of Arable Plants

If our starting point is that conserving arable plant communities is important both for cultural and ecological reasons and that their population trends are closely associated with human activity, to understand trends we need to both quantify changes in species' abundance and diversity and collect information on shifts in agricultural practice. The ideal approach, therefore, is a repeated, systematic vegetation survey focussed on arable fields at a national level coupled with the collection of farm management data. Where there are examples of these studies (in France (Chap. 3), Finland and Denmark (Chap. 4), and Germany (Chap. 5), the data are particularly valuable because population trends can be directly related to changes in management practice. Ideally, these surveys return to the same sample points at intervals over decadal timescales; a challenge made more tractable by the introduction of Global Positioning Systems (GPS). All of the authors in this collection of chapters are strong proponents of such surveys.

Data from a wide scale national botanical survey of this type specifically targeted at arable plants does not exist in the UK; a deficiency also highlighted for other areas of Europe discussed in this section (Chaps. 6 and 7). Rather, to identify trends in arable plant distribution and abundance at a national scale in the UK, we rely on databases from broader vegetation surveys done across all habitats. While lacking the resolution of surveys specifically focussed on weeds and the associated field level management information, these data are still indicative of broad trends in arable floras. The main source of these data in the UK are the records of the Botanical Society of Britain and Ireland (BSBI) reported in the New Atlas of British and Irish Flora (Preston et al. 2002). The Atlas includes a measure of the population change of each species between the 1930s and 1990s. The detail of the calculation of this metric is explained in Telfer et al. (2002); in summary, the proportion of 10 by 10 km squares in which species were recorded in a national botanical survey between 1987 and 1999 were plotted against the proportion of squares in which species were recorded over an earlier period, 1930–1969. The population change index is the residual for an individual species measured against a linear regression model fitted to data for all species from the two time periods. Thus, it is a relative measure of how the distribution of a species has changed between the two surveys in comparison to the rest of the U.K. flora with a negative value indicating relative declines. One of the main conclusions of this analysis was that the group of plants that had suffered the most dramatic declines in their distribution over this period were associated with arable habitats. Examples of species with the most negative population change indices are *Bupleurum rotundifolium*, *Galium tricornutum* and *Scandix pecten-veneris*. The fact that arable plants are a particularly threatened component of the UK flora is also reflected in the Vascular Plant Red Data List for Great Britain that lists five arable plants as recently extinct, seven as Critically Endangered and 14 as Endangered (Storkey et al. 2012), Table 2.1.

Two, more localised, studies in England have been published that report the change in the floras of arable landscapes between the 1950s and 2000's. At this more local scale, the trends in arable plant populations are less dramatic and more subtle. While Potts et al. (2009) documented a marked reduction in arable weed diversity in Sussex after the introduction of herbicides in the 1960s (with an initial 52% fall in species frequency), since the 1970s the abundance and occurrence of annual weeds has remained relatively stable or, for common species increased. Sutcliffe and Kay (2000) also observed many of the rarer arable plant species recorded in a 1960's

Table 2.1 Species on the Vascular Red Data List for Great Britain identified as 'arable plants' (Kate Still pers. com.) as part of a European wide survey (Storkey et al. 2012)

Recently extinct	Critically endangered	Endangered
Arnoseris minima	*Filago gallica*	*Adonis annua*
Bromus interruptus	*Chenopodium urbicum*	*Ajuga chamaepitys*
Bupleurum rotundifolium	*Galeopsis angustifolia*	*Anthemis arvensis*
Caucalis platycarpos	*Galium tricornutum*	*Filago lutescens*
Galeopsis segetum	*Lolium temulentum*	*Filago pyramidata*
	Ranunculus arvensis	*Lithospermum arvense*
	Scandix pecten-veneris	*Lythrum hyssopifolium*
		Scleranthus annuus
		Silene gallica
		Torilis arvensis
		Valerianella dentata
		Valerianella rimosa
		Veronica triphyllos
		Veronica verna

arable field survey in Oxfordshire and Berkshire were still present in a follow-up survey done in 1997. Both studies concluded that although the intensification of agriculture had an impact on arable plant communities, the diversity in the soil seedbank was still sufficient to maintain diversity at this local scale (although it is likely species have been lost from local seed banks since these studies were published). Other countries have also observed positive trends in the frequency and abundance of common species in the last two decades in response to policy measures to support farmland biodiversity (Chap. 4).

3 What Are the Defining Characteristics of an Arable Habitat?

In 1974, the English plant ecologist, Philip Grime, introduced the idea that plants could be thought of having contrasting Plant Ecological Strategies that have evolved in environments with different levels of competition, abiotic stress and disturbance (Grime 1974). Any plant species could be assigned a strategy based on the combination of their functional traits that determined their fitness in different habitats. The functional composition of communities, in terms of the dominance of different strategies was found to shift along gradients of soil fertility and disturbance – the main characteristics of a habitat that determine pant community assembly. 'Competitors' (C) are adapted to undisturbed environments with high soil fertility, 'Stress tolerators' (S) to undisturbed, infertile habitats and 'Ruderals' (R) to highly disturbed, fertile habitats. These three strategies can be visualised as the corners of a 'C-S-R triangle' (Fig. 2.1) and any plant species positioned within that triangle on the

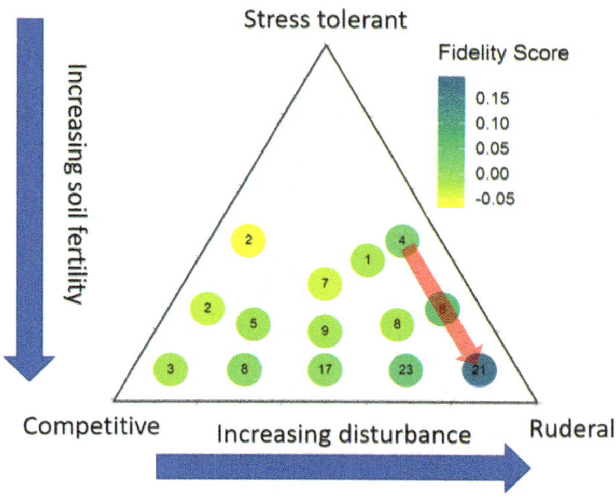

Fig. 2.1 Median fidelity score for common arable plant species showing different ecological strategies according to Grime et al. (2014). Fidelity is a measure of how frequently a species was recorded in arable fields compared to all other habitats; scores range from −1 to +1 with positive (negative) values indicating that the species and the habitat of interest occur more (less) frequently than would be expected by chance. The relative position of the circle indicates the ecological strategy, the colour of the circle represents the median fidelity score for the ecological strategy and the number within the circle is the number of species represented by the ecological strategy. The red arrow indicates the direction of selection pressure as a consequence of increased fertiliser use. (Adapted from Metcalfe et al. 2019)

basis of the traits that determine how well it is adapted to different habitats (Grime et al. 1997). This is a useful framework for defining the important characteristics of arable habitats and understanding the assembly of arable plant communities in response to changes in management. So, how should an arable habitat be defined in this context?

Arable cropping systems are dominated by annual crops meaning soils are generally disturbed at least once a year when crops are established and areas of very low fertility or high abiotic stress (such as arid regions or low pH soils) tend to be avoided as they do not support crop production. Arable habitats are, therefore characterised by frequent disturbance and relatively fertile soils favouring a ruderal plant ecological strategy. Ruderal plants are characterised by traits that enable them to acquire resources quickly (high relative growth rate) and complete their life cycle in between disturbance events (they are generally annuals or 'therophytes' that reproduce exclusively with seed). An analysis of a pool of 118 UK weed species with an ecological strategy listed by Grime confirmed that the majority were concentrated in the 'R' corner of the triangle and that species with this strategy had a higher fidelity to arable fields (more likely to be found in cultivated land than in other habitats), Fig. 2.1. A second study comparing the plant traits of a species pool recorded in French arable fields with a list of species from other open habitats also found that arable plants were more likely to be therophytes (over wintering as seed) and generally flowered earlier with a longer flowering duration allowing them to take advantage of ephemeral opportunities for growth and reproduction (Bourgeois et al. 2019). Within the ruderal strategy, 'segetal' species are sometimes discussed in these chapters as a smaller functional group, defined as annual weeds that share the same life cycle as cereals (autumn germination, fruiting at the end of spring or in summer) and that are of a historical origin (before 1500) as contaminants of cereal seeds (archaeophytes) – see Chap. 3.

4 Changes in Arable Habitats and Response of Arable Plant Communities

The declines of some arable plant species that have now become rare, on a national or continental scale can be related to idiosyncratic effects related to specific weed/crop associations. For example, the fortunes of the parasitic weed of flax, *Cuscutum epilinum*, are determined by the extent to which this crop is cultivated and it has declined across Europe along with the area of flax grown. Similarly, *Lolium temulentum* and *Agrostemma githago* used to be problem weeds in wheat in the UK as their seeds closely resembled wheat grain meaning they were difficult to separate and populations were routinely introduced into fields with farm-saved crop seed. As seed cleaning has improved, these species have subsequently dramatically declined (Firbank and Watkinson 1986). In the following chapters, there are several examples of these species-specific declines. However, the response of weed floras at the community level needs to be understood in the context of broader changes in arable habitats and the main drivers of community assemble discussed in the previous section: intensity and frequency of disturbance and soil fertility.

The timing of field operations (tillage, drilling and harvest) determine the available window for arable plants to germinate, emerge and produce mature seeds. These operations are largely determined by the crops being grown and diversity of rotations. A common theme through the following chapters is the simplification of rotations with a reduction in the diversity of drilling dates and tillage operations. One of the characteristics of the ruderal strategy is a persistent seedbank that allows plants to take advantage of ephemeral opportunities for emergence (although arable plants are very variable in the relative persistence of their seedbanks and for some species, for example *Agrostemma githago*, the fact that their seed is relatively short-lived has contributed to their decline). A diverse crop rotation with a variety of drilling dates, therefore, can support a range of arable plants adapted to emerge and flower at different times of the year. However, if disturbance events become too homogenous over time, the interval between opportunities for some species to emerge and reproduce become too long for them to persist. There are a number of examples of this in the following chapters resulting in the loss of certain species from national arable floras; for example, species adapted to stubbles in Hungary (Chap. 6) and species associated with certain tillage regimes in Spain (Chap. 7). One of the defining changes in arable landscapes in the UK has been the shift from spring sown cereals to narrower rotations based on autumn sown crops (largely driven by a switch from spring to winter barley), this has removed the opportunity for the emergence of obligate spring germinating plants that rely on over winter chilling to break dormancy. This shift in cropping patterns has been implicated in the decline of the populations of several arable plant species in the UK in Table 2.1, including *Galeopsis angustifolia* and *Valerianella dentata* (Wilson and King 2003). Given the reliance of arable floras on regular disturbance, perhaps the factor that has the biggest impact on their communities is land abandonment. This quickly leads to a shift from ruderal pant communities to later successional vegetation with a more competitive ecological strategy. While not a major factor in the UK, it is identified as a growing threat to arable plant communities especially in Eastern Europe and the Mediterranean (Chaps. 6, 7, and 8).

The second driver of plant community assembly is the nutrient status of the soil that selects between stress tolerant and competitive/ruderal ecological strategies (Fig. 2.1). The increased use of fertilisers across Europe associated with the introduction of high yielding crop varieties is identified as a causal factor in the declines of arable plant diversity in all the subsequent chapters. In the UK, the average rate of nitrogen fertiliser applied to winter wheat crops increased from 90 kg/ha N in 1967 to 200 kg/ha N in 1988 (Chalmers et al. 1990) and has stayed high ever since. Along with increases in atmospheric nitrogen deposition associated with pollution from the burning of fossil fuels this has led to the widespread eutrophication of UK habitats leading to the loss of more stress tolerant species (Maskell et al. 2010) and the biotic homogenisation of floras (Smart et al. 2006). A common theme throughout the subsequent chapters is that threatened arable plant communities tend to be found in less productive areas of the landscape – either on less fertile fields that are still farmed extensively or in field boundaries and margins that receive less fertiliser. These species are generally characterised by a 'stress-tolerant/ruderal' strategy – being positioned at the middle of the righthand edge of the CSR triangle (Fig. 2.1). As fertiliser use has increased, the area of these intermediate fertility habitats has decreased selecting against more stress tolerant species that have been outcompeted by nitrophilous species.

One way of visualising this shift in ecological strategies is to compare the Ellenberg indicator values for nitrogen between rare and common species. Ellenberg numbers have been assigned to the European flora indicating preference for different habitats using a nine point scale against seven environmental criteria, including moisture, light and nitrogen, with a higher value indicating that a species is more likely to occur where levels of that resource are high (Hill 1999). If the list of recently extinct, critically endangered and endangered arable plants in Table 2.1 is compared with the species pool of common species used in the Metcalfe et al. (2019) analysis (Fig. 2.1), it becomes clear that species that have suffered declines are characterised by lower Ellenberg-N numbers with a peak at four compared to seven for the common species (Fig. 2.2) – a difference between mesotrophic and eutrophic plant communities. This trend was also observed in a comparison of plant traits between species observed in a local vegetation survey done in 1949–51 and a

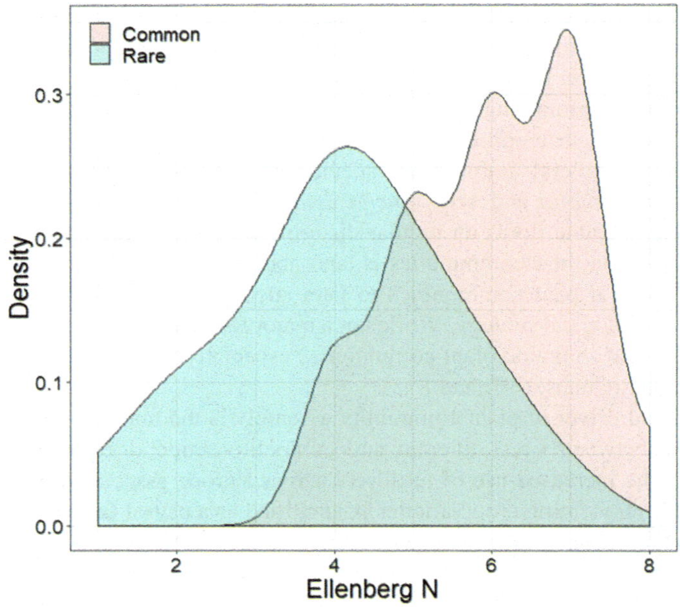

Fig. 2.2 Comparison of Ellenberg-N numbers for a pool of common UK weed species compared to the list of recently extinct, critically endangered and Endangered arable plant species on the UK Red Data List (Table 2.1). A higher number indicates a preference for more fertile environments

follow-up survey in 2004 in an area of intensive agriculture that included arable habitats (Walker et al. 2009). Tall nutrient-demanding species were observed to increase at the expense of small, habitat specialists ('stress tolerators' in Grime's scheme). Similar shifts in Ellenberg numbers are also reported in Chap. 3 on northwestern Europe.

The impact of inorganic fertilisers on arable plant communities is dramatically demonstrated in a long-term experiment on the impact of fertilisers on winter wheat yield at the oldest agricultural research Institute in the world, Rothamsted Research (Hertfordshire, UK). The Broadbalk experiment started in 1843 and originally had long strips of continuous winter wheat that received different amounts and combinations of nutrients either as inorganic fertiliser or farmyard manure, including a gradient of nitrogen inputs from 0 to 288 kg/ha N. Herbicides started to be used on the experiment in 1964 but one section of the experiment has been kept free of herbicides. This section of plots is now characterised by contrasting arable plant communities that are adapted to different levels of soil fertility (Fig. 2.3). The species pool observed on the experiment includes three species on the UK Red Data List (Table 2.1): *Galium tricornutum*, *Ranuculus arvensis* and *Scandix pecten-veneris*. Plots receiving low rates of fertiliser are characterised by short, open crop canopies in comparison to the high crop and weed biomass on plots receiving high rates of nitrogen in combination with phosphorus and potassium (Fig. 2.3).

Fig. 2.3 Photographs from the Broadbalk experiment: plot receiving no fertilisers (*left*) and plot receiving Farmyard manure + 48 kg/ha N inorganic fertiliser (*right*) – neither plot has ever had any herbicides applied. (Photos J. Storkey)

The highest species richness on Broadbalk is observed on the plot with intermediate fertility (receiving P + K+ 96 kg/ha N); non-destructive surveys done annually on the plots record an average of 16 species on this plot compared to 11 on the plots that receives 288 kg/ha N (Moss et al. 2004). The community composition also changes along the nutrient gradient with individual species displaying clear preferences for different levels of soil fertility. When species are grouped on the basis of their functional traits, clear patterns in relative abundance of the groups along the fertility gradient are observed (Fig. 2.4). The group containing common, nitrophilous weeds that are characteristic of intensively managed fields in the UK today (including *Alopecurus myosuroides*, *Papaver rhoeas* and *Tripleurospermum inodorum*) are more frequent on plots receiving high rates of Nitrogen. These species are characterised by small seeds. In contrast, the group containing the three weed species that have become nationally rare (*Ranunculus arvensis*, *Scandix pecten-veneris* and *Galium tricornutum*) are most frequent on plots receiving intermediate rates of nitrogen (96 kg/ha N), declining in frequency at the two extremes of the nitrogen gradient (Storkey et al. 2010). This group of species is characterised by a relatively short stature, large seed and late flowering: a combination of short stature and large seed has been related to increased investment in roots meaning they are better adapted for competition for below ground resources. This combination of three traits has been identified as a 'rare weed trait syndrome' for UK weeds. While there has been some support for this idea elsewhere in Europe (see Chap. 6), a similar Europe wide trait analysis of rare arable plants would be a useful goal for future research.

As well as fertilisers being implicated in the declines in arable plant diversity across Europe, the increased reliance on herbicides as the primary method of weed control is also highlighted as a driver of biodiversity loss by all authors in this section. However, unlike fertilisers, which have a filtering effect on weed communities, species that have declined are not thought to be especially sensitive to herbicides. Rather, herbicides represent an additional, intense disturbance event that effectively 'raises the bar' for all species in terms of the need to grow and

Fig. 2.4 Species on the Broadbalk experiment characteristic of plots with intermediate and high fertility: *Scandix pecten-veneris* (Ellenberg N = 4) (*left*) is most abundant on plots receiving moderate levels of Nitrogen fertiliser (48–96 kg/ha), *Alopecurus myosuroides* (Ellenberg N = 6) (*right*) is dominant on plots with high soil fertility receiving >200 kg N/ha. Photos J. Storkey (*left*) and Rothamsted Research Ltd. (*right*)

reproduce in windows of opportunity between disturbances. There is evidence that this has pushed weeds towards an 'extreme ruderal strategy' (Bourgeois et al. 2019) with successful species being characterised by small seeds, high fecundity (potentially buffering the high mortality rate from herbicides), early flowering and the ability to germinate all year round, facilitating the completion of the life cycle in between herbicide spraying events (Fried et al. 2012) – a phenomena discussed in more detail in Chap. 3. The rare weed trait syndrome described above would be particularly ill suited to tolerating this additional management pressure. Increasing mortality through herbicide use would also be expected to impact arable weed composition through abundance-based mechanisms with species that were less abundant in the landscape before herbicide use being particularly vulnerable. Finally, while it is undoubtedly true that herbicides have had a detrimental effect on arable weed diversity across Europe, Storkey et al. (2012) found that it was variation in fertiliser that explained more variation between countries in the threat status of arable plants.

5 Implications for Conservation

How can an understanding of the assembly of plant communities based on gradients of disturbance and fertility by useful for designing conservation strategies? A first principle must be that regular annual disturbance is a pre-requisite of any conservation strategy and areas managed for arable plants should remain in cultivation systems. Concepts such as land-sparing and re-wilding are, therefore, potentially detrimental to these taxa. Rather, they rely on in-field or in-crop conservation options that tend to be unpopular with farmers (Butler et al. 2009). Secondly, a common feature of the eutrophication of landscapes is the dominance of nitrophilous weeds species, for example in the UK *Alopecurus myosuroides* and *Cirsium arvense*, that quickly out-compete more stress tolerant species that tend to be the focus of conservation efforts. A common feature of the conservation opportunities discussed in the following chapters is, therefore, the focus on less productive areas of the landscape. These could be extensively managed fields on relatively unproductive soils or field margins that receive less agro-chemical inputs. Potential management approaches are reviewed extensively in Albrecht et al. (2016) and fall into two broad categories: (1) managing field margins for arable plants while cropping the rest of the field for yield and (2) establishing arable plant reserves where farmers are compensated for managing whole fields specifically for arable plant diversity. Given the threat from competitive weeds adapted to fertile environments, in both cases, it is important to target conservation measures at habitats that are well suited to arable plant diversity. In the UK, a standardised protocol has been designed to identify these Important Arable Plant Areas (Wilson 2007; Byfield and Wilson 2005). Options for conserving arable plants in field margins are also included in UK subsidised agri-environment schemes; unfertilised conservation headlands (sown with a reduced density of crop (Wagner et al. 2017)) or cultivated uncropped margins are most effective (Walker et al. 2007). Recent work by Metcalfe et al. (2019) has further highlighted the importance of taking the landscape context of these margins into account to reduce the risk of colonisation by competitive species. The importance of these landscape factors on arable plant diversity are discussed in detail in the context of the arable floras of the Iberian Peninsula in Chap. 7 of this section. Perhaps the most ambitious scheme presented in these chapters is the "100 fields for biodiversity" initiative in Germany (Chap. 5) where important arable plant areas are identified and farmers paid to implement the best management for their conservation. The effectiveness of this approach in the UK has been demonstrated through the successful management of arable plant reserves but is currently largely dependent on charitable funding (https://www.plantlife.org.uk/uk/nature-reserves-important-plant-areas/discover-ranscombe). Establishing a network of these sites, not only on a national, but also a continental scale would be a laudable ambition to secure the future of these fascinating, beautiful and scientifically important floras.

References

Albrecht H, Cambecedes J, Lang M, Wagner M (2016) Management options for the conservation of rare arable plants in Europe. Bot Lett 163(4):389–415

Bourgeois B, Munoz F, Fried G, Mahaut L, Armengot L, Denelle P, Storkey J, Gaba S, Violle C (2019) What makes a weed a weed? A large-scale evaluation of arable weeds through a functional lens. Am J Bot 106(1):90–100

Butler SJ, Brooks D, Feber RE, Storkey J, Vickery JA, Norris K (2009) A cross-taxonomic index for quantifying the health of farmland biodiversity. J Appl Ecol 46(6):1154–1162

Byfield A, Wilson PJ (2005) Important arable plant areas:identifying priority sites for arable plant conservation in the United Kingdom. Plantlife International, Salisbury

Chalmers A, Kershaw C, Leech P (1990) Fertilizer use on farm crops in great-Britain – results from the survey of fertilizer practice, 1969–88. Outlook Agric 19(4):269–278

Firbank LG, Watkinson AR (1986) Modeling the population-dynamics of an arable weed and its effects upon crop yield. J Appl Ecol 23(1):147–159

Fried G, Kazakou E, Gaba S (2012) Trajectories of weed communities explained by traits associated with species' response to management practices. Agric Ecosyst Environ 158:147–155

Grime JP (1974) Vegetation classification by reference to srategies. Nature 250(5461):26–31

Grime JP, Thompson K, Hunt R, Hodgson JG, Cornelissen JHC, Rorison IH, Hendry GAF, Ashenden TW, Askew AP, Band SR, Booth RE, Bossard CC, Campbell BD, Cooper JEL, Davison AW, Gupta PL, Hall W, Hand DW, Hannah MA, Hillier SH, Hodkinson DJ, Jalili A, Liu Z, Mackey JML, Matthews N, Mowforth MA, Neal AM, Reader RJ, Reiling K, RossFraser W, Spencer RE, Sutton F, Tasker DE, Thorpe PC, Whitehouse J (1997) Integrated screening validates primary axes of specialisation in plants. Oikos 79(2):259–281

Grime JP, Hodgson JG, Hunt R (2014) Comparative plant ecology: a functional approach to common British species, 2nd edn. Springer, Berlin

Hill M (1999) ECOFACT volume 2: technical annex – Ellenberg's indicator values for British plants. In: UK Government, London, p 46

Jones G, Charles M, Bogaard A, Hodgson JG, Palmer C (2005) The functional ecology of present-day arable weed floras and its applicability for the identification of past crop husbandry. Veg Hist Archaeobotany 14(4):493–504

Marshall EJP, Brown VK, Boatman ND, Lutman PJW, Squire GR, Ward LK (2003) The role of weeds in supporting biological diversity within crop fields. Weed Res 43(2):77–89

Maskell LC, Smart SM, Bullock JM, Thompson K, Stevens CJ (2010) Nitrogen deposition causes widespread loss of species richness in British habitats. Glob Chang Biol 16(2):671–679

Metcalfe H, Hassall KL, Boinot S, Storkey J (2019) The contribution of spatial mass effects to plant diversity in arable fields. J Appl Ecol 56(7):1560–1574

Moss SR, Storkey J, Cussans JW, Perryman SAM, Hewitt MV (2004) The Broadbalk long-term experiment at Rothamsted: what has it told us about weeds? Weed Sci 52(5):864–873

Oerke EC (2006) Crop losses to pests. J Agric Sci 144:31–43

Potts GR, Ewald JA, Aebischer NJ (2009) Long-term changes in the flora of the cereal ecosystem on the Sussex downs, England, focusing on the years 1968–2005. J Appl Ecol 47(1):215–226

Preston CD, Pearman DA, Dines TD (2002) New atlas of the British and Irish flora. Oxford University Press, Oxford

Smart SM, Thompson K, Marrs RH, Le Duc MG, Maskell LC, Firbank LG (2006) Biotic homogenization and changes in species diversity across human-modified ecosystems. Proc R Soc B Biol Sci 273(1601):2659–2665

Storkey J, Moss SR, Cussans JW (2010) Using assembly theory to explain changes in a weed flora in response to agricultural intensification. Weed Sci 58(1):39–46

Storkey J, Meyer S, Still KS, Leuschner C (2012) The impact of agricultural intensification and land-use change on the European arable flora. Proc R Soc B Biol Sci 279(1732):1421–1429

Sutcliffe OL, Kay QON (2000) Changes in the arable flora of central southern England since the 1960s. Biol Conserv 93(1):1–8

Telfer MG, Preston CD, Rothery P (2002) A general method for measuring relative change in range size from biological atlas data. Biol Conserv 107(1):99–109

Vigueira CC, Olsen KM, Caicedo AL (2013) The red queen in the corn: agricultural weeds as models of rapid adaptive evolution. Heredity 110(4):303–311

Wagner M, Bullock JM, Hulmes L, Hulmes S, Pywell RF (2017) Cereal density and N-fertiliser effects on the flora and biodiversity value of arable headlands. Biodivers Conserv 26(1):85–102

Walker KJ, Critchley CNR, Sherwood AJ, Large R, Nuttall P, Hulmes S, Rose R, Mountford JO (2007) The conservation of arable plants on cereal field margins: an assessment of new Agri-environment scheme options in England, UK. Biol Conserv 136(2):260–270

Walker KJ, Preston CD, Boon CR (2009) Fifty years of change in an area of intensive agriculture: plant trait responses to habitat modification and conservation, Bedfordshire, England. Biodivers Conserv 18(13):3597–3613

Wilson PJ (2007) Important arable plant areas – criteria for the assessment of arable sites. Asp Appl Biol 81:183–189

Wilson P, King M (2003) The biology of arable plants. In: Arable plants – a field guide. Hanway Press, London, p 312

Chapter 3
The Status of Arable Plant Habitats in Northwestern Europe

Guillaume Fried

Abstract North-Western Europe is probably the region where the cultivated agro-ecosystem has been modified most rapidly and profoundly with the intensification of agricultural practices during the twentieth century. While the richness and abundance of arable weed species has declined overall, some species historically dependent on cereals (so called 'segetal species') are particularly threatened due to changes in crop rotation, higher sowing densities and fertilisation, more frequent soil tillage and herbicides. Populations of some of these species are often maintained unintentionally; where cultural practices have remained extensive (polyculture-livestock areas), in some crops with an incomplete herbicide spectrum (e.g., oilseed rape), or where practices are compatible with the presence of certain species (for example, geophytes with bulbs in some vineyards). Arable fields under harsh abiotic conditions (very sandy or shallow, stony soils) are also favourable but threatened by abandonment of agriculture. Various initiatives have been taken to protect segetal species: (i) ex situ conservation by harvesting and saving seeds (for future reintroductions), (ii) agri-environmental measures with subsidies paid to farmers to compensate for losses linked to the lack of fertilisation and herbicides in the field or its border, (iii) land acquisition with specifications to ensure the maintenance of certain species. In hindsight, it appears that the best strategy would be to focus on helping to maintain low-input farming that has naturally coexisted with a segetal flora for centuries.

Keywords Segetal flora · Crop rotations · Herbicides · Fertilisers · Agri-environment schemes · Vineyards

G. Fried (✉)
Anses, Laboratoire de la Santé des Végétaux, Unité Entomologie et Plantes Invasives, Montferrier-sur-Lez, France
e-mail: guillaume.fried@anses.fr

C. Hurford et al. (eds.), *The Changing Status of Arable Habitats in Europe*,
https://doi.org/10.1007/978-3-030-59875-4_3

1 The Status of Arable Plant Diversity in Northwestern Europe

A recent compilation of weed survey data indicated that about 1400 plant species have been recorded in cultivated fields in France (Jauzein 1995; Munoz et al. 2017). Approximately 400 of these species have a Mediterranean distribution and are restricted to Southern France, so that the number of arable weed species that occurs in Northwestern Europe (NW Spain, non-Mediterranean France, Belgium, Luxembourg, the Netherlands) can be estimated at around 1000. According to a second rough estimate, about a quarter of the arable weed species in France are still common, including some troublesome weeds, half of the species are weeds of minor importance that are more or less stable, and a quarter of weeds are declining among which most are 'segetal' species, some of which were formerly common (Jauzein 2001). Segetal species ("messicoles" in French) can be defined as annual weeds that share the same life cycle as cereals (autumn germination, fruiting at the end of spring or in summer) and that are of a historical origin (before 1500) as contaminants of cereal seeds (archaeophytes). They occur predominantly on clay-limestone soils and phytosociologists have described them as characteristic of the *Caucalidion* alliance (Tüxen 1950). These segetal species have strong heritage value in popular culture as reflected by their different representations in dialects, imaginary, and symbolic and they have had many food, fodder and medicinal uses in rural areas (Lemonnier 2012).

In Wallonia (southern Belgium), 75 of the 119 segetal species present at the beginning of the twentieth century, i.e. about 60% of them, are now threatened or extinct (Legast et al. 2008). In France, the status of a reference list of 103 segetal species has been assessed as part of the National Action Plan for the conservation of segetal flora (Cambecèdes et al. 2012). Four categories of species were defined according to the trends of their populations determined by expert-knowledge: seven species were considered extinct (*Cuscuta epilinum*, *Cephalaria syriaca*, *Delphinium halteratum*, *Ervilia articulata*, *Lolium remotum*, *Silene cretica*, *Silene linicola*) and two other nearly so (*Camelina alyssum*, *Ridolfia segetum*), 51 taxa of *level 1* are in strong decline and considered under threat, 31 taxa of *level 2* should be monitored, maintaining themselves more or less depending on the region, while 12 taxa of *level 3* are still abundant, at least in some regions. Cambecèdes et al. (2012) compared the number of French departements (i.e., an administrative unit of France, similar to an English county) where these species occurred before the 1970s and after the 1990s. A rate of decline was obtained by dividing the number of departement where the species is considered to have become extinct after the 1990s by the total number of department where the plant had been observed before the 1970s and after the 1990s (Fig. 3.1). This broad scale of assessment allowed sampling periods to be compared despite differences in sampling intensitity and methodology. For 12 *level 1* taxa, the rates of decline between these two periods were more than 70%: *Silene muscipula* (88%), *Nigella nigellastrum* (88%), *Asperula arvensis* (80%), *Turgenia latifolia* (78%), *Polycnemum arvense* (76%), *Polygonum bellardii* (76%),

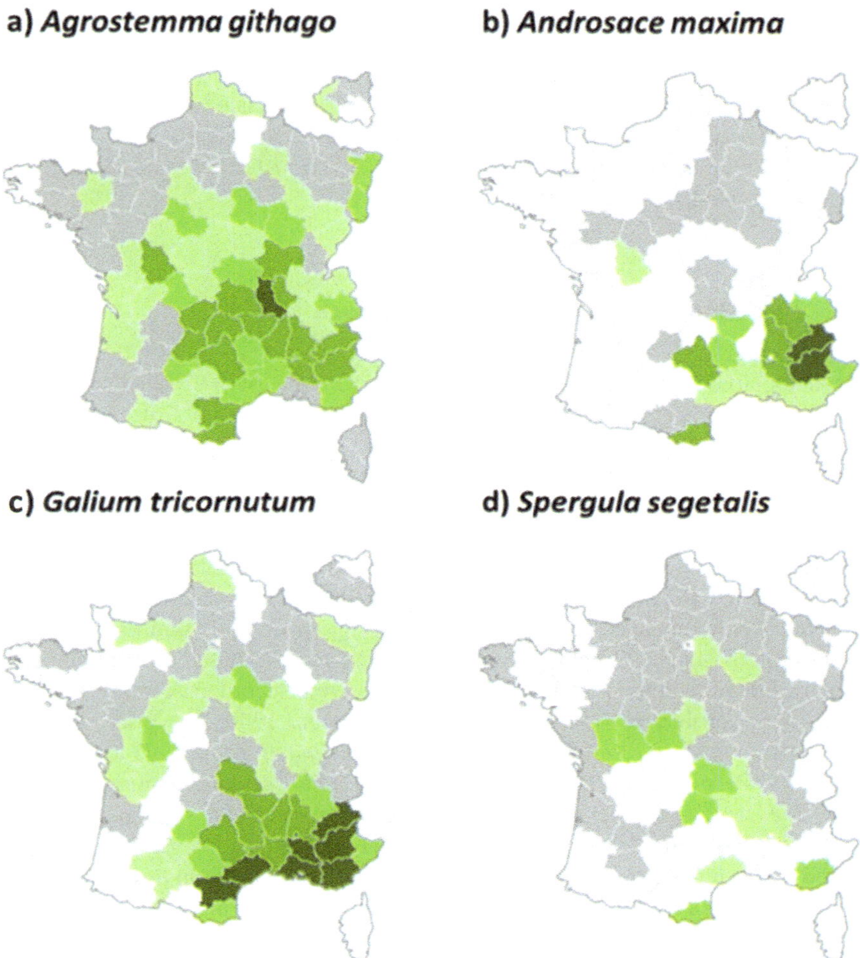

Fig. 3.1 Regression of four segetal species at the scale of France. (Adapted from Cambecèdes et al. (2012)). White: absence of the species, Grey: presence in at least one "commune" before the 1970s. The gradient of green shows the current abundance of the species, from lighter green to the darker green: presence in 1–2 "communes", presence in 3–8 "communes", presence in 9–30 "communes", presence in >30 "communes". Rates of decline were obtained by dividing the number of grey departments (i.e. where the species is considered extinct after 1970s) by the total number of departements that are not white (i.e. where the species was at least historically present). For example, *Androsace maxima* has a regression rate of 17/33 = 52%, while it reaches 43/58 = 74% for *Spergula segetalis*

Hypecoum imberbe (75%), *Spergula segetalis* (74%), *Hypecoum pendulum* (73%), *Vaccaria hispanica* (73%), *Camelina sativa* (73%), *Glaucium corniculatum* (70%). Fourteen species have disappeared from more than half of the departments where they were present before 1970: *Lolium temulentum*, *Nigella arvensis*, *Bifora testiculata*, *Ceratocephala falcata*, *Roemeria hybrida*, *Conringia orientalis*, *Bupleurum subovatum*, *Neslia paniculata*, *Camelina microcarpa*, *Adonis aestivalis*, *Valerianella coronata*, *Androsace maxima* and *Orlaya grandiflora*. Figure 3.1 illustrates the decrease of four segetal species at the scale of France (departments in grey correspond to area where the species is considered extinct). Several regional studies conducted in France over the past decades allow additional detail to be added to this initial assessment.

1.1 The Status of Segetal Flora in Alsace

The first regional example is based on the assessment of a list of 92 segetal species in the Alsace region in the NE of France (Fried 2004, 2009). These 92 species included 61 species of the national list historically present in Alsace as well as 31 species that are specific to arable habitats in Alsace and listed on the Red List of Threatened Species for this region (Vangendt et al. 2014). This study used herbaria data and old literature and floras of the 19th and early twentieth century in order to estimate the frequency of occurrence of the species before 1990. Recent data (after 1990) were gathered by the Botanical Society of Alsace in the Brunfels database and specific surveyed were conducted by Fried (2004) during spring and summer 2004. The comparison of old (<1990) and recent (>1990) records was based on the number of municipalities ("communes") where the species were recorded. Twenty-five taxa (27%) were considered to have become extinct before 1990. The year of the last records of some species was as old as 1900 for *Camelina alyssum*, 1905 for *Legousia hybrida*, 1940 for *Logfia gallica*, and 1950 for *Spergula segetalis*. Thirty taxa (33%) showed a steady decline of more than 75% (*Agrostemma githago*, *Camelina microcarpa*, *Galium tricornutum*); 21 taxa (23%) showed a decline between 33 and 75% (*Buglossoides arvensis*, *Delphinium consolida*, *Ranunculus arvensis*) while only 15 taxa (16%) were stable and/or still present in more than 10% of the surveyed fields in 2004 (*Cyanus segetum*, *Papaver rhoeas*, *Viola arvensis*). Some arable weeds (*Buglossoides arvensis*, *Gagea villosa*, *Malva setigera*), have found refuge in vineyards where weed control is less intensive than in cereal crops (Kazakou et al. 2016) and management is suitable for the conservation of emblematic bulb species such as *Tulipa sylvestris* subsp. *sylvestris* or *Gagea villosa* (Fig. 3.2).

Two other situations were identified as favourable to the maintenance of segetal plants in this region (Fried 2004, 2009). First, livestock areas where farmers often use farm-saved seeds (which allows the maintenance of mimetic species sown with the cereals such as *Bromus secalinus*, Fig. 3.3), and where weeding pressure is less due to an income that depends more on animal products than on crop yields. Tolerance to weeds in crops is also higher among these farmers. Second, areas where abiotic (soil) conditions are "extreme" (sandy or stony soil) which limits yield potential and the benefits or possibilities of intensifying practices (e.g. deep ploughing may not be possible). Draining soils may also limit the effect of herbicides

Fig. 3.2 (**a**) A vineyard with traditional soil cultivation on one inter-row out of two and no management on the row where *Tulipa sylvestris* subsp. *sylvestris* is abundant (Gueberschwihr, Alsace, France). (**b**) *Gagea villosa* and *Muscari neglectum*, two bulbous species that are favoured by superficial and light soil cultivation (Westhalten, Alsace, France). © Guillaume Fried

Fig. 3.3 *Bromus secalinus*, a mimetic species of cereals, still frequent in livestock areas where farmers use farm-saved seeds (Struth, Alsace, France). © Guillaume Fried

with lower persistence on these soils. Finally, these resource-poor situations are more in line with the ecological optimum of many segetal plants with short stature and heavy seeds (Storkey et al. 2010).

1.2 Changes in the Diversity and Composition of Arable Weed Communities in Burgundy

The second regional example comes from a diachronic study based on field-plot surveys at the community-level. In Burgundy, 158 fields were first surveyed between 1968 and 1976 by G. Barralis (Dessaint et al. 2001) and resurveyed in 2005–2006 (Fried et al. 2009b). At the field scale, the average number of species per plot (species richness) decreased from 16.6 (min: 6; max: 40) to 9.3 (min: 1; max: 37), i.e. a loss of 7.3 species per plot (−44%, Fig. 3.4). The total abundance scores (the sum of the abundance scores of all species in the survey) also dropped significantly from an average of 61.5 to 20.2 individuals per m^2 (−67%). Species such as *Buglossoides arvensis*, *Delphinium consolida*, *Legousia speculum-veneris* or *Ranunculus arvensis* that were still present in about one field out of 10 to one field out of four in the 1970s, are now only present in only one field out of 100 (Table 3.1). Species that were already rare in the 1970s (*Agrostemma githago*, *Gagea villosa*, *Nigella arvensis*) were not found any more in any of the sampled fields in the 2000s, although they are still present in a few fields at the regional level (Fried and Cadet 2007). Only a few arable weed species showed a stable regional frequency (e.g. *Alopecurus myosuroides*, *Viola arvensis*) or increased (*Scandix pecten-veneris*) between the two periods. Besides documenting this overall decline, the main lessons of this study were that: (i) even common species such as *Galium aparine* subsp. *aparine*, *Papaver rhoeas* or *Stellaria media* have recorded a significant decline; (ii) some extinct or strongly declining species are better maintained in the edges of fields (*Bunium bulbocastanum*, *Caucalis platycarpos*, *Lathryus tuberosus*, *Veronica triphyllos*); (iii) a positive relationship between weed species richness and functional diversity was found, and it has been highlighted that the current level of weed diversity is now often unable to cover the full range of functions that were provided

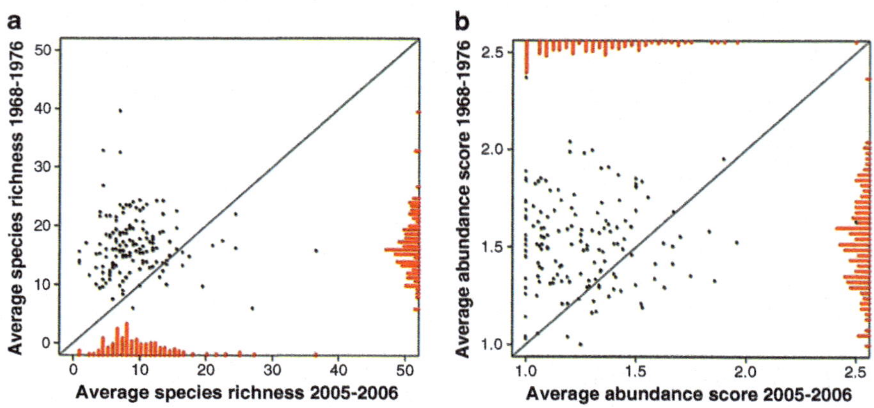

Fig. 3.4 Difference between two surveys (1968/1976 and 2005/2006) of (**a**) the average field species richness and (**b**) the average field abundance scores. The histograms are given on the axes

Table 3.1 Status and presence of rare arable plants in fields surveyed in the 1970s and resurveyed in the 2000s. Fr. and Ab. give the status of the species according to species frequency and species abundance

	Status		Occurrences			
			1970s	2000s		
Species	Fr.	Ab.	Field	Field	Margin	Total
Taxa of list 1						
Adonis flammea	=	=	8	1	0	1
Agrostemma githago	†		1	0	0	0
Camelina microcarpa	=	=	1	1	1	2
Neslia paniculata subsp. *thracica*	†		2	0	0	0
Nigella arvensis	†		3	0	0	0
Taxa of list 2						
Ajuga chamaepytis	=	=	14	1	0	1
Anthemis arvensis subsp. *arvensis*	↘	=	21	1	6	6
Bunium bulbocastanum	† (C.E.)		10	0	3	3
Calepina irregularis	N		0	1	4	4
Caucalis platycarpos	† (C.E.)		2	0	4	4
Cyanus segetum	↘	=	56	27	31	41
Delphinium consolida	↘	↘	53	3	0	3
Gagea villosa	†		2	0	0	0
Legousia hybrida	†		2	0	0	0
Legousia speculum-veneris	↘	↘	58	7	11	16
Papaver argemone	†		8	0	0	0
Ranunculus arvensis	↘	↘	61	2	3	5
Stachys annua	↘	↘	40	4	1	4
Thlaspi arvense	=	=	8	9	0	9
Valerianella spp.	=	↘	70	33	59	65
Taxa of list 3						
Alopecurus myosuroides	=	↘	118	92	109	126
Apera spica-venti	=	=	10	8	2	9
Aphanes arvensis	=	↘	73	35	31	45
Avena fatua	↘	↘	95	22	11	28
Buglossoides arvensis	↘	↘	63	2	10	10
Papaver rhoeas	↘	↘	127	60	83	97
Scandix pecten-veneris	↗	↗	26	26	53	58
Scleranthus annuus	†		13	0	0	0
Spergula arvensis	=	=	16	2	1	2
Viola arvensis	=	↘	145	128	110	139

The comparison is based on a resampling of the 1970s data (757 samples) to take into account the lower number of samples (315) in the 2000s (see Fried et al. (2009b) for details). Legend: † (only found in the 1970s); † (C.E.) (only found in the 1970s but found in crop edges in the 2000s); ↘: significantly decreasing species; = stable species; ↗: significantly increasing species; N: "new" species only found in the 2000s. Presence in the fields is the cumulated number of field (n = 158) where a species was found over the 1970s and 2000s surveys by arable weed communities in the 1970s (Fried et al. 2016). This includes, for example, the timing and amount of resources available to pollinators depending on the flowering phenology and the pollination mode of weed species, with therefore consequences for higher trophic levels (Storkey 2006; Gaba et al. 2016).

2 The Main Drivers of Change in Arable Plant Diversity

Based on the traits and ecology of loser versus winner species (Fried et al. 2009a), the dates of their last record or their changing trend, some hypotheses on the main drivers of changes in weed diversity could be formulated.

Changes in Crop Rotation It is well known that the type of crop strongly influences arable weed communities composition with some species being particularly associated with specific crop species (Fried et al. 2010). Thus, a first driver of changes in weed vegetation was the modification of crop rotation, which has sometimes been drastically altered. The disappearance of flax crops between the end of the nineteenth century and the beginning of the twentieth century led *de facto* to the extinction of mimetic species specialized to this crop. This could explain the early extinction of *Camelina alyssum*, *Cuscuta epilinum*, *Silene linicola* or *Lolium remotum* that were probably only casual species never really established and whose persistence depended on the continuous sowing of contaminated flax seeds.

In Alsace, a large part of the Rhine valley that was once cultivated with winter cereals has undergone profound changes with the introduction of maize that is cropped as an irrigated monoculture thanks to the availability of a plentiful ground water. During the 1980s, maize became the main crop grown in this region reaching 70% of cultivated area in the 2000s. Only a few segetal species, able to germinate and emerge in spring, can persist in this crop (e.g., *Lycopsis arvensis*, *Spergula arvensis*, *Thlaspi arvense*, *Viola arvensis*). Much of this region, which harbours well-draining and basic soils favourable to segetal species, has become unsuitable simply because maize has replaced winter cereals. Conversely, in Burgundy, where acreage of oilseed rape has increased, gaps in the herbicide spectrum in this dicotyledonous crop has allowed the proliferation of certain segetal species (*Adonis annua*, *Camelina microcarpa*, *Caucalis platycarpos*, *Scandix pecten-veneris*) as illustrated by their systematic higher abundance in oilseed rape compared to following cereal crops (Fried & Cadet 2007) – Fig. 3.5.

Amendments, Fertilization and Increased Competition With the Crop In Alsace, as well as in Belgium, many acidiphilic segetal species became extinct during the first half of the twentieth century (Meerts 1997; Fried 2009): *Linaria arvensis* (1901 in Alsace, 1909 in Belgium), *Filago arvensis* (1915 in Belgium), *Logfia gallica* (1940 in Alsace), etc. This is consistent with the concomitant appearance of the first mineral fertilizers and the steep increase in crop fertilization during this period. New varieties of cereals have also been progressively adopted and sown at much higher densities. The greater fertility of the arable habitat has increased the asymmetry of

Fig. 3.5 *Adonis annua* (*left*) and *Calepina irregularis* (*right*) in the margins of an oilseed rape crop (Prenois, Burgundy, France). © Guillaume Fried

competition with greater depletion of resources generated by cultivated species and certain nitrogen demanding weeds, to the detriment of heliophilic species of small size or with low seed potential. This effect further increased during the second half of the XXth century as shown by the significant increase in the mean Ellenberg indicator values for nitrogen at the community-level in the surveyed fields in Burgundy between the 1970s and the 2000s (Fried et al. 2009b).

Herbicides Regular and intensive use of herbicides since the 1960s has greatly reduced the abundance of arable weeds and the size of the weed seed bank in cultivated fields (Jauzein 2001; Fried et al. 2009b). Prolific producers of small seeds with high dormancy (*Sonchus* spp., *Stellaria media*) can regularly compensate the loss due to herbicides spray whereas species producing few large seeds with low dormancy (*Agrostemma githago*, *Turgenia latifolia*, *Vaccaria hispanica,* Fig. 3.6) have rapidly declined in abundance, to the point of local extinction. By comparing weed densities in winter wheat between normally treated areas and neighbouring control areas without herbicides, Fried et al. (2014) computed an index of control (IC) for each weed species, summarizing the average differences of their densities between treated and herbicide free areas. This index ranges between 1 (species only present in herbicides free area) to -1 (species only present in herbicide treated area). They showed that the sixteen segetal species present in the dataset had a significantly greater decrease in density between control and treated areas (IC = 0.90 +/− 0.14) than the sixteen most common species (IC = 0.67 +/− 0.11) which could illustrate the impact of herbicides on this already declining species. Several segetal species were only (*Buglossoides arvensis*, *Caucalis platycarpos*, *Myagrum perfoliatum*, *Petroselinum segetum*, *Stachys annua*) or almost only (*Cyanus segetum*, *Ranunculus arvensis*, *Spergula arvensis*) recorded in herbicides free area while they were always absent in the neighbouring treated area. There may be an 'abundance-based' mechanism in that if herbicides reduce all populations by 95%, declining species that are already less abundant are proportionately less likely to be found in the treated quadrat. However, it shows how the intensive use of herbicides could explain the local disappearance of segetal plant populations already weakened by other changes in the agrosystem.

Fig. 3.6 Examples of segetal species that were formerly largely distributed in Northwestern Europe and that have declined during the twentieth century. *Glebionis segetum* (*top left*), *Nigella arvensis* (*top right*), *Papaver hybridum* (*bottom left*), *Vaccaria hispanica* (*bottom right*). © Guillaume Fried

3 What Conservation Measures Are Being Taken?

Various complementary strategies have been implemented over the past 30 years to conserve the heritage of segetal species. Our examples are taken from France but most are valid for the rest of north-western Europe.

3.1 Regulatory Protection

A first option is to protect the species by law. This kind of regulation usually protects species from human activities such as destruction, cutting, uprooting, removal or sale. Despite segetal species being one of the most threatened groups of plants in France, only 10 segetal species are protected at the national level (*Cephalaria syriaca*, *Delphinium verdunense*, *Gagea villosa*, *Tulipa agenensis*, *Tulipa clusiana*, *Tulipa gesneriana*, *Tulipa radii*, *Tulipa sylvestris* subsp. *sylvestris*, *Nigella gallica*, *Nigella nigellastrum*) as well as 18 taxa at the regional level (Cambecèdes et al. 2012). Most of these species tend to be visually attractive and vulnerable to being collected. However, this way of protection is not optimal as the regulation order stipulates that the prohibitions "*are not applicable to the day-to-day operation of rural land on field plots usually cultivated*". Consequently, the regulations do not prohibit a farmer from carrying out the exploitation of the crop with his conventional practices (including herbicide use) because of the presence of protected segetal plants. On the other hand, they prohibit any operation outside the "current operations", in particular the carrying out of heavy modification of the arable field, which are subject to derogation. It could, therefore, prevent the urbanization of a field plot rich in segetal species. This protection can also be useful when some segetal plants survive outside cultivated crops.

3.2 Ex-situ Conservation

Ex situ conservation is the process of protecting endangered plant species outside their natural habitat, for example in seed banks or cultivated collections. It provides an alternative and complementary method for preventing immediate extinction. It is therefore particularly suited for the most endangered segetal species surviving in only a few locations. By harvesting seeds in the wild since the 1990s, the network of French national botanical conservatories has set up the *ex-situ* conservation of 77 taxa out of the 103 listed segetal species. However, 12 taxa among the most threatened are not yet conserved in any botanical reserves and should be the priority target of future harvests (Cambecèdes et al. 2012). In most cases, the seeds are of wild origin except for a few taxa that have been extinct in the wild for a long time that were harvested in botanical gardens (e.g., *Silene linicola*). One of the most effective ways to preserve plant diversity through *ex situ* conservation is by storage in seed banks, which allows the conservation of large amounts of genetic material in a small space and, under suitable conditions, for a long time with minimum risk of genetic damage. A strong limitation of the *ex situ* conservation approach is the lack of adaptation of the individuals cultivated in botanical gardens and conserved in seed banks to the changing cropping and environmental conditions. This strategy is therefore generally linked to reintroduction opportunities in favourable areas.

3.3 Reintroductions of Segetal Species

Another complementary approach to *ex situ* conservation programs, is to harvest segetal species locally, cultivate and multiply them on dedicated field plots and quickly reintroduce them to volunteer farmers. In various European countries, there have been recent efforts in this direction to re-establish new populations of threatened arable species in regularly managed fields or their margins, assumed to provide suitable habitat conditions for successful establishment and survival (Krick 2011; Cambecèdes et al. 2012; Lang et al. 2016). In France, this has resulted in the development of the "vraies messicoles" quality trademark (that could be translated as "true segetal species") in order to distinguish and encourage the use of seeds of rare arable plants propagated from local source populations (Cambecèdes et al. 2015). Such sourcing of seed from within the same region is considered vital for ensuring adaptation to local conditions and for maintaining intraspecific genetic variation at larger spatial scales (Keller et al. 2000). In Belgium, this is part of the measure "landscaped strips of wildflowers" where three species (*Cyanus segetum*, *Glebionis segetum* and *Papaver rhoeas*) are sown with cereals on strips at the edges of fields by volunteer farmers, who commit themselves for five years and receive in return EUR 1250 per strip to compensate for yield losses on the flowered strips (Piqueray et al. 2012).

3.4 Agri-environmental Schemes

Introducing segetal species to volunteer farmers, sometimes where these species were not historically present, may rightly seem artificial. Stace and Crawley (2015) noted that fields where segetal plants have been voluntarily sown have an artificial appearance that does not resemble the physiognomy of fields traditionally hosting a high diversity of segetal flora (Fig. 3.7). To address this issue, some conservation strategies first identify areas with the highest concentrations of segetal species in order to concentrate resources where it is worthwhile. Some environmental organisations or conservation bodies focussed on natural areas have purchased cultivated plots for the protection of their segetal flora. Land control can ensure the preservation of plots with an exceptionally high floristic richness or containing a population of rare species. For example, in SW France, the *Conservatoire Régional d'Espaces Naturels d'Aquitaine* has acquired an orchard plot to preserve *Tulipa agenensis*. Sometimes, farmers remain owners of the field but agree to sign an agreement defining the management methods and the conservation objective on the plot. Thus, in Champlitte (NE France), a farmer has an agreement with the *Conservatoire Régional d'Espaces Naturels de Franche-Comté* in order to set up farming practices allowing the maintenance of *Gagea villosa* and many other segetal species including *Nigella arvensis*.

Land acquisition is the surest way to control and guide management, but it is costly for environmental organisations and difficult to implement at the national level. Financial incentives to adopt favourable practices have been possible under the second pillar of the Common Agricultural Policy. During the 2000s, several agri-environmental schemes (AES) had potential for the conservation of segetal species. The AES "1402A: Maintenance of segetal plants and biodiversity in general" has been used in 29 contracts covering 116 ha in France. The specifications of this AES prohibited pesticide treatments, fertilisation and harvesting on at least 10% of the plot. Other AES potentially favourable to segetal plants included the establishment of crops of faunistic or floristic interest (1403A), the absence of herbicide treatment (PHYTO02), the maintenance of organic farming in phytosanitary problem areas, the creation and maintenance of a canopy of flora for threatened birds (bustard or other lowland birds) (Couver07). However, on the whole, these measures have been

Fig. 3.7 Examples of fields managed with traditional practices, rich in segetal species in the south of the Massif central, Millau, France. General view of the field with *Delphinium orientale* (*top*) and close up of a portion of the community with *Adonis flammea* and *Bupleurum rotundifolium* (*bottom*) © Guillaume Fried

little used or poorly explained and poorly understood by farmers with little awareness or motivation for the conservation of the segetal flora (Cambecèdes et al. 2012).

In France, the latest version of the "*Plan national d'actions en faveur des plants messicoles*" (Cambecèdes et al. 2012) aims to develop indicators based on species richness and rarity to identify areas where such AES would be most suited. Based on these results, a new AES should be developed to encourage farmers to maintain traditional low-intensity farming practices that has sustained segetal species to the present day. Similarly in Wallonia (southern Belgium), one of the AES (MC8: 'Strips adapted for segetal plant conservation') is targeted at fields presenting a high potential for segetal plant conservation (Lemoine et al. 2018). The selection of these fields relies on expert judgment, based on recent observations of threatened species in the close vicinity (ca. 500 m radius). The idea of these agri-environment measures focusing on areas of existing diversity of rare arable plants is to maintain extensive farming that could be threatened by either intensification or abandonment.

4 Conclusions

In Northwestern Europe as elsewhere on the continent, a strong decline of arable weeds has been observed during the second half of the twentieth century. Quantitative data show that the species richness at the scale of cultivated plots has decreased on average by 50%, while more than 50 species have a rate of decline of more than 75% at the scale of France. Around 10 species are considered extinct in France, but locally this figure is higher and increases towards the north of the territory under consideration. Changes in crop rotation, with the early disappearance of specific crops (flax), and the more recent increase in the acreage of unsuitable summer-sown crops (especially maize) has been an important driver of the decline of the segetal flora in many regions. Increasing inputs of agro-chemicals within cereal fields together with landscape simplification has reduced the abundance of weed species within the fields, their ability to use open marginal areas in the landscape as a refuge, and therefore to replenish the seed banks by dispersal. Increased disturbances, including more intensive weeding with deep tillage or herbicides, and increased competition with the crops, has filtered out many traditional segetal species and selected a restricted list of weeds able to cope with or to avoid these filters (Bourgeois et al. 2019). Areas where a high segetal flora diversity persists correspond mostly to areas where harsh environmental conditions prevail (sandy acidic soils or conversely, dry stony calcareous fields) and prevent agricultural intensification or reduce its negative effects. Mixed crop-livestock farms are also very favourable to the segetal flora as the weeding pressure is generally lower and the weeds are sometimes perceived favourably and even grazed by the herd in the stubble after the harvest (Lemonnier 2012). In fact, the highest level of diversity of segetal flora in France is found when these two factors (harsh environments and livestock) coincide, for example in the south of the Massif central (Grands Causses), in the Luberon or the southern Pre-Alps (Albrecht et al. 2016). Conservation efforts should focus on helping to maintain this type of low-input farming that has *naturally* coexisted with a segetal flora for centuries. Although many segetal species are 'flagship species' with a high cultural value and merit conservation actions *per se* (e.g., *Adonis* spp., *Cyanus segetum*), with the current transitioning to agro-ecology, conservation strategies should also be broader in scope by supporting practices that are compatible with a higher diversity and abundance of common weeds, which may also benefit supporting fauna (insects and invertebrates) (Gaba et al. 2016). In turn, a change in the perception of weeds by the farming community may facilitate the acceptance of measures for segetal species.

References

Albrecht H, Cambecèdes J, Lang M, Wagner M (2016) Management options for the conservation of rare arable plants in Europe. Bot Lett 163:389–415

Bourgeois B, Munoz F, Fried G, Mahaut L, Armengot L, Denelle P, Storkey J, Gaba S, Violle C (2019) What makes a weed a weed? A large-scale evaluation of arable weeds through a functional lens. Am J Bot 106:90–100

Cambecèdes J, Largier G, Lombard A (2012) Plan national d'actions en faveur des plantes messicoles. Conservatoire botanique national des Pyrénées et de Midi-Pyrénées–Fédération des Conservatoires botaniques nationaux–Ministère de l'Écologie, du Développement durable et de l'Énergie

Cambecèdes J, Malaval S, Bischoff A, Provendier D, Boutaud M, Gourvil J, Largier G, Dutrève B (2015) "Vraies Messicoles": a label for rarearable plants produced from the wild for functional and ornamental purposes as a way to preserve biodiversity. In: Proceedings 17th EWRS-Symposium, p. 245. EWRS, Montpellier

Dessaint F, Chadoeuf R, Barralis G (2001) Diversity of weed communities of annual crops in côte-d'Or, France. Biotechnol Agron Soc Environ 5:91–98

Fried G (2004) La flore messicole en Alsace. Etat des lieux, analyses des capacités de maintien et perspectives de conservation. Mémoire de fin d'études pour l'obtention du diplôme d'ingénieur de l'agriculture et de l'environnement, Enesad, Dijon

Fried G (2009) Les plantes messicoles et les plantes remarquables des cultures en Alsace. Atlas écologique et floristique. Société Botanique d'Alsace, DIREN Alsace, Conseil Général du Bas-Rhin, Strasbourg

Fried G, Cadet E (2007) Le lieu-dit "Beuchail" à Fleurey-sur-Ouche (21) : un site remarquable pour la flore messicole. Le Monde des Plantes 493:19–23

Fried G, Chauvel B, Reboud X (2009a) A functional analysis of large-scale temporal shifts from 1970 to 2000 in weed assemblages of sunflower crops in France. J Veg Sci 20:49–58

Fried G, Petit S, Dessaint F, Reboud X (2009b) Arable weed decline in northern France: crop edges as refugia for weed conservation? Biol Conserv 142:238–243

Fried G, Petit S, Reboud X (2010) A specialist-generalist classification of the arable flora and its response to changes in agricultural practices. BMC Ecol 10:1–11

Fried G, Chauvel B, Gaba S, Jullien J, Pointereau P, Rodriguez A, Reboud X (2014) Flore du blé d'hiver : les facteurs qui influencent sa composition et sa diversité. Phytoma – La santé des végétaux 679:43–47

Fried G, Dessaint F, Reboud X (2016) Local and regional changes in taxonomic and functional diversity of arable weed communities in Burgundy (France) between the 1970s and the 2000s. Bot Lett 163:1–13

Gaba S, Reboud X, Fried G (2016) Agroecology and conservation of weed diversity in agricultural lands. In: Taylor & Francis

Jauzein P (1995) Flore des champs cultivés. Sopra-INRA, Paris

Jauzein P (2001) L'appauvrissement floristique des champs cultivés. In: Les Dossiers de l'environnement de l'INRA, pp 65–78

Kazakou E, Fried G, Richarte J, Gimenez O, Violle C, Metay A (2016) A plant trait-based response-and-effect framework to assess vineyard inter-row soil management. Bot Lett 163:1–16

Keller M, Kollmann J, Edwards P (2000) Genetic introgression from distant provenances reduces fitness in local weed populations. J Appl Ecol 37:647–659

Krick F (2011) Unkraut vergeht nicht -stimmt nicht! LZ Rheinland 32:48–49

Lang M, Prestele J, Fischer C, Kollmann J, Albrecht H (2016) Reintroduction of rare arable plants by seed transfer. What are the optimal sowing rates? Ecol Evol 6:5506–5516

Legast M, Mahy G, Bodson B (2008) Les messicoles, fleurs des moissons. Région Wallone, Namur

Lemoine C, Sérusiaux E, Mahy G, Piqueray J (2018) Agro-environmental scheme for segetal plant conservation in Wallonia (Belgium): an assessment in conventional and organic fields. BASE 22:35–44

Lemonnier S (2012) L'aventure est dans les blés. Redécouvrir les plantes messicoles, les sauvageonnes des moissons. Savoirs de Terroirs

Meerts PJ (1997) La régression des plantes messicoles en Belgique. Faut-il sauver les mauvaises herbes: Actes du Colloque sur la régression de la flore messicole (ed by, pp. 49–55

Munoz F, Fried G, Armengot L, Bourgeois B, Bretagnolle V, Chadoeuf J, Mahaut L, Plumejeaud C, Storkey J, Violle C, Gaba S (2017) Database of weeds in cultivation fields of France and UK, with ecological and biogeographical information In: Zenodo, https://doi.org/10.5281/zenodo.1112342

Piqueray J, Gilliaux V, Mahy G (2012) La conservation des messicoles dans le cadre du programme agro-environnemental Wallon. Parcs et Réserves 67:18–22

Stace CA, Crawley MJ (2015) Alien plants. Harper Collins, London

Storkey J (2006) A functional group approach to the management of UK arable weeds to support biological diversity. Weed Res 46:513–522

Storkey J, Moss SR, Cussans JW (2010) Using assembly theory to explain changes in a weed flora in response to agricultural intensification. Weed Sci 58:39–46

Tüxen R (1950) Grundriβ einer Systematik der nitrophilen Unkrautgesellschaften in der eurosibirischen Region Europas. Mitt Flor Soz Arb Gem, NF 2:94–175

Vangendt J, Berchtold JP, Jacob JC, Holveck P, Hoff M, Pierne A, Reduron JP, Boeuf R, Combroux I, Heitzler P, Treiber R (2014) La Liste rouge de la Flore vasculaire menacée en Alsace. CBA, SBA, ODONAT

Chapter 4
The Status of Arable Plant Habitats in Scandinavian Countries

Terho Hyvönen, Lars Andersson, and Christian Andreasen

Abstract Scandinavia is situated in the northern part of Europe where climate conditions limit both agricultural production and weed diversity. Scandinavian countries differ from each other in terms of climate conditions, land use intensity and agricultural production. In Finland and Denmark, the status of arable plant populations has been recorded by regular weed surveys covering several decades. In both countries, a tremendous decline in weed abundance was documented between 1960s and 1980s caused by the significant intensification of the cropping practices. The decline was attributed to the higher rate of nitrogen fertiliser application, more effective fertilisation methods, the increased use of crop monocultures and the application of herbicides. During the following decades, a slight increase in weed abundances was recorded in both countries probably mainly caused by stricter regulation of herbicide use decided nationally and by the EU. In the 1990s, agri-environmental support schemes to facilitate environmental issues were introduced, which resulted in, among other things, increasing area of organic farming. Organic farming has enhanced weed species diversity. However, conservation of rare arable weeds has not been a primary focus in Scandinavian countries and there are no specific conservation methods to protect rare plant species on arable land.

Keywords Denmark · Sweden · Norway · Finland · Cropping patterns · Weed diversity · Organic farming · European Union

T. Hyvönen (✉)
Natural Resources Institute Finland (Luke), Jokioinen, Finland
e-mail: terho.hyvonen@luke.fi

L. Andersson
Swedish University of Agricultural Sciences, Uppsala, Sweden
e-mail: lars.andersson@slu.se

C. Andreasen
University of Copenhagen, Taastrup, Denmark
e-mail: can@plen.ku.dk

C. Hurford et al. (eds.), *The Changing Status of Arable Habitats in Europe*,
https://doi.org/10.1007/978-3-030-59875-4_4

1 Introduction

Scandinavia is situated in the northern part of Europe where climatic conditions limit both agricultural production and weed diversity. However, Scandinavian countries (here Denmark, Sweden, Norway and Finland) also differ from each other in terms of climate conditions, land use intensity and agricultural production. The arable land area in Denmark, Sweden and Finland is similar, ca 2.2–2.5 million ha, but lower in Norway (0.8 million ha). Denmark has the most favorable climatic conditions for agriculture where the arable land constitutes about 56% of the total land area (https://tradingeconomics.com). In the other Scandinavian countries, the share of arable land varies between 2.2 and 7.6%. In these countries, the arable area is concentrated in the southern regions while other parts are either field-forest mosaic or entirely forested areas.

Weed flora composition and weed infestation are affected directly and indirectly by climatic factors, soil characteristics, and land use (Andreasen and Skovgaard 2009). In all countries the production of annual crops is dominated by cereals, ranging from 36% of arable land in Norway to 60% in Denmark. Grass clover leys are also an important part of the arable landscape covering 20, 30 and 45% of arable land in Denmark, Finland and Sweden, respectively. The proportion of leys increases toward the north in the two latter countries. Climatic conditions and soil characteristics have resulted in large differences in land use; while Danish cereal production is dominated by autumn sown cereals, spring barley and spring oat are the most common annual crops in Finland. In Sweden, with a mid-position, spring barley and winter wheat are the major annual crops. The proportion of organically managed land has increased substantially over the last decades. In 2018, the proportion of organic land (including under conversion) was 4.7, 9.8, 13.1 and 20.3% in Norway, Denmark, Finland and Sweden, respectively, as compared to an average of 7.5% in the European Union as a whole (Eurostat 2020).

As a probable consequence of the interaction between climate and land use, and the high proportion of organic farming, the use of herbicides is lower in the three Scandinavian countries than in EU 28 as a whole. The average sale of herbicides, including preharvest desiccation agents and moss killers, during the period 2013–2017 ranged from 0.5 kg ha^{-1} (Finland) to 0.8 kg ha^{-1} (Denmark) as compared to, for example, 1.4 kg ha^{-1} in Germany and 1.1 kg ha^{-1} across all the EU28 (Eurostat 2020).

Weed diversity declines along a south-north climate gradient in Europe (Glemnitz et al. 2006; Hyvönen et al. 2011) and in Scandinavia, the distribution of several weed species common in Central Europe (e.g., *Amaranthus retroflexus* and *Echinochloa crus-galli*) is limited by the climate (Hyvönen et al. 2012) or lack of crop species, that cover the ground late in the growing season such as beets and maize (Hyvönen and Ramula 2014). However, frequent reports from advisors and farmers indicate that *E. crus-galli* is a fast-increasing weed problem in southern Sweden, Denmark and Norway (L. Andersson (own observation), Andreasen et al. 1992; Andreasen and Streibig 2011; VKM 2016).

2 The Status of Arable Plant Diversity in Scandinavian Countries

In Finland and Denmark, the status of arable plant populations has been recorded by regular weed surveys covering several decades (Mukula et al. 1969; Erviö and Salonen 1987; Haas and Streibig 1982; Andreasen et al. 1996, 2018; Salonen et al. 2001, 2011; Andreasen and Stryhn 2008, 2012). The major aim of these weed surveys has been to record changes in the abundance of harmful weed species in terms of crop production. However, they also provide useful information on changes in weed species diversity. In both countries, dramatic declines in weed abundance were documented between the 1960s and 1980s (Andreasen et al. 1996; Erviö and Salonen 1987) caused by the significant intensification of the cropping practices. During the following decades, a slight increase in weed abundances was recorded in both countries (Andreasen and Stryhn 2008, 2012; Salonen et al. 2001, 2011) probably mainly caused by stricter regulation of herbicide use decided nationally and by the EU.

The total numbers of weed species recorded in the weed surveys in the two countries are not directly comparable due to different sample sizes and sampling methods. However, in both countries, the total number of weed species is relatively high compared to the number of species that are targeted for weed control. In the two latest weed surveys of Finnish spring cereal fields the total numbers of weed species recorded were 160 and 148 (from 690 and 595 sampled fields), respectively (Salonen et al. 2001, 2011). In the countrywide survey in Denmark in 1987–1989, 199 weed species were found in nine crops in 213 fields (Andreasen 1990), while 224 species were found in 11 crops during 2001–2004 distributed on 240 fields (Andreasen and Stryhn 2008). These data do not indicate that the total number of weed species has decreased on a country level as different species occur in different regions. However, the number of species in a single field was observed to have declined.

Average species numbers per sampled area are easier to compare between the two countries. In Denmark, long-term changes in average species numbers have followed the trend of weed abundance: declining between the 1960s and the 1980s followed by a slight increase, but not fully recovering by the 2000s. In cereal fields (winter wheat, rye and spring barley), the species numbers in a random selected circular sample plot of 0.1 m^2, varied between 5.8 and 6.9 in the 1960s, 2.1 and 2.9 in the 1980s and 3.4 and 4.9 in the 2000s (Andreasen and Stryhn 2008). Average species numbers of grass leys were lower than in cereal fields in all decades (e.g., 3.4 species in the 1960s) and did not differ between the two latest weed surveys (1.5 species in both). In these surveys, the average number of species was calculated based on the 67 weed species observed across all the sampled plots. In the Finnish weed surveys, determination of average number of weed species was based on records of all species, but it is available only from the two latest weed surveys (Salonen et al. 2001, 2011). The comparison between the weed surveys showed a decline in the average number of weed species (per 10×0.1 m^2) both in conventionally (16 vs. 12 species) and in organically managed (24 vs. 21 species) fields between the 1990s and 2000s. Conventionally grown fields included both fields sprayed and unsprayed with herbicides during the growth season. This increased the variation of average species number among conventional fields (e.g., variation of 3–26 species in the 2000s). This is an indication of the impact of weed control measures on weed species numbers.

In both countries, seed banks of weeds have also been studied (Salonen et al. 2011; Andreasen et al. 2018). Surveys of the same fields in Denmark, showed that the mean soil seed bank per m^2 in the uppermost 20 cm of the ploughing layer was estimated to be 19,390 seeds in 1964, 10,120 in 1989 and 20,455 in 2014. Thus, the size of seed bank had recovered to the level of 1960s in the 2010s. However, a few common species made up most of the soil seed bank in all years (e.g., *Poa annua*, *Juncus bufonius* and *Capsella bursa-pastoris*) but diversity has declined over time. Relatively few species have been able to adapt to modern cultivation practice reflecting the narrowing of the ecological niche for weeds in arable fields. In Finland, the number of seeds were 1684 per m^2 when samples were taken at the depth of 5 cm. The majority of the seeds were either *Chenopodium album* (66.2%) or *Spergula arvensis* (10.6%) Finnish samples were taken on spring cereal fields while Danish ones represented several crop species.

Weed species abundance and diversity affects the higher trophic levels using weeds as food sources. Based on such interactions, Hyvönen and Huusela-Veistola (2008) developed an indicator for farmland birds, pollinators and phytophagous insects. The indicator showed the decline in the weed abundance between 1960s and 1980s to have been harmful for all species groups in terms of decline of food sources. However, the increase of weed abundance between 1980s and 1990s benefitted more farmland birds and phytophagous insects than pollinators. This was due to lower abundance of weed species important for pollinators such as *Galeopsis speciosa* and *Galeopsis bifida*. Recognition of decline in the arable biodiversity has led to research on the measures for enhancing agrobiodiversity. For pollinators, rotational fallows (Kuussaari et al. 2011) and long-term set-asides (Alanen et al. 2011) as well as wildflower strips (Korpela et al. 2013) have been found to be most effective. Long-term set asides also provide important food sources for farmland birds. (Hyvönen and Huusela-Veistola 2011). In Finland, these measures have already been introduced in an agri-environmental support scheme.

3 The Specific Drivers of Change

Intensification of agricultural practices has negatively affected weed abundance and species diversity. In both Denmark and Finland, a tremendous decline in weed abundance took place between the1960s and 1980s. The decline was attributed to the higher rate of nitrogen fertilizer application, more effective fertilization methods, the increased use of crop monocultures and the application of herbicides. Large-scale application of herbicides was probably the most important single factor responsible for the decline in weed abundance (Andreasen and Streibig 2011). Sub-surface draining also became more common which affected the landscape structure in terms of increased field size, reducing field margins and hedges. Furthermore, the landscape became dominated by a few crop species with a simplified crop rotation system. These landscape scale impacts reduced habitat heterogeneity and the richness of the regional species pool that is able to colonise cultivated fields.

After decades of intensification, concerns about the environmental problems related to use of pesticides and fertilizers were raised, which led to political initiatives to reduce their use. For example, a Danish action plan required a 50% reduction in pesticide use by 1997 compared with 1981–1985. Both the quantity of ingredient sold and the spraying intensity were reduced (Haas and Streibig 1994). The average amount of active ingredient of herbicide used per hectare declined from 1.5 to 1.0 kg, while the spraying intensity declined from 1.5 to 1.3 applications per hectare (Anon 2006).

In 1995, Finland and Sweden joined the EU, which affected agricultural policy in many ways. One of the consequences was an introduction of agri-environmental support schemes to facilitate environmental issues. In Finland and Sweden, enrollment in agri-environmental schemes has been high (https://ec.europa.eu/eurostat/statistics-explained/index.php/Agri-environmental_indicator_-_commitments). This resulted in, among other things, an increasing area of organic farming, which in 2017 constituted 19.2% of the total utilized agricultural area in Sweden and 11.4% in Finland (https://ec.europa.eu/eurostat/statistics-explained/index.php/Organic_farming_statistics). In Denmark, conversion from conventional to organic farming has been politically and economically encouraged in order to obtain more sustainable agricultural practices, and the area of registered organic land increased from 0.2% in 1989 to 5.6% in 2004 (Anon 2005). Organic farming generally enhances weed species diversity (Hyvönen et al. 2003; Andreasen and Andresen 2011).

A decrease in the profitability of agriculture has led to an increase in the farm and field sizes. For example, in Denmark, the average farm size increased from 33.1 ha to 53.7 ha from 1987–89 to 2001–04 (Anon 2006). In 2018, 22% of the farms were larger than 100 ha and the number of farms was reduced from 81,267 in 1989 to 34,114 in 2018 (Anon 1990, 2020). In Sweden, the number of agricultural holdings decreased from 96,000 in 1990 to 63,000 in 2016, and the average farm size increased from 20 to 41 ha during the same period. There is a tendency towards greater pesticide use per unit area on large farms (Jensen 2003), and the sale of herbicides has increased in Sweden since the 1990’s. The average sale during the period 2011–2017 was 18% higher than the period 1991–1995 (KEMI 2020). In Finland, the use of glyphosate has increased due to direct drilling (Salonen et al. 2011).

4 Conservation of Rare Plant Species

There has been little consideration given to the conservation of rare arable weeds in Scandinavian countries and there are no specific conservation methods to protect rare plant species on arable land. However, there are several weed species whose conservation status has been identified as being of concern (Table 4.1). Many of these species occurred on agricultural fields before the introduction of herbicides in the 1950's were introduced (Suominen 1986; Svensson and Wigren 1986). Some of these weed species are associated with cultivation of flax (*Camelina alyssum*, *Cuscuta epilinum* and *Spergula arvensis* subsp. *maxima*) or rye (e.g., *Bromus secalinus*). Cropping practices of these crops have changed (e.g., combine harvesting, more competitive cultivars) or the cropping area has declined (flax) tremendously which caused a decline in the populations. *Centaurea cyanus* is also a weed species associated with rye which declined in abudance during earlier

Table 4.1 Red-listed plant species occurring in arable habitats and their conservation status in Scandinavian countries

	Finland[a]	Sweden[b]	Norway[c]	Denmark[d]
Primarily or solely on arable habitat				
Agrostemma githago	RE	CR	NKR	NA
Anthemis arvensis	NT	NT	LC	NA
Bromus secalinus	VU	EN	NE	NA
Camelina alyssum	RE	RE	NE	NA
Cuscuta epilinum	RE	RE	NE	NA
Delphinium consolida	EN	NT	NE	–
Fumaria vaillantii	CR	NT	NE	NA
Galeopsis ladanum	NT	NT	EN	NA
Lolium remotum	RE	RE	NE	NA
Odontites vernus	RE	NT	SE	LC
Papaver dubium	NT	LC	NA	NA
Spergula arvensis subsp. *maxima*	RE	RE	RE	–
Vicia villosa subsp. *villosa*	RE	NA	NA	NA
Primarily on other habitats				
Buglossoides arvensis	EN	NE	CR	–
Epilobium lamyi	EN	LC	–	LC
Hyoscyamus niger	NT	NT	EN	NA
Oxybasis urbica	RE	RE	NE	–
Urtica urens	NT	NT	VU	NA

The categories of conservation status: *RE* Regionally Extinct, *CR* Critically Endangered, *EN* Endangered, *VU* Vulnerable, *NT* Near Threatened, *LC* Least Concern, *NA* Not Applicable, *NE* Not Evaluated, *SE* Severe Risk, *NKR* No Known Risk - = not found from the database
[a]Ryttäri et al. (2019)
[b]Artfakta från ArtDatabanken
[c]Artsdatabanken (Norwegian Biodiversity Information Centre)
[d]Wind and Pihl (2004)

decades in Finland and Sweden (Suominen 1986; Svensson and Wigren 1986). The likely reasons for the decline are denser crop stands and the use of herbicides. Many of the threatened species have already gone extinct. The highest number of extinct species is found in Finland (Table 4.1). However, the information on the status of rare weed populations of arable fields is often insufficient to evaluate their status. This is true especially in Denmark where only two of the species have been systematically evaluated.

References

Alanen E-L, Hyvönen T, Lindgren S, Härmä O, Kuussaari M (2011) Differential responses of bumblebees and diurnal Lepidoptera to vegetation succession in long-term set-aside. J Appl Ecol 48:1251–1259

Andreasen C (1990) The occurrence of weed species in Danish arable fields. PhD thesis, Department of Agricultural Sciences, The Royal Veterinary and Agricultural University, Frederiksberg, Copenhagen. 125 pp

Andreasen C, Andresen LC (2011) Managing farmland flora to promote biodiversity in Europe. CAB Reviews: Perspectives in Agriculture, Veterinary Science, Nutrition and Natural Resources 6, No. 047

Andreasen C, Skovgaard IM (2009) Crop and soil factors of importance for the distribution of plant species on arable fields. Agric Ecosyst Environ 133:61–67

Andreasen C, Streibig JC (2011) Evaluation of changes in weed flora in arable fields of Nordic countries – based on Danish long-term surveys. Weed Res 51:214–226

Andreasen C, Stryhn H (2008) Increasing weed flora on Danish arable fields and its importance for biodiversity. Weed Res 48:1–9

Andreasen C, Stryhn H (2012) Increasing frequency of weed species in Danish beet, pea and winter barley fields. Crop Prot 36:11–17

Andreasen C, Jensen JE, Haas H (1992) Grøn Skærmaks (*Setaria víridis*) og Hanespore (Echinochloa sp.): Nye græsukrudtsarter i Danmark. (*Setaria viridis* and *Echinochloa*: New grass weed species in Denmark) 9th Danish Plant Protection Conference/Weeds. Statens Planteavlsforsøg, beretn S 2178–1992. 49–60

Andreasen C, Stryhn H, Streibig JC (1996) Decline of the flora in Danish arable fields. J Appl Ecol 33:619–626

Andreasen C, Jensen HA, Jensen SM (2018) Decreasing diversity in the soil seed bank after 50 years in Danish arable fields. Agric Ecosyst Environ 259:61–71

Anon (1990) Agricultural Statistics 1990. Danmarks Statistik ISSN 0070-3559

Anon (2005) Statistik over økologiske jordbrugsbedrifter 2004. Autorisation og produktion. Juli 2005, 2. Plant Directorate. Ministry of Food, Agriculture and Fishery, Denmark

Anon (2006) Agricultural Statistics 2006. Danmarks Statistik. ISSN 1602-4550

Anon (2020) Danmarks Statistik. https://www.dst.dk/. Accessed 6 Jan 2020

Erviö L-R, Salonen J (1987) Changes in the weed population of spring cereals in Finland. Ann Agric Fenniae 26:210–226

Eurostat (2020). https://ec.europa.eu/eurostat/statistics-explained/index.php/Organic_farming_statistics

Glemnitz M, Radics L, Hoffmann J, Czimber G (2006) Weed species richness and species composition of different arable field types – A comparative analysis along a climate gradient from south to north Europe. J Plant Dis Prot Sonderheft 20:577–586

Haas H, Streibig JC (1982) Changing patterns of weed distribution as a result of herbicide use and other agronomic factors. In: Lebaron HM, Gressel J (eds) Herbicide resistance in plants. Wiley, New York, pp 57–79
Haas H, Streibig JC (1994) Policy and research strategies to reduce herbicide use in northern Europe. Rice pest science and management. In: Teng PS, Heong KL, Moody K (eds) Selected papers from the International Rice-Research Conference. IRRI, Los Ban˜os/Laguna, pp 245–254
Hyvönen T, Huusela-Veistola E (2008) Arable weeds as indicators of agricultural intensity – A case study from Finland. Biol Conserv 141:2857–2864
Hyvönen T, Huusela-Veistola E (2011) Impact of seed mixture and mowing on food abundance for farmland birds in set-asides. Agric Ecosyst Environ 143:20–27
Hyvönen T, Ramula S (2014) Crop–weed competition rather than temperature limits the population establishment of two annual C4 weeds at the edge of their northern range. Weed Res 54:245–255
Hyvönen T, Ketoja E, Salonen J (2003) Changes in the abundance of weeds in spring cereal fields in Finland. Weed Res 43:348–356
Hyvönen T, Glemnitz M, Radics L, Hoffmann J (2011) Impact of climate and land use type on the distribution of Finnish casual arable weeds in Europe. Weed Res 51:201–208
Hyvönen T, Luoto M, Uotila P (2012) Assessment of weed establishment risk in a changing European climate. Agric Food Sci 21:348–360
Jensen JE (2003) Behandlingsintensitet 2000, 2001 og 2002 (*Treatment intensity* in 2000, 2001 and 2002). DJF-rapportnr. Danske Planteværnskonference, Danmarks JordbrugsForskning, Foulum, Danmark, 88–2003:20, 143–153
KEMI (2020). https://kemi.se/kemikaliestatistik/statistik-om-bekampningsmedel
Korpela E-L, Hyvönen T, Lindgren S, Kuussaari M (2013) Can pollination services, species diversity and conservation be simultaneously promoted by sown wildflower strips on farmland? Agric Ecosyst Environ 179:18–24
Kuussaari M, Hyvönen T, Härmä O (2011) Pollinator insects benefit from rotational fallows. Agric Ecosyst Environ 143:28–36
Mukula J, Raatikainen M, Lallukka R, Raatikainen T (1969) Composition of weed flora in spring cereals in Finland. Ann Agric Fenniae 8:59–109
Ryttäri T, Reinikainen M, Hæggström C-A, Hakalisto S, Hallman J, Kanerva T, Kulmala P, Lampinen J, Piirainen M, Rautiainen V-P, Rintanen T, Vainio O (2019) Vascular plants. In: Hyvärinen E, Juslén A, Kemppainen E, Uddström A, Liukko U-M (eds) The 2019 red list of Finnish species. Ministry of the Environment & Finnish Environment Institute, Helsinki, pp 263-–312
Salonen J, Hyvönen T, Jalli H (2001) Weeds in spring cereal fields in Finland – a third survey. Agric Food Sci Finland 10:347–364
Salonen J, Hyvönen T, Jalli H (2011) Composition of weed flora in spring cereals in Finland – a fourth survey. Agric Food Sci 20:245–261
Suominen J (1986) The vanishing element of Finnish field weeds. Memoranda Soc. Fauna Flora Fennica 62:79–82. (In Finnish)
Svensson R, Wigren M (1986) Observations on the decline of some farmland weeds. Memoranda Soc Fauna Flora Fennica 62:63–67
VKM (2016). Risk assessment of cockspur grass (Echinochloa crus-galli). Scientific Opinion of the Panel on Plant Health of the Norwegian Scientific Committee for Food Safety, ISBN: 978-82-8259-213-0, Oslo, Norway. https://vkm.no/download/18.2994e95b15cc54507161c7ae/1498204990105/c2542ec415.pdf. Accessed 6 Jan 2020
Wind P, Pihl S (eds) (2004) The Danish Red List – The National Environmental Research Institute, Aarhus University. redlist.dmu.dk (updated April 2010)

Chapter 5
The Status of Arable Plant Habitats in Central Europe

Stefan Meyer

Abstract In Central Europe arable land is the most dominant land use type. This, often intensively managed habitat, is home to about 400 plant species. However, the last decades have seen a dramatic decline in species richness and, in some cases the complete disappearance of arable plant communities. A multitude of comparative studies show that, in the field interiors, the total species number pooled over all samples has declined by approximately one third since the middle of the twentieth century. In the 1950s the first appeals for the protection of these species and their populations were published. Following the establishment of conservation measures in field margins in the late 1970s as a forerunner of the first agri-environment schemes (AES) in the EU, important arable plant areas have been set up. So-called field flora reserves or conservation fields are particularly important for the protection of the rare and highly endangered segetal plants and are managed in accordance with the guidelines agreed with the farmer. As well as the beneficial practices of organic farming, extensification schemes and the use of fallow land or set-aside, rare and endangered arable plants are now being re-introduced on arable land with care being taken to maintain genetic provenance.

Keywords Marginal habitats · Fertilisers · Herbicides · Crop rotation · 100 fields for diversity · Extensification schemes · Conservation guidelines

1 The Change and Impoverishment of the Segetal Flora

As early as 1915, the pharmacist Adolf Andree from Lower Saxony complained that *'one has to travel far to see colourful fields dense with poppies and cornflowers [...]; arable weeds are now rare here'*.

S. Meyer (✉)
Albrecht-von-Haller-Institute for Plant Sciences, Plant Ecology and Ecosystems Research, University of Göttingen, Göttingen, Germany
e-mail: smeyer1@gwdg.de

C. Hurford et al. (eds.), *The Changing Status of Arable Habitats in Europe*,
https://doi.org/10.1007/978-3-030-59875-4_5

Arable habitats currently cover about one third of the surface area of Central Europe, and alongside forests are one of the most important ecosystems in the region. This, often intensively managed, habitat is home to about 400 plant species, i.e. slightly more than 10% of the Central European vascular plant flora. Of these, about 150 species are obligate 'segetal' plants, requiring arable habitats to persist. Many of them have evolved under the pressure of land use practices over the course of the roughly 7000-year history of Central European agriculture, and now possess traits adapted to this man-made habitat. Despite the considerable changes in agricultural management during its development from the Middle Ages to the end of the nineteenth century, the arable flora remained relatively stable throughout this period. The three-field-crop rotation with its alternation of winter cereals - spring cereals - fallow was the dominant land-use form in medieval and early-modern Central Europe. Even the widespread shift to the improved three-field-crop rotation (fodder or root crop cultivation instead of fallow land) in the nineteenth century continued to provide the conditions for a species-rich arable flora. The often-quoted statement of Buchli (1936) '*In no other land use system do we find such an abundant and rich segetal flora … as in the improved three-field-crop rotation*' expresses this. Palaeobotanical studies, herbal books and floras, also suggest that the species spectrum did not undergo any fundamental changes until relatively recently, despite the addition of several neophytes especially from the nineteenth century onwards.

The first signs of noticeable declines in the arable flora were at the end of the nineteenth century. This concerned mainly speirochore species (unintentionally spread with the harvested grain) and species with large diaspores, which were disadvantaged by the development of better seed cleaning techniques (e.g. *Adonis* species or Corn cockle, *Agrostemma githago*). Others, like epizoochore species that had been described as 'common, widely distributed' as late as the middle of the nineteenth century also soon became rarities. Bogenhard (1850) mentions the Greater Bur-parsley (*Turgenia latifolia*, Fig. 5.1) and White laceflower (*Orlaya grandiflora*) as field weeds in Thuringia *'which appear in places as a real plague, as a plague of the fields, … known and hated by the farmer*'. Today, they are unknown even to many botanists in Central Europe.

The species spectrum found in arable fields has also shifted with the habitat conditions. In the economically difficult war and post-war years, in which even marginal sites were still cultivated, indicator species of lime and acid conditions were still abundant. In the Thuringian limestone areas, a rich flowering display of arable plants could still be widely experienced in several fields in the 1950s (Hilbig 2007a, Fig. 5.2). But as the area of intensive agriculture with fertile soils and intermediate environmental conditions (not too wet or dry, no extremes in chemical composition etc.) increased in the early twentieth century, so did the dominance of habitat generalist, highly competitive, less herbicide-sensitive species and those adapted to nutrient-rich habitats (Hanf 1937). Numerous floristic and vegetation studies, especially those repeating historical surveys of the segetal vegetation (see the overview in Albrecht 1995; Richner et al. 2014) have highlighted the massive decline and, in some cases the complete disappearance of, the arable plant communities of shallow chalk soils, nutrient-poor acid sandy soils and the wet and

Fig. 5.1 Only three recent arable populations of the Greater bur-parsley (*Turgenia latifolia*) are known in Germany. In Switzerland, Austria and Czechia this plant is either already extinct or only occurs in a few populations. © Stefan Meyer

temporarily flooded sites (Fig. 5.3). In Western Pomerania, the Lamb's lettuce community (Teesdalio-Arnoseridetum) now occurs '*on less than 1% of the arable land on which occurrence would be conceivable according to substrate and climatic conditions*' (Litterski et al. 2005, Fig. 5.4). In contrast to the historical data, arable plant stands tend no longer to be differentiated according to substrate or altitude. Assigning of modern relevés to phytosociological communities is now impossible, and even diagnostic species of higher-level syntaxa are lacking. Character species of the plant communities that were still frequent in the 1920–1950s are now very rare (Meyer et al. 2015; Richner et al. 2017).

Quantitatively, comparative studies show that in the field interiors, total species number pooled over all samples declined by approximately one third since the 1950s/60s and, in some regions like on the Swabian Alb, by up to two thirds (Gerhards et al. 2013). On the plot level (100 m^2), mean species losses were around 45–80% (Kohlbrecher et al. 2012; Meyer et al. 2013a, 2015; Dedek and Wesche 2017; Richner et al. 2017). The cover of segetal vegetation in the field interior is now often less than 3%, in some areas of north-western Germany only 0.5%. Thus, despite early warnings by Tüxen (1962) pointing to the serious destruction of the arable plant communities in Central Europe, it is clear that a widespread collapse of phytodiversity on arable land has taken place in the twentieth century. Consequently, the Red Lists of endangered, lost and extinct plant species have become populated with species of the segetal flora, which is often the group with the highest risk levels.

In summary, several factors have contributed to the great changes in the arable flora over recent decades (Albrecht et al. 2016). The main causes of the decline,

Fig. 5.2 The Kyffhäuser Mountains in Thuringia (Germany) are home to one of the most species-rich and colourful limestone segetal communities in Central Europe. © Stefan Meyer

Fig. 5.3 The Toad rush (*Juncus sphaerocarpus*) used to be found rarely in temporarily flooded arable fields. © Stefan Meyer

Fig. 5.4 The Orange lily (*Lilium bulbiferum* ssp. *croceum*), a former characteristic species of the Teesdalio-Arnoseridetum is very rare today, and is considered to be the queen of the sandy fields in Northern Germany and the Netherlands. © Stefan Meyer

homogenization and endangerment of the segetal flora in Central Europe are the widespread use of herbicides in agriculture, high fertilisation rates resulting in dense crop stands, the abandonment of low-yielding sites or their conversion into grassland, and the disappearance of arable farming in mountain regions (Kohler et al. 2011; Reidl 2015). Furthermore, crop rotations have been simplified, site environmental conditions have been homogenised by soil improvement measures, seed cleaning has been perfected, historical or special crop types (such as emmer) with their adapted species are no longer cultivated and the immediate ploughing of the stubble after harvesting prevents the reproduction of late flowering species (e.g. Gottwald et al. 2018; Seifert et al. 2014; Richner et al. 2014; Meyer et al. 2013a; Storkey et al. 2012).

2 Efforts to Conserve the Segetal Flora

> '*... one could now also preserve a field with the old cereal varieties and the former arable plant communities in a museum-like manner...*' (Waldis-Meyer 1978).

The sharp decline in the occurrence of numerous segetal species and species-rich arable plant communities, especially on shallow chalk and acid sandy soils, has led to discussions among conservationists about the protection of these species and their populations. The first calls for the conservation of the segetal flora date back to the 1950s/1960s. The phytosociologist Robert Gradmann wrote as early as 1950 in his very comprehensive work on the flora of the Swabian Alb that '*the flower-decorated cornfields have almost disappeared from our cultural landscape, and soon we will have to set up small protected areas where three-crop rotation farming is carried out with poorly cleaned seeds*' (Gradmann 1950). Militzer (1960) recommended '*to maintain some small fields on marginal soils with low-intensity cultivation. In this way, the species-rich segetal flora … can be preserved in a few examples and protected as important arable plant areas*'. In Germany, the thesis on the protection of arable plants (Hilbig 1986), the Karlstadt position paper (van Elsen et al. 2005), the project '100 fields for diversity' (Meyer and Leuschner 2015) and the review of management options for the conservation of rare arable plants by Albrecht et al. (2016) have made significant contributions to the protection of arable plants and to the requirements for conservation measures of various kinds. Since 2007, an annual 'field conference on the protection of segetal plants has been held in Germany in June, which is increasingly attended by segetal flora specialists from abroad. Consequently, some years ago a bibliography on the 'protection of arable plants' was published by Meyer et al. (2013b). The introduction to this publication outlines the definition, the genesis and the habitat preferences of arable plants in Central Europe, as well as describing the strong decrease of numerous arable plants in recent decades and efforts and conservation measures to protect rare and endangered arable plant species. In the bibliography, all known publications dealing with arable plant conservation are listed. The majority are German texts, but references from other European countries are also included. Altogether, the bibliography cites more than 1700 texts that deal with conservation measures for arable plants and their locations, with special measures for protection and re-establishing of species, with extensification programmes, their successes and their effect on agrobiodiversity. Every citation also includes the country or region of the data source and up to seven additional key words.

Other Central European countries have also seen a sharp decline in the number of arable plants recorded in cultivated fields in recent decades, a trend that continues and is documented in numerous publications (e.g. Richner et al. 2017; Legast et al. 2008; Schopp-Guth et al. 2006; Traxler et al. 2005; Warcholińska 2004). In the former Czechoslovakia, a conference on the protection of plant genetic resources was held around 1980, at which the protection of the arable flora was also discussed (Skalický 1981). Since the 1980s, and increasingly since the 1990s, efforts to protect the phytodiversity in agroecosystems through the conservation of species-rich and typical communities of marginal arable habitats have begun, e.g. for Switzerland (Waldis-Meyer 1986; Waldis 1987) and Poland (Warcholińska 1986). Numerous publications are available on conservation programs in Switzerland, Poland, and Luxembourg (listed in Meyer et al. 2013b). Field flora reserves have been established in Poland (Siciński 1986; Warcholińska 1986), Luxemburg (Lenerz et al. 2017), Austria (Schmid 2008) and Switzerland (Birrer et al. 2018).

For lower Austria, Holzner (1978) had already compiled a list of potential arable protection areas at the end of the 1970s. In the area surrounding Vienna, conservation fields were established where a species-rich segetal vegetation was present (Schmid 2008). A current binational Interreg-project 'Wild and Cultivated' in

Fig. 5.5 Alternatively cultivated vineyard in the district of Kitzingen (Bavaria, Germany) with a large population of Wild tulip (*Tulipa sylvestris*) © B. Blümlein

several biosphere reserves of Bavaria and Austria is dedicated to the propagation of seeds of the regional segetal flora (www.wildundkultiviert.at). One of the project's main goals is the re-establishment of rare and endangered arable plants and their communities. In Switzerland (Halder 1982) special species protection programmes were aimed at preserving the cornflower (*Cyanus segetum*). Brunner and Gigon (2001) described the conservation of bulb-geophytes in Swiss vineyards, which are also highly endangered in Central Europe. Vineyard flora reserves have also been established for the very endangered Geranio-Allietum vinealis community with its magnificent spring geophytes of the genera *Allium*, *Gagea*, *Muscari*, *Ornithogalum* and *Tulipa* and the numerous spring ephemera species (Ehmke 2001; Meyer and Leuschner 2015) (Fig. 5.5).

As long ago as the 1970s, Waldis-Meyer (1978) called for the preservation of examples of the small fields of mountain farmers in the Swiss Valais. Almost 30 years later, this idea was realised and the first plots of land for the conservation of the diverse segetal flora were acquired by Pro Natura Valais (Stipa 2008) (Fig. 5.6). In addition, the 'Resource project for the conservation and promotion of the endangered Swiss arable flora' was implemented in Switzerland between 2012–2018 to preserve existing arable plant communities. By the end of the project in seven cantons, a total of 79 ha of plots were contracted and 60 indicator species of the segetal flora as well as rare arable mosses were secured (Birrer et al. 2018).

Fig. 5.6 Conservation field with rye and saffron cultivation on a south-west facing slope in a mountainous region near Mund/Upper Valais (Switzerland) © Stefan Meyer

Building on this, a project co-ordinated by the HotSpots association in the cantons of Schaffhausen and Zurich is currently underway to promote rare arable plants on suitable arable land. There, at least 20 fields with autochthonous populations of arable plants or that are suitable for reintroduction are to be secured and managed appropriately.

3 Measures and Concepts for the Conservation of Arable Flora

'Several possibilities are emerging for creating and maintaining reserves for relics of anthropogenic flora and vegetation. Since the species affected by this are particularly closely linked to the development of land use by man and thus to the history of man himself, all possibilities for the protection of these species and biotopes should be used'. (Willerding 1986)

3.1 Important Arable Plant Areas, Field Flora Reserves and Conservation Fields

Fields, which have been identified as being particularly important for the protection of the rare and highly endangered segetal plants and included in conservation schemes must be managed in accordance with the guidelines agreed with the farmer. In Baden-Württemberg they were called a field flora reserve ('Feldflorareservat'), a term that was later used by other authors. Hilbig (1985) introduced the term 'Schutzacker' (conservation field). In Poland such conservation fields are called 'agroreserwaty' (agroreserves) (Siciński 1986).

One of the first conservation sites for arable plants in Germany was a field flora reserve on the Swabian Alb (Baden-Württemberg). This has been in existence since 1970 and serves to preserve and demonstrate the diversity of the species-rich field flora of calcareous soils (Rodi and Schill 1982). In the former German Democratic Republic (GDR), intensive research had been conducted since the 1950s on the distribution and environmental requirements of segetal plants, their occurrence in plant communities and on the changes caused by intensive agriculture with large-scale cultivation. According to Hilbig (1978) it was time to '*grant special species protection to the declining arable plants, mostly species of extreme locations*'. Krausch (1978) had already written an '*obituary for an arable plant*' for the corn cockle (*Agrostemma githago*), which was still '*frequent and widespread in cereal fields . . . from the plains to the foothills of the Alps . . . a feared, unwelcome weed*' at the beginning of the twentieth century (Hegi 1910). For knowledge transfer and coordination, and in preparation for the creation of further areas for the protection of the arable flora, a working group 'segetal plant conservation' was founded in 1984 in the former GDR. By the late 1980s, about 25 field flora reserves had been created and some were secured as natural monuments or by integration into a larger nature reserve (Illig 1990; Hilbig 2007b). Various field flora reserves were also established in the federal states of western Germany (for details see Meyer et al. 2013b).

Despite these efforts, the lack of general success in ensuring the long-term conservation of highly endangered arable species led to the development of the '100 fields for diversity' project (Meyer and Leuschner 2015). The aim of this project is to set up a nation-wide network of conservation fields ('Schutzacker'). These arable fields have outstanding plant species inventories and are protected in the long term (up to 25 years or open ended) by appropriate contractual agreements or legal requirements.

The following guidelines are given for the appropriate management of these conservation fields:

- no or only little, mostly organic fertilisation
- no pesticides
- a high proportion of cereals in the rotation, in particular winter cereals, seed preferably uncleaned from the field itself
- late stubble tillage, no catch crops
- only shallow ploughing, no subsoil loosening

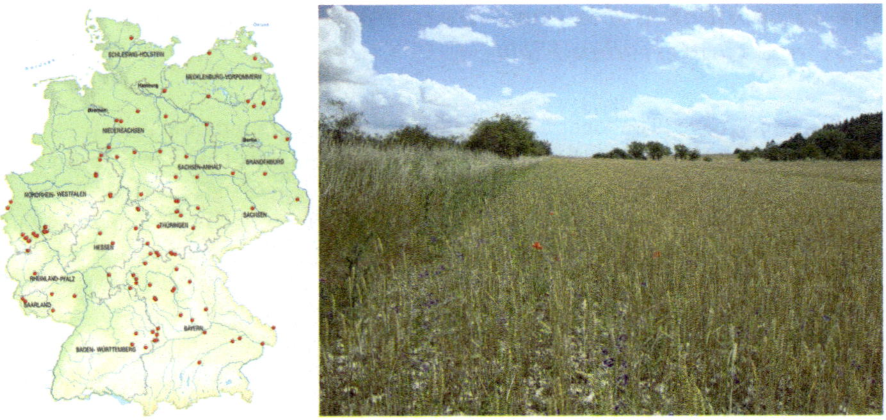

Fig. 5.7 Locations of the 112 conservation fields for arable flora in Germany (left) and conservation field for the segetal flora in the northern foothills of the Harz Mountains (Saxony-Anhalt, Germany) (right). © Stefan Meyer

- no draining or irrigation
- inclusion of old crop types typical for the area
- seeds and plant material of rare arable plant species from the immediate vicinity that have not been identified in the conservation field but are present in the area may be brought in
- the plant species community must be documented at the time of the establishment of the conservation field, as well as its changes over the years, the introduction of species and their development, and the cultivation regime.

Based on a nationwide screening of the floristically most valuable sites for arable vegetation in Germany, 112 important arable plant areas with a total area of around 475 ha had, by 2015, been secured with appropriate low-intensity agricultural practices for at least 10 years (often up to 20–25 years or open ended) by contract. The conservation fields are spread over the whole area of Germany on limestone, sandy or loamy soils with arable communities harbouring many rare and endangered species (Fig. 5.7). A key element is the commitment by the farmer to low-intensity farming practices targeted at promoting endangered species. For this purpose, various instruments for achieving the long-term financial support of conservation management, such as ecological compensation measures and agri-environmental schemes (AES) are used.

3.2 Field Margin Programs

A project creating herbicide-free field margins in the Limestone Eifel begun by Professor Wolfgang Schumacher (University of Bonn) and colleagues in the period 1978–1982 represented a great boost to the protection of segetal plants by integrating the concept of biodiversity conservation into mainstream agriculture (Schumacher 1979). This project was a forerunner of the first agri-environment schemes (AES) in the EU. In the second half of the 1980s and 1990s, the ministries and federal states responsible for agriculture set up special protection programmes for segetal plants and provided funds to support the farmers involved. The field margin programmes attempt to preserve and promote rare arable plants by omitting the chemical weed control methods commonly used today in the headlands and margins of fields (Fig. 5.8). With otherwise the same cultivation, a zone of 3–12(−24) metres of the cultivated area remains unsprayed. Farmers who voluntarily participate in the project receive compensation for reduced yields. In order to promote rare species and to prevent harvesting difficulties caused by the occurrence of problematic weeds, a reduction in fertilization and, on sandy soils, a limitation of lime applications are also recommended. With comparatively little financial outlay, unsprayed field margins can make a major contribution to the medium-term conservation of rare plant species and their animal communities if the areas are carefully selected and targeted based on the presence of endangered species. The effectiveness of the field margin programmes is therefore highly dependent on the commitment of knowledgeable experts on site, who also provide technical support to the participating farmers. These colourful flowering field margins have also increased awareness among the public of segetal flora protection with the help of interpretation boards (Karkow and Gronemann 2005).

Fig. 5.8 Colourful field margin in the district of Goettingen (Lower Saxony, Germany) with flowering poppies. © R. Urner

3.3 Organic Farming

Early comments on the '*impact of alternative land management on vegetation*' are found in Meisel (1978, 1979), who noted '*a positive effect of organic farming*', which was reflected in greater biodiversity on arable land and to some extent also in grassland. In the meantime, numerous comparative studies of biodiversity on organically and conventionally farmed fields, have been published by (among others) van Elsen (2000) and Frieben et al. (2012) for Germany and, more generally by Stein-Bachinger et al. (2020). In the Czech Republic, an analysis of 290 records revealed the abundance of endangered arable species in organically managed fields was 4.5 times higher than in conventional farming (Kolářová et al. 2013). Common to all comparative studies is that - depending on the intensity of cultivation - usually at least 2–3 times more segetal species are found in organically managed fields, ranging from only slightly higher to ten times higher. On an organic farm in the federal state Brandenburg, a total of 21 Red List species of segetal flora were identified (Gottwald 2010, Fig. 5.9). However, even in organic systems, management intensity plays a role, and intensive harrowing or cultivation of grass–clover leys in such systems may be as harmful to threatened species as herbicide application is in conventional systems. Such practices should therefore be limited if management objectives for a given field include preservation of rare arable plants (Albrecht et al. 2016).

3.4 Extensification Schemes and Fallow Land or Set-Aside

Extensification schemes can also contribute to varying degrees to the maintenance of a basic level of arable plant diversity especially on arable land with low productivity. If there is a sufficient soil seed bank, and depending on the location, a few years of extensive cultivation following previous low-input conventional farming is often sufficient to rejuvenate species-rich segetal stands (Litterski and Jörns 2004). A widespread and effective extensification measure is annual rotational fallowing (Hilbig 1998). Here, the favourable effects of the fallow year for the rarer arable plants are maximised when cereal stubble is left over winter: immediate cultivating before the one-year fallow period prevents many positive effects. Which species occur depends on the site conditions, the seed bank and the seed rain from neighbouring areas. Fallow land can, especially on marginal areas, lead to species-rich, colourful vegetation (Fig. 5.10) and to the occurrence of rare, uncompetitive segetal species in the first years. Leaving the field undisturbed for several years, however, leads to successional processes from the initial annual segetal species to increasing numbers of perennial species (Manthey 2003) to ruderal perennial meadows, after which grasses tend to dominate and young tree and shrub saplings start to develop. To promote the competitively weak arable flora, at most a two (to three) year fallow period is suitable. A permanent fallow will eventually kill off a species-rich segetal community. Ritschel-Kandel (1988) wrote in the 1980s of the

Fig. 5.9 Organic cereal cultivation with Field black cumin (*Nigella arvensis*) near Brodowin (Brandenburg, Germany). © Stefan Meyer

Fig. 5.10 Annual fallow fields: on a limestone field on the Franconian Alb (Bavaria, Germany) (*left*) © Stefan Meyer and a colourful annual fallow on a sandy field in the biosphere reserve 'Schorfheide-Chorin' (Brandenburg, Germany) (*right*) © F. Gottwald

'catastrophic effects on the protection of endangered arable plants of set-aside'. This phenomenon is mainly due to the fact that farmers used their least productive fields (i.e. remote, sloping, low-yielding, small or stony) for set-aside, which were usually those still supporting a species-rich segetal flora. Especially in such areas, the preservation of arable land with low management intensity is essential for the protection of arable plants and must also be financially supported. In contrast, the extensification or set-aside of highly productive arable land generally does not lead to the occurrence of rare arable plants as the seed banks have been greatly thinned

out after decades of intensive cultivation and the seedbank community that will be dominated by common, nitrophilous species. In the interest of preserving a level of arable plant diversity, the following applies to fallow land in general even if this may, however, be in conflict with the protection of animal species on the fallow:

- extensively farmed arable land is better than undisturbed fallow land,
- rotational fallow is better than permanent fallow,
- fallow land is better than seeding,
- the vegetation developing on the fallow land is allowed to grow, it is not flailed and usually develops into a dense sward or it is kept short by flailing (the latter is better for less competitive arable plants) and
- the removal of biomass is better than leaving it lying on the fallow land, as it decreases the nutrient availability and thus promotes less competitive arable plants.

More recently, so-called 'production-integrated compensation' measures have been introduced to support biodiversity conservation on arable land in Germany. These measures are financed by developers as ecological compensation for building on farmland (e.g. roads, housing, industrial complexes, wind energy plants) and can be used to promote endangered arable vegetation and other species of arable land through low-intensity production (e.g. skylarks or hamsters) (Druckenbrod and Meyer 2013; Druckenbrod and Beckmann 2018).

3.5 *Reintroduction of Rare and Endangered Arable Plants*

For about a decade now, the re-establishment of rare and endangered wild arable plants has been tested in Germany on selected sites. Initially, the focus was on projects on organically farmed fields, such as 'Field flower strips for the integration of autochthonous segetal plants in organically cultivated arable land' (Hotze et al. 2009) or through 'Reintroduction of rare and endangered arable wild plants on organic farms' (Wiesinger et al. 2015; Lang et al. 2016a, b). Field trials showed that early autumn sowing and low competition from crops yielded best results for three rare winter annual species of limestone arable sites (Forking larkspur, *Consolida regalis*, Large Venus's-looking-glass, *Legousia speculum-veneris* or Field gromwell, *Lithospermum arvense*). For the successful establishment of these rare arable plants, sowing without or with reduced density of winter cereals, such as spelt or rye, by mid-October at the latest is recommended. Sowing rates of 100 seeds/m^2 for *C. regalis* and *L. arvense*, and 50 seeds m^2 for *L. speculum-veneris* are recommended to achieve successful establishment with negligible crop yield losses (Lang et al. 2016b). Grass-clover leys and summer crops like peas allowed little or no emergence of the target species, although they survived to some extent in the seed bank. Seed mixtures were then sown in separate plots and the neighbouring cereal stand and the transfer of topsoil from species-rich areas was tested. In the year of sowing, a proportion of the introduced species developed into adult plants in both methods. This tended to be less successful when competing with cereals. In the following years, some species were again detected in cereal crops; most of the seeds reached deeper soil layers during tillage and enriched the seed bank (Lang et al. 2016a).

According to project team (Bayerische Kulturlandstiftung 2018) the following recommendations should be followed for the reintroduction of rare arable plants:

- there is species-specific variation, with highest success in reintroduction of *L. speculum-veneris*, followed by *C. regalis* and *L. arvense*.
- species lacking a long-term seed bank depend on suitable crop rotations that allow sufficient seed production each year. For winter annuals this means crop rotations containing high percentages of winter cereals, preferably without mechanical weeding.
- seed dispersal by field management is several meters per year. Thus, seeds should be sown on many plots spread over large parts of a reintroduction field.

Currently, several projects (e.g. Lang et al. 2018; Muchow and Fortmann 2019) are trying to establish an economically viable system for producing seeds of regional genotypes with certified provenance. This is especially important for endangered segetal flora species to reintroduce them to areas where they have largely died out and maintain viable populations to preserve them in the long term.

Several essential criteria have been developed for seed collection and sowing. Considering all herbaceous plants in Germany, 22 regions with similar gene pools have been defined. Seed is allowed to be propagated and distributed within these regions, but not between them, in order to prevent the homogenisation of the gene pool. It is important to note that these seed transfer zones are not species-specific, but are, to simplify practical implementation, equally applied to all species. To supply the market with regional seeds adapted to wider range of environmental conditions, the German system requires seeds to be collected from at least five large populations across a region (Prasse et al. 2010). The seed collection should be carried out in a sustainable way - only populations over 100 individuals should be harvested. In order to maintain the genetic diversity of the donor population, the seed collection should be distributed over at least 50 plants per population (Lang et al. 2018). The propagated seeds are available for restoration projects in a given seed zone, or are used for re-establishing another generation of cultivation. This procedure can be repeated for up to five generations, afterwards the seed production must start from a new wild collection (for details see Prasse et al. 2010). To balance the value of local adaptation with the need for future adaptation potential, Bucharova et al. (2019) propose this way of regional provenancing as a compromise strategy. Here, seeds are sourced from multiple populations within the same natural region as the target locality and mixed prior to use. The mixing of seeds will increase the genetic diversity necessary for future adaptation, while restricting seed origins to a regional scale will maintain regional adaptation and reduce the risk of unintended effects on other biota. This approach is feasible in practice and has recently been implemented in Bavaria and North Rhine Westphalia (Lang et al. 2018; Muchow and Fortmann 2019).

References

Albrecht H (1995) Changes in the arable weed flora of Germany during the last five decades. In: Proceedings of the 9th EWRS-symposium 'Challenges for weed science in a changing Europe' in Budapest, 28th – 30th of March 1995:41–48

Albrecht H, Cambecèdes J, Lang M, Wagner M (2016) Management options for the conservation of rare arable plants in Europe. Bot Lett 163:389–415

Bayerische Kulturlandstiftung (2018) Zwischenbericht 2018 über den Stand der Umsetzung im Projekt: Ackerwildkräuter für Bayerns Kulturlandschaft – Produktionsintegrierte Förderung seltener und gefährdeter Ackerwildkrautarten. Project Report, 185 p

Birrer S, Doppler V, Uebersax A (2018) Erhaltung und Förderung gefährdeter Schweizer Ackerbegleitflora – Schlussbericht 2017. Manuscript, 24 p

Bogenhard C (1850) Taschenbuch der Flora von Jena. Leipzig, 483 p

Bucharova A, Bossdorf O, Hölzel N, Kollmann J, Prasse R, Durka W (2019) Mix and match: regional admixture provenancing strikes a balance among different seed-sourcing strategies for ecological restoration. Conserv Genet 20(1):7–17

Buchli M (1936) Ökologie der Ackerunkräuter der Nordostschweiz. Beiträge zur geobotanischen Landesaufnahme der Schweiz 19: 354 p

Brunner AC, Gigon A (2001) Erhaltung und Förderung attraktiver Zwiebelpflanzen in Rebbergen der Nordostschweiz. Schweizerische Zeitschrift für Obst- und Weinbau 5:102–105

Dedek M, Wesche K (2017) Die Segetalflora der Oberlausitz im Wandel – heutige Situation im Vergleich zu historischen Daten von Max Militzer. Berichte der Naturforschenden Gesellschaft der Oberlausitz 25:83–106

Druckenbrod C, Beckmann V (2018) Production-integrated compensation in environmental offsets – a review of a German offset practice. Sustainability 10:4161

Druckenbrod C, Meyer S (2013) Produktionsintegrierte Kompensation (PIK) in Thüringen – landwirtschaftliche Akzeptanz und naturschutzfachliche Aufwertung von Ackerflächen. Landschpfl Natschutz Thüring 50(1):31–35

Ehmke W (2001) Das Artenhilfsprogramm für Ackerwildkräuter im Rheingau und Westtaunus - Teil 1: Grundlagen. Jahrbücher des Vereins für Naturkunde im Herzogtum Nassau 122:59–81

Frieben B, Prolingheuer U, Wildung M, Meyerhoff E (2012) Aufwertung der Agrarlandschaft durch ökologischen Landbau - Eine Möglichkeit der produktionsintegrierten Kompensation? (Teil 1). Natschutz Landschplan 44(4):108–114

Gerhards R, Dieterich M, Schumacher M (2013) Rückgang von Ackerunkräutern in Baden-Württemberg – ein Vergleich von vegetationskundlichen Erhebungen in den Jahren 1948/49, 1975–1978 und 2011 im Raum Mehrstetten – Empfehlungen für Landwirtschaft und Naturschutz. Gesunde Pflanz 65:151–160

Gottwald F (2010) Segetalflora. In: Stein-Bachinger K, Fuchs S, Gottwald F (eds.). Naturschutzfachliche Optimierung des Ökologischen Landbaus 'Naturschutzhof Brodowin'. Naturschutz und Biologische Vielfalt 90:98–105

Gottwald F, Peil J, Stein-Bachinger K, Flade M, Meyer S (2018) Wege im Ackerwildkrautschutz: Schutzäcker, Schonstreifen und ökologische Landwirtschaft -Erfahrungen und Perspektiven in Nordostdeutschland. Naturschutz und Biologische Vielfalt 164:233–253

Gradmann R (1950) Das Pflanzenleben der Schwäbischen Alb. Band 2, 4. Auflage. Stuttgart, 407 p

Halder U (1982) Aktion Kornblume. Eine Chance für die Natur. Informationsblatt des Schweizerischen Bund für Naturschutz (SBN) und Schweizer Heimatschutz (SHS), Zürich

Hanf M (1937) Pflanzengesellschaften des Ackerbodens (Untersuchungen über die Unkrautvegetation in der Umgebung von Halle/Saale). Pflanzenbau 13:449–476 & Pflanzenbau 14:29–48

Hegi G (1910) Illustrierte Flora von Mitteleuropa 3. Wien

Hilbig W (1978) Zur Veränderung der Segetalflora. In: Kulturbund der DDR, ZfA Botanik (ed.). Tagungsbericht von der II. Zentralen Tagung für Botanik - Florenwandel und Florenschutz - 23. - 24. April 1977 in Karl-Marx-Stadt, pp 68–75

Hilbig W (1985) Aufgaben und Ziele des Schutzes von Ackerwildpflanzen im Rahmen des Arten- und Biotopschutzes. Arch Natschutz Landschforsch 25:101–108

Hilbig W (1986) Bericht über eine Arbeitsexkursion der AG Ackerwildpflanzenschutz. Mitteilungen der Biologischen Gesellschaft der DDR 3/1986:22–24

Hilbig W (1998) Auswirkungen unterschiedlicher Intensitäten im Ackerbau auf die Ackerwildpflanzen (Segetalvegetation). Bodenkultur und Pflanzenbau, Schriftenreihe der Bayerischen Landesanstalt 2(1):23–39

Hilbig W (2007a) Die Haftdolden-Gesellschaft der Kalkäcker in Thüringen - Erinnerungen an einstige Blütenpracht. Landschpfl Natschutz Thüring 44(2):76–83

Hilbig W (2007b) Die Arbeitsgruppe "Ackerwildpflanzenschutz" in der Biologischen Gesellschaft der DDR. Studienarchiv Umweltgeschichte 12:3–15

Holzner W (1978) Ackerunkrautschutzgebiete in Niederösterreich. Manuscript, 4 p

Hotze C, van Elsen T, Haase T, Hess J, Otto M (2009) Ackerwildkraut-Blühstreifen zur Integration autochthoner Ackerwildkräuter in ökologisch bewirtschaftete Ackerflächen. In: Beiträge zur 10. Wissenschaftstagung Ökologischer Landbau in Zürich (Band 1), pp 426–429

Illig H (1990) Feldflora-Reservate als neue Form von Naturschutzgebieten. Abhandlungen und Berichte Naturkundemuseum Görlitz 64(1):99–101

Karkow K, Gronemann S (2005) Akzeptanz und Zahlungsbereitschaft bei den Besuchern der Ackerlandschaft. In: Hampicke U, Litterski B, Wichtmann W (eds) Ackerlandschaften. Nachhaltigkeit und Naturschutz auf ertragsschwachen Standorten. Springer, Berlin/Heidelberg, pp 115–128

Kohlbrecher C, Wesche K, Hilbig W, Leuschner C, Meyer S (2012) Veränderungen in der Segetalflora am Kyffhäusergebirge in den letzten 50 Jahren. Landschpfl Natschutz Thüring 49 (1):1–9

Kohler F, Vandenberghe C, Imstepf R, Gillet F (2011) Restoration of threatened arable weed communities in abandoned mountainous crop fields. Restor Ecol 19(101):62–69

Kolářová M, Tyšer L, Soukup J (2013) Impact of site conditions and farming practices on the occurrence of rare and endangered weeds on arable land in the Czech Republic. Weed Res 53:489–498

Krausch H-D (1978) Rade, Rade, rot … Nachruf auf ein Ackerunkraut. Rathenower Heimatkalender 22:69–74

Lang M, Prestele J, Wiesinger K, Kollmann J, Albrecht H (2018) Reintroduction of rare arable plants: seed production, soil seed banks and dispersal three years after sowing. Restor Ecol 26:170–178

Lang M, Albrecht H, Kollmann J, van Elsen T, Gärtner A, Hotze C, Fink S, Becher T, Lenerz M, Wiesinger K, Prestele J (2016a) Nature conservation by organic farming: Restoration of local populations of rare and endangered arable field plants on organic farms in Germany (in German). Final report, 159 p

Lang M, Prestele J, Fischer C, Kollmann J, Albrecht H (2016b) Reintroduction of rare arable plants by seed transfer. What are the optimal sowing rates? Ecol Evol 6:5506–5516

Legast M, Mahy G, Bodson B (2008) Les Messicoles Fleurs des Moissons. Coll Agri 1:124

Lenerz M, van Elsen T, Schneider S (2017) Ackerwildkrautschutz in Luxemburg durch Schutzäcker, Feldflorareservate und Ökologische Landwirtschaft. In: Wolfrum S, Heuwinkel W, Reents H J et al. (eds.) Ökologischen Landbau weiterdenken – Verantwortung übernehmen – Vertrauen stärken. Beiträge der 14. Wissenschaftstagung Ökologischer Landbau, Freising-Weihenstephan, 7.-10. März 2017, Verlag Dr. Köster, Berlin, pp 798–801

Litterski B, Jörns S (2004) Der Einfluss extensiven Anbaus von Winterroggen auf die Segetalflora und -vegetation - Eine Untersuchung in Nordostdeutschland. Z Pflanzenkr Pflanzenschutz, Sonderheft 19:65–72

Litterski B, Jörns S, Grabow M, Manthey M (2005) Extensiv bewirtschaftete Sandstandorte aus vegetationsökologischer Sicht. In: Hampicke U, Litterski B, Wichtmann W (eds) Ackerlandschaften. Nachhaltigkeit und Naturschutz auf ertragsschwachen Standorten. Springer, Berlin/Heidelberg, pp 191–206

Manthey M (2003) Vegetationsökologie der Äcker und Ackerbrachen Mecklenburg-Vorpommerns. Diss Bot 373: 209 p

Meisel K (1978/1979) Auswirkungen alternativer Landbewirtschaftung auf die Vegetation. Jahresbericht der Bundesforschungsanstalt für Naturschutz und Landschaftsökologie 1978/1979, Bonn

Meyer S, Bergmeier E, Becker T, Wesche K, Krause B, Leuschner C (2015) Detecting long-term losses at the plant community level – arable fields in Germany revisited. Appl Veg Sci 18 (3):432–442

Meyer S, Wesche K, Krause B, Leuschner C (2013a) Dramatic losses of specialist arable plants in Central Germany since the 1950s/60s – a cross-regional analysis. Divers Distrib 19 (9):1175–1187

Meyer S, Hilbig W, Steffen K, Schuch S (2013b) Ackerwildkrautschutz. Eine Bibliographie BfN-Skripten 351: 222 p

Meyer S, Leuschner C (eds.) (2015) 100 Äcker für die Vielfalt - Initiativen zur Förderung der Ackerwildkrautflora in Deutschland. Universitätsverlag Göttingen, 351 p

Militzer M (1960) Über die Verbreitung von Ackerunkräutern in Sachsen. Berichte der Arbeitsgemeinschaft sächsischer Botaniker N.F. 2:113–133

Muchow T, Fortmann L (2019) Konzepte zur Erhaltung gefährdeter Ackerwildkräuter auf Naturraumebene. Erfahrungen des Modellprojekts "Unkraut vergeht nicht – stimmt nicht!". Nat Landsch 94:202–210

Prasse R, Kunzmann D, Schröder R (2010) Development and practical implementation of minimal requirements for the verification of origin of native seeds of herbaceous plants (in German). In: Cooperation with Verband Deutscher Wildsamen- und Wildpflanzenproduzenten. DBU, reference no. 23931

Reidl D (2015) Ackerbau im Paznaun und in der Silvretta-Region (Österreich/Schweiz) - Segetalflora, Kulturtechnik, Geschichte - Dokumentation einer im Verschwinden begriffenen Landwirtschaftstradition. MSc. thesis University of Innsbruck, 144 p

Richner N, Linder HP, Holderegger R, Walter T (2014) Reviewing change in the arable flora of Europe: a meta-analysis. Weed Res 55(1):1–13

Richner N, Holderegger R, Linder HP, Walter T (2017) Dramatic decline in the Swiss arable flora since the 1920s. Agric Ecosyst Environ 241:179–192

Ritschel-Kandel G (1988) Die Bedeutung der extensiven Ackernutzung für den Arten- und Biotopschutz in Unterfranken. Schriftenreihe des Bayerischen Landesamtes für Umweltschutz 84:207–218

Rodi D, Schill G (1982) Feldflorareservat Beutenlay. In: Stadt Münsingen (ed.) Münsingen - Geschichte, Landschaft, Kultur. Sigmaringen, pp 659–672

Schmid RL (2008) Brachenmanagement in der Wiener Agrarlandschaft. Diploma thesis, University of Vienna, Austria, 160 p

Schopp-Guth A, Junck C, Sowa F, Naumann S, Schoos F (2006) Erfassung der seltenen Ackerwildkrautarten im Gebiet der Dogger-Hochflächen im Südwesten Luxemburgs. Bull Soc Nat luxemb 107:105–110

Schumacher W (1979) Untersuchungen zur Erhaltung seltener und gefährdeter Ackerwildkräuter durch extensive Bewirtschaftungsmaßnahmen. Verhandlungen der Gesellschaft für Ökologie 9:75–76

Seifert C, Leuschner C, Meyer S, Culmsee H (2014) Inter-relationships between crop type, management intensity and light transmissivity in annual crop systems and their effect on farmland plant diversity. Agric Ecosys Environ 195:173–182

Siciński JT (1986) Agro-reserves (Agricultural nature reserves) (in Polish) In: Summary of the 47th Meeting of the Polish Botanical Society Łódz 1986, p 19

Skalický V (1981) The questions of decline and weed extinction (in Czech). In: Holub J (ed.) The vanishing flora and protection of the gene pool in Czechoslovakia (in Czech). Studie Čsav 20:83–88

Stein-Bachinger K, Gottwald F, Haub A, Schmidt E (2020) To what extent does organic farming promote species richness and abundance in temperate climates. Organic Agric: 12 p

Stipa – Beratungsbüro Für Natur, Landschaft (2008) Kulturlandschaft Brentjong - Maßnahmen Ackerbegleitflora 2008: Vorprojekt. Projektbericht i. A. von Pro Natura Wallis und Gemeinde Leuk. Manuscript, 12 p

Storkey J, Meyer S, Still KS, Leuschner C (2012) The impact of agricultural intensification and land use change on the European arable flora. Proc R Soc B 279:1421–1429

Traxler A, Minarz E, Höttinger H, Pennerstorfer J, Schmatzberger A, Banko G, Placer K, Hadrobolec M, Gaugitsch H (2005) Biodiversitäts-Hotspots der Agrarlandschaft als Eckpfeiler für Risikoabschätzung und Monitoring von GVO. Report for the Umweltbundesamt, 159 p

Tüxen R (1962) Gedanken zur Zerstörung der mitteleuropäischen Ackerbiozönosen. Mitteilungen der Floristisch-soziologischen Arbeitsgemeinschaft, N.F. 9:60–61

van Elsen T (2000) Species diversity as a task for organic agriculture in Europe. Agric Ecosyst Environ 77:101–109

Van Elsen T, Berg M, Drenckhahn D, Dunkel FG, Eggers T, Garve E, Kaiser B, Marquart H, Pilotek D, Rodi D, Wicke G (2005) Karlstadter Positionspapier zum Schutz der Ackerwildkräuter. Natschutz Landschplan 9:284–286

Waldis R (1987) Unkrautvegetation im Wallis. Pflanzensoziologische und chorologische Untersuchungen. Beiträge zur geobotanischen Landesaufnahme der Schweiz 63: 348 p

Waldis-Meyer R (1978) Die Verarmung der Unkrautflora und einige Gedanken zu ihrer Erhaltung (mit besonderer Berücksichtigung des Wallis). Mitteilungen des Vereins für Forstliche Standortskunde und Forstpflanzenzüchtung 26:70–71

Waldis-Meyer R (1986) Über die Situation der Segetalflora in der Schweiz und Bestrebungen zu ihrer Erhaltung. Aus Liebe zur Natur 4:63–69

Warcholińska AU (1986) Anthropogenic endangerment of segetal weeds *Polycnemum arvense* L., *Herniaria hirsuta* L. and *H. glabra* L. and problems of their protection (in polish). Acta Universitatis Łódziensis – Folia Botanica 4:53–84

Warcholińska AU (2004) The segetal flora of Central Poland (in polish). Acta Agrobot 57 (1/2):327–352

Wiesinger K, Lang M, van Elsen T, Albrecht H, Prestele J, Kollmann J (2015) Wiederansiedlung seltener und gefährdeter Ackerwildkräuter im Biobetrieb. Booklet, 25 p. URL: http://www.lfl.bayern.de/mam/cms07/schwerpunkte/dateien/praxisbrosch%C3%BCre_ackerwildkraut.pdf

Willerding U (1986) Zur Geschichte der Ackerunkräuter Mitteleuropas. Göttinger Schriften zu Vor- und Frühgeschichte 22: 382 p

Chapter 6
The Status of Arable Plant Habitats in Eastern Europe

Gyula Pinke

Abstract Today large parts of Eastern Europe can be considered as strikingly species-poor "agrarian-deserts". Nevertheless, the region also retains relatively large areas of species-rich farmland. Changes in the weed flora in this region, with special regard to the disappearing weed species, is the subject of relatively small numbers of international scientific studies compared to the western part of Europe. The average weed species number per plot seems to have declined less in eastern than in western countries since the Second World War. However, by the turn of the Millennium the number of threatened weed species had increased considerably, which is apparent in the recently updated national weed red lists. Many studies indicate that lower farming intensity and diversified farming systems at higher altitudes provided better conditions for the occurrence of rare species and greater diversity than intensively farmed lowlands. Unfortunately, only a few traditionally managed small fields remain in extreme habitats, and they are continuously being abandoned. Regrettably, Eastern Europe mostly lacks any conservation initiatives which directly target the preservation of rare and threatened arable weeds, consequently further declines are anticipated.

Keywords Collectivisation · Land abandonment · Stubbles · Extensive farming · Marginal habitats · Urbanisation · Agricultural intensification

1 Setting the Scene

The current status of arable plant diversity in Eastern Europe is a product of two contrasting patterns. In this region, the agricultural intensification which dramatically changed the European weed flora after the Second World War (Storkey et al. 2012; Richner et al. 2015; Albrecht et al. 2016), was also associated with forced

G. Pinke (✉)
Faculty of Agricultural and Food Sciences, Széchenyi István University, Mosonmagyaróvár, Hungary
e-mail: pinke.gyula@sze.hu

C. Hurford et al. (eds.), *The Changing Status of Arable Habitats in Europe*,
https://doi.org/10.1007/978-3-030-59875-4_6

collectivization (Constantin and Bauerkämper 2013). In the newly consolidated, large fields of big cooperatives, industrialized methods of plant production quickly led to the rapid decline of weed diversity (Pinke and Pál 2005). At the same time, some countries like Poland and the former Yugoslavia escaped collectivisation and, in more or less remote Eastern European regions, traditional farming systems still remain. Consequently, although today large parts of Eastern Europe can be considered as strikingly species-poor "agrarian-deserts" with weed infestations of, at most, a few troublesome weeds, nevertheless, according to Sutcliffe et al. (2015) it also retains relatively large areas of species-rich farmland of international importance for weed conservation; fields in any transitional state between these two extremities also exist.

Changes in the weed flora in Eastern Europe, with special regard to the disappearing weed species, is the subject of relatively small numbers of international scientific studies compared to the western part of Europe. However, in Poland where farms managed to resist collectivization but generally could not avoid intensification, the response of weed communities gained the interest of many botanists. A decreasing trend in weed diversity was first reported by Kornas (1961) who highlighted the disappearance of specialized flax weeds (e.g. *Lolium remotum*, *Silene linicola*, *Camelina alyssum*, *Spergula maxima* and *Cuscuta epilinum*) in a mountainous district in southern Poland ten years after traditional flax growing had ceased. In the following decades these species were also no longer found within intensively cultivated flax crops and considered to be extinct in the entire country (Heller 2010). The main causes of broader declines in arable plants were roughly the same as elsewhere in Europe (Storkey et al. 2012) – Sicinski and Sieradzki (2010) identified the following factors as being particularly responsible for the changes in Poland: herbicides, mineral fertilizers, seed cleaning, simplification of crop rotation and intensive soil cultivation. According to Lazarski (2015) the introduction of new, more competitive, high-yielding varieties of cereals is an additional threat to rare weeds. High-yielding crops in conjunction with the use of mineral fertilizers result in higher crop densities which increases the shading of lower vegetation layers. This has resulted in species that are poor competitors for light, such as *Scandix pecten-veneris* (Fig. 6.1), disappearing from crop habitats. Skrajna et al. (2012) highlighted that on acidic soils, the number of sites of occurrence as well as population abundance of *Illecebrum verticillatum* and *Radiola linoides* have also drastically decreased after the turn of the Millennium in Eastern Poland. This retreat and the decline of other acidophilous species were associated with increasing nitrophilous species caused by the increase of intensive maize production in the region.

Early stubble ploughing has also been a decisive factor responsible for the declines of farmland biodiversity observed in some Eastern European cropping systems. Formerly, weeds on stubbles were grazed by livestock and also provided resources for bees, representing a high nature value type of farming. Since the 1950s, however, farms have tended to plough the stubble as soon as possible after the harvesting of cereals. As a consequence, the range and population size of weeds typical of stubble fields has reduced significantly, because these species fail to flower and set seeds. An example of one such typical stubble-weed species is *Stachys annua*

Fig. 6.1 *Scandix pecten-veneris*, an example of a weed that has dramatically declined in Poland partly owing to the increased competitiveness of crop canopies. (Photo G. Pinke)

(Fig. 6.2), which before agricultural intensification in Hungary frequently covered stubble fields in dominant stands during late summer and early autumn (Fig. 6.2). This species is an outstanding nectar-producer and it was one of the most important honey plants in Hungary providing beekeepers with profitable stubble-honey until the 1950s (Fig. 6.2). Studies revealed that where remnants of this habitat remain, it provides significant food sources for pollinating insects and farmland birds (Pinke and Pál 2009). Even though stubble weed communities are clearly of high conservation priority for biodiversity, the decline in the area it covers has dramatically accelerated in the last two decades, mainly due to the large expansion of the invasive competitive weed, *Ambrosia artemisiifolia* (Fig. 6.3). On the one hand this species is invading and overwhelming more and more habitats of *Stachys annua*. On the other hand, the allergenic pollen of this invasive plant has resulted in greater emphasis being placed on the importance of early ploughing of stubbles. Authorities can even impose financial penalties in case farmers fail to plough stubbles at the right time. The impact of this invasive plant is, therefore, impacting biodiversity through indirect effects on management as well as direct biotic effects.

Fig. 6.2 *Stachys annua* (*top left*), a species that has declined in Hungary because of the loss of stubbles and is a particularly valuable source of resource for pollinators (*top right*). (*Bottom*): Stubble field covered with *Stachys annua* (2005, Kálóz, Hungary). (Photos G. Pinke)

2 Historical Changes in Weed Communities in Eastern Europe

According to Richner et al. (2015) the average weed species number per plot seems to have declined less in eastern than in western countries since the Second World War. However, by the turn of the Millennium the number of threatened weed species had increased considerably in the region, which is apparent in the recently updated national weed red lists. The continental part of Croatia has 78 endangered weed species (Hulina 2005), while Slovakia has 150 (Eliáš et al. 2007), Poland 105 (Bomanowska 2010) and Hungary 149 (Pinke et al. 2011) threatened weed species, respectively. As reported by these lists, specialized flax weeds can be regarded extinct in all these countries, but mainly because of the modern seed

Fig. 6.3 Stubble field covered with *Stachys annua* with patch-like infestation of *Ambrosia artemisiifolia* (2011, Mosonudvar, Hungary). (Photo G. Pinke)

Fig. 6.4 Examples of weed species that have declined in Eastern Europe for different reasons discussed in the text: *Adonis flammea* (*left*) and *Papaver argemone* (*top middle*) have been selected against by the increased use of fertilisers and, for *P. argemone*, liming of more acidic soils. *Agrostemma githago* (*middle bottom*) has declined because of more efficient seed cleaning and *Lathyrus aphaca* (*right*) is threatened by expanding urbanisation in Poland. (Photos G. Pinke)

cleaning technologies specialised cereal weeds, like *Agrostemma githago* and *Bromus secalinus* (Fig. 6.4) are also among the most symbolic endangered weeds, with varying conservation status across the countries. Weeds, typical for acidic soils, like *Arnoseris minima* and *Papaver argemone* (Fig. 6.4) have become vulnerable mainly because of liming and the application of fertilizers. One of the most threatened and fascinating arable weed communities of this region is the Caucalidion alliance, with species like *Adonis flammea* (Fig. 6.4), *Caucalis platycarpos*, *Turgenia latifolia*, *Scandix pecten-veneris*, *Vaccaria hispanica*, *Bupleurum rotundifolium* and *Galium tricornutum*. These communities are distributed from the Balkans up to Poland, tending to become more specialized to the north, where they occupy ever more calcareous, warmer and drier habitats and become especially susceptible to the methods of modern agriculture (Pinke 2004; Wegrzynek and Nowak 2010, Silc et al. 2014). After 1950, the number of occurrences of this weed community decreased significantly, but surprisingly in the 1990s, some new sites for these weed species were recorded in the Czech Republic. This trend is presumably associated with a reduction in the application of herbicides and fertilizers after the socio-economic changes in 1989 (Lososová 2003).

A similar phenomenon has also been observed in Hungary, partly because of the chaotic changes in farming practices following the collapse of many large cooperatives in the post-communist period after 1990. The distribution of one of the most emblematic endangered weeds, *Agrostemma githago* significantly retreated after the 1950s; this once common arable weed had only 22 known localities in Hungary at the end of the 1980s. However, between 1995 and 2005 it was reported from 366 Hungarian localities. Although, it should also be noted that at this time field botanical research was also significantly boosted. Therefore, it is difficult to assess whether the temporary reduction in the intensity of farming or the more thorough surveys were to a greater extent responsible for this increase in observations (Pinke et al. 2006). Most of the new occurrences were discovered in extensively cultivated small fields from the hilly and sandy regions of Hungary, which mostly avoided the attention of botanists in the former decades. Unfortunately, there are no records of the share of permanent low intensity farming habitats, which already existed before the 1990s, and were not the temporary products of the economic crisis in the early 1990s that allowed the possible beneficial effects of changes in farming practice on arable weeds to be quantified with certainty.

3 Important Habitats for Arable Plant Conservation in Eastern Europe and Future Threats

Arable mosaics combined with grasslands and other semi-natural habitats usually result in the greatest arable plant diversity in Eastern Europe, particularly in hilly regions (Kovács-Hostyánszki et al. 2011; Loos et al. 2015). Mountainous areas tend to preserve traditional farming practices in contrast with the great plains, where crop production has been easier to intensify. Many studies indicate that lower farming intensity and diversified farming systems at higher altitudes provided better conditions for the occurrence of rare species and greater diversity than intensively farmed

Fig. 6.5 Example of a traditional farming landscape (2007, Zagradi, Albania) that can still be found in parts of Eastern Europe and are particularly important for conserving arable plant communities. (Photos G. Pinke)

lowlands. This trend has been reported by Vrbnicanin et al. (2009) in Serbia, by Kolarová et al. (2013) in the Czech Republic, by Nagy et al. (2018) in Romania and by Májeková et al. (2019) in Slovakia. Barina et al. (2017) reported that, in Albania, both at higher and lower altitudes small cultivated fields are still typical (Fig. 6.5) and weeds that have become threatened at a European scale, like *Agrostemma githago*, *Vaccaria pyramidata*, *Scandix pecten-veneris* and *Adonis flammea* are still relatively frequent. Further eastwards outside Europe, closer to the centre of their distribution area, many of these weed species can still occur even in large populations (Nowak et al. 2014).

In extensive, or in other words in the "traditional low-intensity farming systems" intensification of arable land use is usually unprofitable due to extreme soil conditions or difficult topography. In these habitats the absence or low use of agrochemicals benefits a wide range of flora and fauna throughout the food chain (Beaufoy et al. 1994). To validate this assumption, the floristic composition and conservation value of the weed vegetation of winter cereal fields on base-rich soils in western Hungary was compared between extensively and intensively managed fields. The findings showed that total weed cover was, on average, six times higher in extensively compared with intensively managed fields (Pinke and Gunton 2014). In addition, Red List and insect-pollinated plant species occurred more frequently in extensively managed fields (Pinke et al. 2009). Community responses to arable intensification using functional trait analyses were also characterized. The response of weeds to intensification in these cereal fields was best described by a functional classification based on species' flowering duration, maximum height and seed weight. One of the groups particularly associated with the extensive fields is characterized by medium stature, the largest seeds and the shortest flowering duration. This could be identified as a particular conservation concern and includes many

Fig. 6.6 Two species of *Melampyrum, M. arvense* (*left*) and *M. barbatum* (*right*) that are only found in extensively managed fields and display a rare weed traits syndrome of large seed, intermediate stature and short flowering duration. (Photos G. Pinke)

of the most threatened and declining arable weed species in Hungary (e.g. *Agrostemma githago*, *Caucalis platycarpos*, *Galium tricornutum*, *Ranunculus arvensis*, *Turgenia latifolia*, *Melampyrum arvense* and *M. barbatum* (Fig. 6.6)). The medium stature reflects competitive exclusion of weeds of similar height to cereal crops in intensive fields, seed size is negatively correlated with persistence in the seed bank and low seed production, and short flowering period also represents a short phenology making this species unable to re-emerge after herbicide applications (Pinke and Gunton 2014). This shows some congruence with the rare weed trait syndrome of short stature, large seeds and late flowering proposed by Storkey et al. (2010) – see also Chap. 2 in this volume.

These results show that extensively managed fields have greater conservation value compared with intensively managed fields. Therefore, extensively managed arable habitats should have a high priority for biodiversity conservation. Unfortunately, only a few traditionally managed small fields remain in extreme habitats (Fig. 6.7), and they are continuously being abandoned. Extensive fields are usually owned by old farmers, who grow cereals to feed their own livestock. There is no market orientation in their farm business and they do not receive any financial support for this environment-friendly cropping. After the demise of this older generation, who are the last members of the traditional peasant culture, there will be no younger generation continuing this traditional lifestyle. Extensive fields are, therefore, likely to be abandoned (Fig. 6.8), intensified or built up. This is a dramatically accelerating trend; more than 90% of the extensive fields, which were sampled for the study of Pinke et al. (2009) in 2007, were lost in the following decade in North-Western Hungary. Expanding Hungarian towns are also

Fig. 6.7 Last remnants of extensively managed fields in Western Hungary with rich arable floras including abundant populations of *Centaurea cyanus* in Öskü taken in 2007 (*top left*) and 2011 (*top right*), *Melampyrum barbatum* and *Anthemis austriaca* (*bottom left*) and *Centaurea cyanus*, *Agrostemma githago* and *Anthemis austriaca* (*bottom right*) taken in Öskü 2019. (Photos G. Pinke)

Fig. 6.8 Abandoned extensive fields in Western Hungary (2019, Öskü) (*left*) and an expanding town incorporating once extensively managed small fields in its vicinity (*Anthemis austriaca* at the building site; 2017, Mosonmagyaróvár, Hungary) (*right*). (Photos G. Pinke)

increasingly incorporating small fields in their vicinity (Fig. 6.8), while around villages once extensively managed fields turn to be more and more intensified or abandoned. A considerable part of the Eastern European region is likely to be involved in this land use trend with negative impacts on arable plant habitats. For instance, Nobis et al. (2011) reported that although extensively managed fields provided suitable habitats for the threatened *Lathyrus aphaca* (Fig. 6.4), a species

Fig. 6.9 Oil pumpkin field with massive populations of *Chenopodium album* and *Persicaria lapathifolia* (2015, Vasszentmihály, Hungary). (Photo G. Pinke)

of Mediterranean-Irano-Turanian origin around the Polish city of Kraków; the progressive development of the city, and especially by numerous building investments as well as the discontinuation of cultivation threaten its survival. Mehmeti et al. (2009) noted that the abandonment of cultivation, especially on shallow calcareous soils threaten the existence of rare weed species in Kosovo. According to Öllerer (2013) small-sized subsistence household farms in Romania are continuously being abandoned due to economic constraints. Furthermore, Batáry et al. (2015) highlighted that Eastern European agriculturally marginal areas, where the productivity of land is limited by biophysical or socioeconomic constraints and are currently home to the highest concentrations of biodiversity and host the largest populations of threatened species are also under the highest pressure from agricultural intensification and abandonment. Storkey et al. (2012) also pointed out that the factor most commonly identified as causing national declines in arable floras in Eastern Europe was the abandonment of marginal land.

There are some examples of specialised cropping systems that remain important refugia for threatened plant species. Before the intensification of rice production, flooded paddy fields in the region provided habitat for a great diversity of wetland species, which have also drastically diminished since the 1960s. Nevertheless, studies suggest that waterlogged arable fields, including temporary pools and rice paddy fields even today can harbour many endangered plants (e.g. *Elatine triandra*, *E. alsinastrum*, *E. hungarica*, *Lindernia procumbens*, *Alisma gramineum*) and high species diversity, especially when the fields are cultivated less intensively (Lukács et al. 2013; Pinke et al. 2014). The weed vegetation of oil pumpkin fields in the region can be also remarkable, because it is often cultivated in "eco-friendly" ways. Even in conventional farms growers usually rely on mechanical weed management rather than herbicides. They regularly do not control weeds from midsummer, partly because they consider them beneficial, as they can provide some shelter from wind and heliosis for the ripening pumpkin fruits (Pinke et al. 2018). Although most of the weed flora are recruited from the commonest plants in these fields, they can possibly supply farmland birds with important seed food sources and cover (Fig. 6.9).

4 The Future of Arable Plant Habitats in Eastern Europe

Unfortunately, Eastern Europe lacks any conservation initiatives which directly target the preservation of rare and threatened arable weeds. Even in Poland, where scientists have payed attention to this issue for a long time (Dostatny 2004, Sicinski and Sieradzki 2010), the concepts of arable conservation areas in national and landscape parks, as well in botanical gardens and open air museums have still remained largely unimplemented proposals. In Hungary, the applied agri-environment schemes do contain certain elements that indirectly could be beneficial for the weed flora, although not directly targeted at these taxa. They include subsidies for long lasting green fallow as well as the bee-foraging margins and farmland bird protection zones (Sztahura and Rezneki 2015). Although these programs are benign in terms of arable plants, they are not focussing on the last remnants of the vanishing extensively managed fields, thus they fail to account for and preserve the most important "biodiversity hotspots" of rare and threatened arable weeds. Among the applied conservation options reviewed by Albrecht et al. (2016) the most appropriate way to save these habitats in Eastern Europe would be to create new programs similar to the German project "100 Fields for Diversity", in which a sustainable network of sanctuaries for permanent conservation of endangered arable weed species is established through long lasting maintenance of extensive farming practices (Meyer et al. 2010) – see Chap. 5 in this volume. Finally, it should be noted again that agricultural biodiversity and its conservation strategies in Eastern Europe is also relatively under-represented in the international literature (Sutcliffe et al. 2015). In addition, the adopted western European type agri-environment schemes seem to be generally ineffective in this region and there is a great need for better locally adapted strategies (Babai et al. 2015; Sutcliffe et al. 2015; Reif and Hanzelka 2016; Spulerová et al. 2017). Sutcliffe et al. (2015) also emphasised that the biodiversity benefits for Europe of existing low-intensity farmland, particularly in the eastern states, should be harnessed before they are lost as they represent important biodiversity hotspots at the European scale. Instead of waiting for species-rich farmland to further decline, targeted research and monitoring to create locally appropriate conservation strategies for these habitats is needed now.

Acknowledgments The work of Gyula Pinke was supported by the EFOP-3.6.3-VEKOP-16-2017-00008 "Innovative Scientific Institutions in Domestic Agricultural Higher Education" project. The project is co-financed by the European Union and the European Social Fund.

References

Albrecht H, Cambecèdes J, Lang M, Wagner M (2016) Management options for the conservation of rare arable plants in Europe. Bot Lett 163:389–415

Babai D, Tóth A, Szentirmai I et al (2015) Do conservation and agri-environmental regulations effectively support traditional small-scale farming in East-Central European cultural landscapes? Biodivers Conserv 24:3305–3327

Barina Z, Mullaj A, Pifkó D et al (2017) Distribution atlas of vascular plants in Albania. Hungarian Natural History Museum, Budapest

Batáry P, Dicks LV, Kleijn D, Sutherland WJ (2015) The role of agri-environment schemes in conservation and environmental management. Conserv Biol 29:1006–1016

Beaufoy G, Baldock D, Clark J (1994) The nature of farming: low intensity farming systems in nine European countries. Institute for European Environmental Policy, London

Bomanowska A (2010) Threat to arable weeds in Poland in the light of national and regional red lists. Plant Breed Seed Sci 61:55–74

Constantin I, Bauerkämper A (2013) The collectivization of agriculture in communist Eastern Europe: comparison and entanglements. CEU Press, Budapest

Dostatny DF (2004) Preservation of weed diversity in protected areas. Bull Bot Gard 13:79–83

Eliáš P, Eliáš P, Baranec T (2007) The new red list of Slovak endangered weeds. In: Proceedings of traditional agroecosystems – 1st international conference. Threatened Weedy Plant Species. Nitra, Slovakia, pp 23–28

Heller K (2010) "Flax specialists" – weed species extinct in Poland? Plant Breed Seed Sci 61:36–40

Hulina N (2005) List of threatened weeds in the continental part of Croatia and their possible conservation. Agric Conspec Sci 70:37–41

Kolarová M, Tyser L, Soukup J (2013) Impact of site conditions and farming practices on the occurrence of rare and endangered weeds on arable land in the Czech Republic. Weed Res 53:489–498

Kornas J (1961) The extinction of the association *Sperguleto-Lolietum remoti* in flax cultures in the Gorce (Polish Western Carpathian Mountains). Bulletin De L'Academie Polonaise des Sciences, Ser Sci Biol 9:37–40

Kovács-Hostyánszki A, Batáry P, Báldi A, Harnos A (2011) Interaction of local and landscape features in the conservation of Hungarian arable weed diversity. Appl Veg Sci 14:40–48

Lazarski G (2015) *Scandix pecten-veneris* L. (Apiaceae) in the Małopolska Upland (S Poland) – regional changes in the distribution and population resources of declining weed species. Biodivers Res Conserv 40:13–20

Loos J, Turtureanu PD, von Wehrden H et al (2015) Plant diversity in a changing agricultural landscape mosaic in Southern Transylvania (Romania). Agric Ecosyst Environ 199:350–357

Lososová Z (2003) Estimating past distribution of vanishing weed vegetation in South Moravia. Preslia 75:71–79

Lukács BA, Sramkó G, Molnár A (2013) Plant diversity and conservation value of continental temporary pools. Biol Conserv 158:393–400

Májeková J, Zaliberová M, Skodová I (2019) Weed vegetation of arable land in Slovakia: diversity and species composition. Acta Soc Bot Pol 88:3637

Mehmeti A, Demaj A, Waldhardt R (2009) Plant species richness and composition in the arable land of Kosovo. Landsc Online 11:1–29

Meyer S, Wesche K, Leuschner C et al (2010) A new conservation strategy for arable weed vegetation in Germany – the project "100 fields for biodiversity". Plant Breed Seed Sci 61:25–34

Nagy K, Lengyel A, Kovács A et al (2018) Weed species composition of small-scale farmlands bears a strong crop-related and environmental signature. Weed Res 58:46–56

Nobis A, Nobis M, Urbisz A (2011) *Lathyrus aphaca* L.: the distribution, habitats and remarks on the status of the species in Poland. Acta Soc Bot Pol 80:237–244

Nowak A, Nowak S, Nobis M, Nobis A (2014) A report on the conservation status of segetal weeds in Tajikistan. Weed Res 54:635–648

Öllerer K (2013) On the spatio-temporal approaches towards conservation of extensively managed rural landscapes in Central-Eastern Europe. J Land Ecol 6:32–46

Pinke G (2004) Last occurrence of *Caucalion* species in North-Western Hungary. J Plant Dis Prot Spec Issue 19:73–82

Pinke G, Gunton RM (2014) Refining rare weed trait syndromes along arable intensification gradients. J Veg Sci 25:978–989

Pinke G, Pál R (2005) Gyomnövényeink eredete, termőhelye és védelme. Alexandra Kiadó, Pécs

Pinke G, Pál R (2009) Floristic composition and conservation value of the stubble-field weed community, dominated by *Stachys annua* in western Hungary. Biologia 64:279–291

Pinke G, Pál R, Schmidt D et al (2006) The current occurrences of *Agrostemma githago* in Hungary. Magyar Gyomkutatás és Technológia 7:63–81

Pinke G, Pál R, Botta-Dukát Z, Chytry M (2009) Weed vegetation and its conservation value in three management systems of Hungarian winter cereals on base-rich soils. Weed Res 49:544–551

Pinke G, Király G, Barina Z et al (2011) Assessment of endangered synanthropic plants of Hungary with special attention to arable weeds. Plant Biosyst 145:426–435

Pinke G, Csiky J, Mesterházy A et al (2014) The impact of management on weeds and aquatic plant communities in Hungarian rice crops. Weed Res 54:388–397

Pinke G, Karácsony P, Czúcz B, Botta-Dukát Z (2018) When herbicides don't really matter: weed species composition of oil pumpkin (*Cucurbita pepo* L.) fields in Hungary. Crop Prot 110:236–244

Reif J, Hanzelka J (2016) Grassland winners and arable land losers: the effects of post-totalitarian land use changes on long-term population trends of farmland birds. Agric Ecosyst Environ 232:208–217

Richner N, Holderegger R, Linder HP, Walter T (2015) Reviewing change in the arable flora of Europe: a meta-analysis. Weed Res 55:1–13

Sicinski J, Sieradzki J (2010) Protection of segetal flora and vegetation in Poland (historical outline). Plant Breed Seed Sci 61:123–131

Silc U, Lososová Z, Vrbnicanin S (2014) Weeds shift from generalist to specialist: narrowing of ecological niches along a north-south gradient. Preslia 86:35–46

Skrajna T, Kubicka H, Rzymowska Z (2012) *Illecebrum verticillatum* L. – endangered species in agrocenoses of eastern Poland: assessment of ecological and genetic indicators for protection goal. Pol J Ecol 60:577–589

Spulerová J, Bezák P, Dobrovodská M et al (2017) Traditional agricultural landscapes in Slovakia: why should we preserve them? Landsc Res 42:891–903

Storkey J, Moss SR, Cussans JW (2010) Using assembly theory to explain changes in a weed flora in response to agricultural intensification. Weed Sci 58:39–46

Storkey J, Meyer S, Still KS, Leuschner C (2012) The impact of agricultural intensification and land-use change on the European arable flora. Proc R Soc B Biol Sci 279:1421–1429

Sutcliffe LME, Batáry P, Kormann U et al (2015) Harnessing the biodiversity value of Central and Eastern European farmland. Divers Distrib 21:722–730

Sztahura E, Rezneki R (2015) Agrár-környezet-gazdálkodás. Nemzeti Agrárgazdasági Kamara, Budapest

Vrbnicanin S, Stevanovic ZD, Radovanov KJ, Uludag A (2009) Weed vegetation of small grain crops in Serbia: environmental and human impacts. Turk J Agric For 33:325–337

Wegrzynek B, Nowak T (2010) Rare and endangered segetal weed species in the Silesian Upland (s Poland) recorded in the last twenty years. Plant Breed Seed Sci 61:75–84

Chapter 7
The Arable Flora of Mediterranean Agricultural Systems in the Iberian Peninsula: Current Status, Threats and Perspectives

Jordi Recasens, Alejandro Juárez-Escario, Bàrbara Baraibar, and Xavier O. Solé-Senan

Abstract In the last decades, the diversity of the arable flora from the Mediterranean farming systems of the Iberian Peninsula has declined significantly as a result of agricultural intensification and changes in soil management, land use and landscape structure. Some of the arable plants that belong to these segetal communities are representatives of chorological elements with high ecological relevance including: Mediterraneo-Irano-Turanian element, Mediterranean arid element, Ibero-Maghrebian element and Iberian endemism element. The conservation of this specific, highly ecologically significant flora depends on the preservation of those specific habitats where these species are still present. This involves actions at both field and landscape scales. Such measures include the promotion of low-intensity organic farming practices (since they increase richness and diversity of arable plants), the preservation of well-structured margin systems (since they act as a shelter for these species) and the fostering of those landscape structures, both at compositional and configurational level, that favour the conservation of arable plants. From a European perspective, Mediterranean agricultural systems still constitute a significant reservoir of diversity of the arable flora and every effort must be made to preserve them.

Keywords Spain · Portugal · Rare Arable Plants (RAP) · Field margins · Landscape heterogeneity · Fertilisers · Herbicides

---

J. Recasens (✉) · A. Juárez-Escario · B. Baraibar · X. O. Solé-Senan
Research Group of Weed Science and Plant Ecology, Department of Horticulture, Botany and Landscaping, University of Lleida, Lleida, Catalonia, Spain
e-mail: jordi.recasens@udl.cat

C. Hurford et al. (eds.), *The Changing Status of Arable Habitats in Europe*,
https://doi.org/10.1007/978-3-030-59875-4_7

1 Features of Agricultural Systems in the Western Mediterranean Area

Agriculture in a vast area of the Iberian Peninsula is conditioned by the Mediterranean climate, which is characterised by mild and relatively wet winters, few frosts, hot summers, sometimes very dry, and irregular and stormy rainfall. Following the classification of Trewartha modified by Fernández-Suberviola (2016), the different types and subtypes of Mediterranean climate (Warm Temperate climate Csa and Csb, and Temperate climate Dsa and Dsb) (Fig. 7.1), dominate most of the dry areas of the Iberian Peninsula where the agricultural systems are limited to traditional crops such as winter cereals, vineyards, olives, almonds, and occasionally also to dry pastures. Other crops such as orchards, alfalfa, maize, rice or vegetables are found in irrigated areas, some of them representing historical land use (more than 100 years) and others more recent transformations. In the rainfed areas of Spain, winter cereals are the main crop, whereas in Portugal, the surface dedicated to winter cereals has declined significantly over the last 30 years being substituted by olive groves. In 2018, in Spain, barley represented a 38.7% of the total grain production (with more than 2.5 million ha) followed by soft wheat (29%) (1.7 million ha) and durum wheat (6.7%) (3.7 million ha) (MAPA 2019). In Portugal, in 2018, winter cereals overall only represented around 50,000 ha (close to a 17% of total grain production) (INE 2020).

Traditionally, the prevalent cropping system in the region was called "año y vez" that is a cereal – fallow rotation, with tillage before cereal sowing (Fig. 7.2). This system allowed the land to rest for a year, which was a practical necessity due to the limited possibilities of adding organic or chemical fertilisers. Nowadays, the fallow phase of the rotation is disappearing unless farmers are forced by European policy, and the cereal monocrop is becoming increasingly frequent.

In a similar way to other parts of the world, the advent of the green revolution in the 1960s resulted in the intensification of the production systems of Spain and Portugal. Fertiliser and herbicide use started to increase gradually until their use

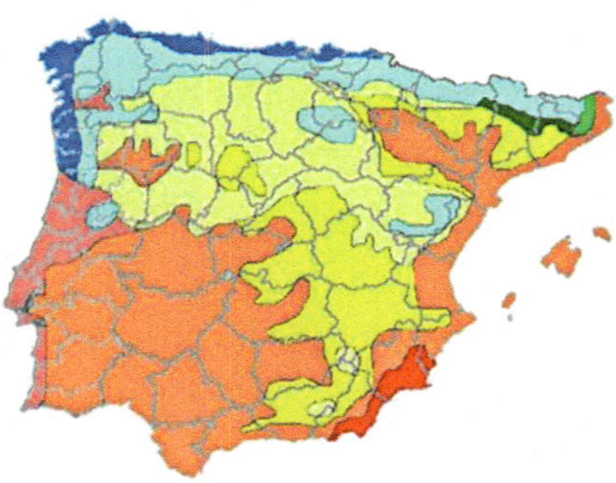

Cfa		Warm temperate rainy climate of hot summers
Cfb		Mild warm rainy climate of mild summers
Csa		Warm temperate climate of dry and hot summers; typical Mediterranean climate
Csb		Warm temperate climate of dry and mild summers: Mediterranean oceanic climate
Dsa		Temperate climate of dry and hot summers (Mediterranean Continental Climate)
Dsb		Temperate climate of dry and fresh summers (Climate sub-Mediterranean Continental)
Doa		Mild oceanic climate of hot summers
Doa		Temperate oceanic climate of fresh summers
BSh		Arid Climate

Fig. 7.1 Climates of Spain and Portugal (excluding Atlantic islands) according to the classification of Trewartha and modified by Fernández-Suberviola (2016)

Fig. 7.2 Landscape from Teruel province, where the traditional system called "año y vez" (cereal – fallow rotation) is still implemented. (Photo J. Recasens)

peaked around the 1980s and 1990s. At the same time, rotations started to be simplified and the fallow years were replaced by continuous cropping, which was only possible by relying on fertilisers and pesticides. Agricultural intensification also brought changes at the landscape scale, where field sizes gradually increased at the expenses of field margins and edges, which are now very scarce and poor. This homogenization of the landscape is still an ongoing process, especially in Spain, and is mainly motivated by the expansion of the irrigated areas and the associated programs of land consolidation. Access to irrigation has caused shifts in crop rotations and an increase in the use of agricultural inputs throughout the Mediterranean area.

In the last couple of decades, soil management and especially tillage intensity has also changed in Spain and Portugal. Conservation agriculture and no-till systems have rapidly increased in rainfed areas and, even if adoption is still very variable across regions, for winter cereals it now represents more than 600,000 ha (close to 10% of the area of this crop in Spain) (AEAC 2020). Conservation agriculture in rainfed systems where rainfall is scarce, largely unpredictable and frequently stormy has multiple agronomic and environmental benefits such as a better soil water retention and more water availability, a reduction in erosion and soil evaporation, as well as economic benefits such as fuel and time savings (Cantero-Martínez et al. 2007). However, the role of tillage as a primary weed management tool has been replaced by an increasing use of herbicides, which has caused a shift in weed community composition and a higher incidence of pernicious weeds (Alvaro Fuentes et al. 2007; Arrúe et al. 2007). Furthermore, the increasing reliance on herbicides as the only weed management tool has resulted in an increase of herbicide resistant weeds in winter cereals (SEMh 2020).

2 Current Status of Arable Floras in Spain and Portugal

Several authors have highlighted a decline of the arable flora throughout the Mediterranean region during the last century (Wilson and King 2003; Baessler and Klotz 2006; Storkey et al. 2012). As a result of the decline, some species are currently threatened or endangered whereas a small number of species have become dominant and well adapted to new management strategies (Rotchés-Ribalta 2015b; Chamorro et al. 2016). This community shift has great ecological relevance because arable plant communities in the Mediterranean area, such as in the Iberian Peninsula, support higher plant diversity than those in Central and Northern Europe (José-María et al. 2010; Solé-Senan et al. 2014; Sirami et al. 2019). A particular set of arable plant species, also called 'segetal' species, have suffered the most drastic population declines in recent years, to the point of becoming rare arable plants (RAP) or even locally extinct in many countries. However, it is still possible to find localised areas where populations of these species are still abundant (Solé-Senan et al. 2014; Torra et al. 2020) – Fig. 7.3.

The lack of specific weed surveys at the national scale makes it difficult to estimate the response of the arable flora to the changes in arable habitats in recent decades, and even more to evaluate the status of rare, vulnerable or probably locally extinct of arable plants. The last edition of the Red List of Spanish Vascular Flora (Bañares et al. 2010) and the practically finished Red List of Vascular Flora of Continental Portugal (SPB – PHYTOS 2020) do not include any arable plants. Furthermore, those Red Lists published at regional level, for example in Andalusia (Cabezudo et al. 2005), Catalonia (Sáez et al. 2010), or Aragón (Alcántara 2007), do not consider any arable plants either. There are two explanations for this. On the one hand, Red Lists are generally compiled by expert botanists focused on the study of natural systems, and probably with scarce information of ecological processes inherent to agricultural systems. On the other hand, the species included as arable plants belong to a traditional group of species considered as noxious weeds, requiring a philosophical shift in thinking when considering them as vulnerable or having a conservation interest.

Despite the lower number of published works, the decline of plant diversity of the arable plant communities in the Mediterranean area seems to be consistent with the declines observed in central and northwest Europe (Sutcliffe and Kay 2000; Baessler and Klotz 2006; Van Calster et al. 2008; Fried et al. 2009). In the case of arable habitats of the Iberian Peninsula only studies done at regional or at local scales are available. Guillerm and Maillet (1982) accounted for near 550 weed species in

Fig. 7.3 Crop edge in a cereal field in the Jiloca county (Teruel) with *Agrostemma githago*, *Consolida orientalis* and *Bupleurum rotundifolium*. (Photo J. Recasens)

winter cereals in Spain and reported significant modifications in the floristic composition of segetal flora compared to the surveys made by Braun-Blanquet and Bolòs 25 years before (Braun-Blanquet and Bolòs 1957). Recasens and Conesa (2004) described the status of arable weeds in the Ebro Valley confirming that a large percentage of these species are endangered or in a critical situation. Cirujeda et al. (2011), in a survey carried out in Aragon comparing the arable flora recorded in 1976 and 30 years later, found a significant decrease in the mean number of arable plant species, with three key results: (a) a striking increase of grass weeds, (b) a high decrease in the average species richness per field, declining from 9 to 3, and (c) a decrease in the frequency of many weed species. Similarly, Chamorro et al. (2016), in a comparative survey carried out in arable fields in Central Catalonia, reported a loss of 47% of species from 1953 to 2007, being even more dramatic for the considered subsets -according to Bolòs and Vigo (1984)- of segetal and rare species (75% and 87%, respectively). Furthermore, Ramôa et al. (2013) identified 264 different species in cereal fields in Alentejo (Portugal) but, in general, with a low level of frequency with only three grass species present in more than half of surveyed fields.

The arable flora from the western Mediterranean region includes species from different chorological groups (Bolòs and Vigo 1984). Some species with a wide distribution, included in the pluri-regional or cosmopolitan group (such as

Androsace maxima, Ceratocephala falcata, Agrostemma githago, Vaccaria hispanica, Papaver argemone, Turgenia latifolia, Adonis annua, Adonis aestivalis or *Asperula arvensis*) are rare or endangered, similar to other European countries (Wilson and King 2003; Storkey et al. 2012). But the arable fields of the Iberian Peninsula also host a specific flora that is also present in the surrounding therophitic wastelands, with a more restricted chorology and, in consequence, with a greater ecological significance (Recasens and Conesa 2004) – Fig. 7.4. These species are: (a) species from the Mediterraneo-Irano-Turanian element as *Cerastium perfoliatum, Garidella nigellastrum, Hypecoum pendulum, Roemeria hybrida, Silene conoidea, Centaurea depressa, Consolida orientalis* and *Velezia rigida*; (b) species belonging to the Mediterranean arid element: *Consolida pubescens, Euphorbia helioscopia* ssp. *helioscopodioides* and *Nonea echioides*; (c) species from the Ibero-Maghrebian element: *Biscutella auriculata, Delphinium gracile, Diplotaxis virgata, Erysimum incanum, Nonea micrantha, Scandix australis* ssp. *microcarpa* and *Silybum eburneum*; and also (d) Iberian endemisms: *Sisymbrium assoanum*, and *Valerianella multidentata.*

Before the advent of herbicides, some of these arable weeds were abundant in winter cereals of Spain and Portugal, such as *Centaurea cyanus, Vaccaria hispanica, Adonis flammea* and *Agrostemma githago.* The latter was specifically problematic due to the toxicity that their seeds could cause if milled with the wheat grain (Guillerm and Maillet 1982). This species was one of the first to be targeted by the first herbicides applications carried out in the middle of the twentieth century (Fig. 7.5).

3 The Role of Field Margins

Field margins perform important ecological functions in agroecosystems providing suitable habitat for wildlife and connecting habitats across the landscape. Particularly in conventional managed fields, margins tend to support higher overall plant species richness (Marshall 1989; Romero et al. 2008) as well as a higher presence of rare arable plants (Wilson and Aebischer 1995; Fried et al. 2009) compared to the field interior (Fig. 7.6). In field margins arable plants benefit from increased light availability and reduced management intensity (Kleijn and Van der Voort 1997). In Mediterranean arable fields, margins are mainly composed of a boundary strip and the adjacent field edge (Romero et al. 2008; Solé-Senan et al. 2014) creating a gradient of plant diversity from the inner field to the boundary (see Fig. 7.7). In the field edge, the lower level of soil disturbance, fertilisation and efficacy of herbicide application determine the richness and diversity of arable plants, whereas in the boundary the presence of arable flora is largely determined by the boundary structure itself. Cirujeda et al. (2019) describe five types of boundaries in different regions of Spain, ranging from structurally simple (flat, narrow, dominated by annual species) to structurally complex (with the presence of a bank, more than 3 m wide and dominated by perennials and trees). These authors pointed out that the presence of

Fig. 7.4 Some of the arable plants that belong to chorological elements that are particularly characteristic of Mediterranean arable fields. From *left* to *right* and from *top* to *bottom*: *Garidella nigelastrum, Velezia rigida, Silene conoidea, Hypecoum pendulum, Roemeria hybrida, Consolida orientalis, Consolida pubescens, Delphinium gracile, Biscutella auriculata, Centaurea depressa, Nonea micrantha, Cerastium perfoliatum, Sisymbrium assoanum, Valerianella multidentata* and *Euphorbia helioscopia* ssp. *helioscopodioides*. (Photos from J. Recasens except *G. nigellastrum, V. multidentata* and *E. helioscopia* ssp. *helioscopodioides* from J. A. Conesa and *V. rigida* from J. Pedrol)

Fig. 7.5 Historical picture of a farmer prepared to apply herbicide 2,4-D against *Agrostemma githago* in cereal fields in North of Lleida province in 1967. (Photo from A. Bigatà)

Fig. 7.6 Field edge and boundary of an organic farming in Valle del Jarque (Teruel) with presence of *Vaccaria hispanica*, *Consolida orientalis*, *Bupleurum rotundifolium*, *Turgenia latifolia* and *Centaurea depressa* in the field edge. (Photo from J. Recasens)

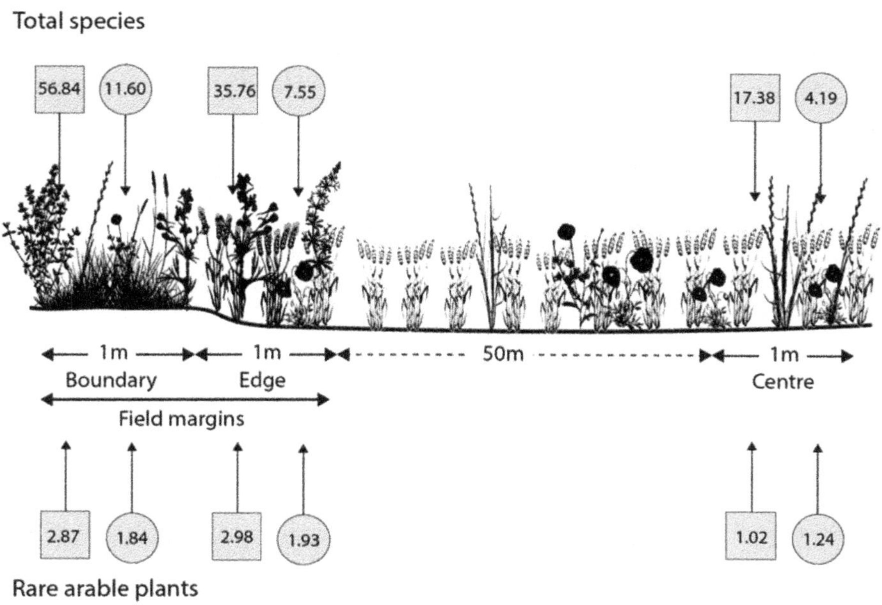

Fig. 7.7 Mean values of species richness (squares) and Shannon diversity -Shannon entropy index sensu Jost (2007)- (circles) per field for both total species and rare arable plants depending on the field position. Results are computed from the species recorded in a survey of 90 arable fields in Catalonia. (Adapted from Solé-Senan et al. 2014)

RAP was more common in boundaries with a bank with <60% of woody and evergreen perennials or in those with vegetation strips next to stonewalls compared to flat and narrow boundaries with >60% of woody and evergreen perennials or with a bank and trees with >60% of wood and evergreen perennials. But in either case their presence was also conditioned by the use of fertilizers and herbicides in the field that also affected the boundary.

Although a gradual reduction of species richness and diversity of arable plants has been observed in cereal fields of the Iberian Peninsula from boundaries to edges to inner fields, weaker differences have been found for RAP between the two outer positions of the field (Fig. 7.7) (José-María et al. 2010; Solé-Senan et al. 2014). From a landscape perspective, despite it being suggested that landscape heterogeneity compensates for the negative effects of farming intensification on the arable plants species richness and diversity (Tscharntke et al. 2005), such compensation is greatest at the boundary and decreases towards the field centre, reflecting the increasing impact of farming practises (Wilson and Aebischer 1995).

4 Causes of the Decline of Arable Plants in Western Mediterranean Area

Across Europe, and years before agriculture intensification, at the end of the 19th and the beginning of the twentieth century, two factors started to affect arable floras and cause its decline. One of them was the improvement of specific seed cleaning devices in conjunction with the development of threshing machines and combine harvesters that encouraged farmers to move away from using their own seed from the previous years' crop and to purchase seed from commercial producers (Hilbig 1982; Albrecht et al. 2016). The other factor was the virtual disappearance of certain traditional crops, such as flax, that hosted specialized arable plants that were unknown in any other habitat (Hilbig 1982; Barret 1983). But since the twentieth century, the intensification of agriculture with its known changes to arable farming systems (Robinson 2002) and the consequent homogenization of the landscape (Benton et al. 2003) have been two of the main factors driving arable plant declines.

4.1 *Intensification of Agriculture*

Agricultural intensification can be technically defined as an increase in agricultural production per unit of inputs (which may be labour, land, time, seed, fertilizer or pesticides) (FAO 2004). Generally, this concept is related to the increase in the use of pesticides and fertilisers, simplification of rotations, and transformation of marginal land into grassland or forestry (Kleijn and Van der Voort 1997; Robinson 2002). Several researchers have highlighted the important role of herbicides in the decline of arable plant diversity in arable fields (Rydberg and Milberg 2000; Hyvönen and Salonen 2002). Nonetheless, Egan et al. (2014) indicated that rarity *per se* may not necessarily be the result of herbicide intensity. This conclusion was reached from a comparative study involving pairs of closely related and common arable species, where herbicide sensitivity tended to be similar in both members of a given species pair. Similarly, in Mediterranean conditions, Rotchés-Ribalta et al. (2015a) observed that rarity by itself is not directly linked to herbicide sensitivity and pointed out that even if herbicides have been important drivers of plant diversity loss, they are unlikely to be the sole determinant in explaining the overall decline of the rare weed species. Ulber et al. (2010) even suggested that applying selective herbicides could be an option for managing weed diversity, by maintaining beneficial or rare weed species occurring at moderate densities while simultaneously controlling highly competitive weeds.

Storkey et al. (2012) pointed out that factors other than herbicides have been decisive in reducing the populations of some arable plants that have become rare. The interaction of different agricultural management practices, such as fertiliser application and mechanisation may have also played a significant role. Fertilised fields are detrimental to rare arable plants (RAP) due to the high competition between crops and weeds for the same resources (Ponce et al. 2011). Rotchés-Ribalta et al. (2016) highlighted that RAP are more affected by the indirect competitive effect of the cereal crop than by the dose of both organic and mineral fertilisers applied. In any case, these studies on the effect of herbicides and fertilization on arable plants have been carried out on specific target species, and with specific herbicides. Future work should increase the number of species tested since they may have unequal sensitivity to different herbicides and different competitive ability to capture mineral resources.

Other studies include more factors that help explain the causes of rarity in the arable flora. In a review, Albrecht et al. (2016) pointed out that phenological development can be a significant filter for some arable species in winter cereals; if germination takes place early and at low temperatures (characteristic for winter-annual species), those species can be more exposed to early herbicide applications. In contrast, spring and summer-annual species, that require higher germination temperatures, may be better able to avoid herbicide exposure, and can benefit from reduced competition. Torra et al. (2018) observed the influence of annual tillage and seed burial time on the emergence rate of 30 RAPs collected in Mediterranean arable fields. Most of the RAPs analysed, particularly those with high persistence in the soil, had induced secondary dormancy in the absence of tillage, highlighting the importance of tillage in promoting RAP emergence, and seed bank replenishment. Therefore, the absence of soil tillage may represent a driver of seed bank decline for those species with secondary dormancy processes. These results suggest that the increase of direct drilling (or "no till") in Mediterranean areas may also contributing to the decline of RAP.

In another study, Torra et al. (2020) observed that weed emergence patterns of RAP are primarily influenced by temperature and secondarily by soil moisture, reflecting a strong effect of seasonal weather variations, but also highlighting species-specific temperature and moisture germination requirements. In this work, the authors divided the same 30 RAP into four distinct groups according to their relative emergence levels during the germination season from autumn to spring (Table 7.1): (i) species almost only germinating and emerging in spring, (ii) species with emergence in winter and spring, (iii) species emerging mainly in autumn with some winter emergence, and (iv) species that emerged both in autumn and winter. This research confirmed that species in the genus Galeopsis are obligate spring-germinating species and require high temperatures (Milberg et al. 2000; Karlsson et al. 2006); that the Brassicaceae species as *Thlaspi arvense*, *Camelina microcarpa* and *Neslia paniculata* mainly emerge in autumn and winter (Royo-Esnal et al. 2015a, b, 2019); that species belonging to the Papaveraceae family show important fluxes in autumn and winter, like *Papaver argemone*, which show cooler germination requirements compared to other species from this family (Torra et al. 2016; Karlsson and Milberg 2007); and that Consolida species can germinate mainly in autumn winter, while Delphinium species and Nigella gallica also germinated in spring (Torra et al. 2015). These results highlight the importance of phenology as a filter for some species, especially if germination takes place early and at low temperatures, leaving them more exposed to the action of herbicides. In any case,

Table 7.1 Four distinct groups of arable plants proposed by Torra et al. (2020) according to their relative emergence levels during the germination season from autumn to spring

Groups	Emergence patterns	RAP species
Group 1	Spring	*Galeopsis ladanum*
Group 2	Mainly winter	*Adonis flammea, Hypecoum pendulum, Legousia hybrida, Nigella gallica Turgenia latifolia*
Group 3	Mainly autumn	*Agrostemma githago, Androsace maxima, Biscutella auriculata, Camelina microcarpa, Cerastium perfoliatum, Neslia paniculata, Papaver dubium, Silene conoidea*
Group 4	Autumn-winter	*Adonis aestivalis, Asperula arvensis, Bifora radians, Bupleurum rotundifolium, Conringia orientalis, Consolida orientalis, Consolida pubescens, Delphinium gracile, Delphinium halteratum, Iberis amara, Lathyrus aphaca, Papaver argemone, Ranunculus arvensis, Roemeria hybrida, Thlaspi arvense, Vaccaria hispanica*

the authors are aware that these results need an extensive verification before they can be extrapolated to the entire of Spain or Europe.

4.2 Loss of Landscape Heterogeneity

Mediterranean agricultural landscapes have been traditionally characterized by a heterogeneous mosaic of habitats (e.g. arable fields, perennial crops such as almond and olive trees, abandoned fields, semi-natural and natural vegetation). The structural heterogeneity of the agricultural landscape can be assessed in terms of its composition and configuration (Fahrig et al. 2011). While compositional heterogeneity focuses exclusively on the diversity of cover types, configurational heterogeneity refers to the number, the type and the arrangement of patches and habitats within a landscape. Increased compositional heterogeneity promotes more niches to support more species (Sirami et al. 2019), whereas increased configurational heterogeneity improves the mosaic structure and the connectivity of patches along boundaries and edges. Thus, the denser the network of boundaries and edges, the higher the spatial mass effect of species colonisation of fields (Solé-Senan et al. 2018).

Agricultural landscape homogenization in the Mediterranean area has been happening over recent decades and it is still an ongoing process. Landscape simplification has occurred mainly through land concentration causing field enlargement and the loss and fragmentation of natural habitats and field margins (Fig. 7.8). Simplifying agricultural landscape heterogeneity has triggered not only a substantial reduction of plant diversity (José-María et al. 2010; Armengot et al. 2011; Bassa et al. 2012; Solé-Senan et al. 2014, 2018; Cirujeda et al. 2019), but also of functional diversity of the local arable plant communities (Solé-Senan et al. 2017).

A landscape with high compositional heterogeneity, in terms of a larger percentage of natural vegetation, enhances the diversity of arable plant communities within-

Fig. 7.8 Agricultural landscape in Cubells (Lleida province), without (left) and with land consolidation (right). (Photos J. Recasens)

fields due to neighbourhood effects and the fluxes of seeds over short distances from adjacent habitats (José-María et al. 2010; Solé-Senan et al. 2014; Rotchés-Ribalta et al. 2015b). For RAP species, the highest positive effect of compositional heterogeneity occurs in the field edges as they represent an intermediate level of disturbance compared to both the boundaries and the field centre (Fried et al. 2009; José-María et al. 2010; Solé-Senan et al. 2014, 2017). Furthermore, crop edges may increase neighbourhood effects with boundaries (Devlaeminck et al. 2005; Bassa et al. 2012) leading to boundaries being additional local reservoirs for rare species. However, when natural vegetation is substituted for newly cultivated arable land, it is often first colonised by wind-dispersed, competitive species that can travel long distances (Solé-Senan et al. 2017). These species can outcompete less aggressive arable plants and RAP and limit their spread. Conversely, when land abandonment occurs, there is an initial increase of species richness and abundance of arable plants (Plieninger et al. 2014). However, over a longer time period, the process of secondary succession results in communities dominated by perennial species displacing annual species (Bassa et al. 2012; Solé-Senan et al. 2018; Cirujeda et al. 2019). When this process takes place, RAP are expected to disappear due to the absence of soil disturbance favouring seed burial and seed incorporation into the seed-bank (Solé-Senan et al. 2014).

Increasing configurational heterogeneity can also promote arable plant species diversity. Some studies show that the highest values of perimeter/area ratio of arable fields are related to lower species richness of RAP due to the low ability of these species to spread at medium and long distances (Rotchés-Ribalta et al. 2015b). However, other studies have found that increasing perimeter/area ratio may benefit plant diversity as much as semi-natural cover (Solé-Senan et al. 2014; Alignier et al. 2020). The authors suggest that fields with high values of perimeter/area ratio are more elongated, convoluted and have dendritic edges which act as a source of diversity from where species can spread to the whole field. Unfortunately, the ongoing process of landscape simplification and land consolidation is resulting in the loss of fields with high perimeter/area ratios and favouring regular and large fields where spread of arable species is more challenging.

5 Opportunities for Arable Plant Conservation

The conservation status of arable plants and particularly of RAP is increasingly raising concerns in Europe. These species are indicators of good ecological equilibrium and sustainable management of the fields, and it is suggested that they contribute to maintaining ecosystem functions (Storkey and Westbury 2007). Maintaining and improving habitats where RAP are already present can be achieved through a decrease of agricultural intensification via the implementation of organic farming practices, or by modifying landscape structure, both its composition and/or its configuration. Both strategies have been shown to increase arable plant diversity and also RAP abundance in Mediterranean conditions but their applicability may be highly management-dependent or site- specific.

5.1 Conservation Through Organic Agriculture

Organic farming has been found to increase the richness and diversity of arable species (Cirujeda et al. 2011; Kleijn et al. 2012; Gibson et al. 2007), which has been also observed in North-East Spain (Armengot et al. 2012), even for rare arable and segetal plants (Chamorro et al. 2016). Moreover, in the Mediterranean area it has been shown that organic farming enhances the abundance of those functional groups, such as the insect-pollinated plants, high flowering plants and legumes (Romero et al. 2008), which are key in the delivery of ecosystem services (Ricou et al. 2014) (Figs. 7.9 and 7.10). These trends are similar to other regions of Europe, such as England (Moreby et al. 1994), Denmark (Hald 1999) or Finland (Hyvönen et al. 2003).

Although it is widely acknowledged that organic farming is an important tool for arable plant conservation, it also involves a wide range of management practices that differ in intensity and can affect the potential for the conservation of those plants (Van Elsen 2000). In Mediterranean conditions, the richness of arable plants is affected by land use intensity, based on the agricultural practices performed (i.e. the annual inputs of exogenous nitrogen, weed control intensity, crop diversity, and the origin of the sown seeds) and this intensity can be high even in organic farming systems (Armengot et al. 2011). For example, in organic cropping systems, fertilisation, particularly slurry, had a negative effect on arable plant richness (Rotchés-Ribalta et al. 2015c), and reduced tillage increased monocotyledon and perennial weed infestations (Sans et al. 2011) which not only negatively affects crops yield (Tørresen et al. 2003) but also displaces the specialized annual arable plants by outcompeting them (Fagúndez 2014). On the other hand, a promotion of RAP has been observed in rotations where cereals dominate since the sowing time of the cereals and the germination requirements of those plants are synchronized (Rotchés-Ribalta et al. 2015c).

Fig. 7.9 Crop edge of an organically farmed field in Valle del Jarque (Teruel) with *Bupleurum rotundifolium*, *Papaver rhoeas* and *Consolida orientalis*. (Photo J. Recasens)

Fig. 7.10 Organically farmed field in Cerdanya county (North of Catalonia), with a large population of *Centaurea cyanus*. (Photo J. Recasens)

The distribution of arable plants in organic winter cereal fields in Spain is more homogeneous from the centre to the edges comparing to conventional fields, suggesting a reduced role for field margins as plant reservoirs (Romero et al. 2008). However, in the case of RAP, Rotchés-Ribalta et al. (2015b) reported that they are mainly located on the edges of organic fields and were related to management based on winter cereal production and an adjusted fertilisation. These results show the importance of selecting the proper management practices to promote the conservation, support and development of a diverse arable flora (Van Elsen 2000). In this sense, management practices could be based on mechanical weeding on field margins, avoiding field ploughing after harvest if RAP are detected and promoting structural elements such as hedges or boundaries.

5.2 Field and Landscape Perspectives

Disentangling the appropriate scale at which conservation measures must be applied is important because the effectiveness of the measures conducted in the agricultural landscapes of the European Union based on AES (Agri-Environmental Schemes) has recently been questioned (Whittingham 2007). Most of the studies conducted in the Mediterranean area suggest that measures aimed at increasing species diversity must be taken not only at the field but also at the landscape scale. Hitherto, field enlargement and landscape simplification have been underestimated drivers of past and ongoing diversity declines in arable plant communities (Sirami et al. 2019), and therefore, landscape management would be an effective and complementary way to promote farmland biodiversity without taking land out of agricultural production. Accordingly, Alignier et al. (2020) recommend adopting manipulation of compositional landscape heterogeneity as a specific, effective management option in future policy measures, to contribute to the conservation of arable plant diversity. Increasing compositional crop heterogeneity may benefit plant diversity at a landscape scale, since the presence of patches subjected to different levels of intensification may shelter a different species pool that will increase diversity from a local perspective (Solé-Senan et al. 2018). However, most of the agricultural Mediterranean landscapes are in dryland and rainfed areas where it is a challenge to promote such crop diversity. Consequently, policy makers must properly target the areas where the conservation measures have the best chance of succeeding.

In the particular case of RAP, several studies have pointed out that conservation measures of plant diversity in the Mediterranean area should be based on targeting those landscape structures in which species pools of RAP have not already been depleted (José-María et al. 2010; Bassa et al. 2012; Solé-Senan et al. 2014). Managing configurational crop heterogeneity in these areas opens new effective and complementary approaches to farmland biodiversity conservation (Solé-Senan et al. 2018; Sirami et al. 2019). The use of a set of landscape metrics may help to determine the structures and landscape on which policy makers should focus. Moreover, José-María et al. (2010) and Solé-Senan et al. (2014) suggest that a

1-m-wide strip is enough to enhance RAP. Therefore, it is crucial to focus low-intensity farming practices at the edges of the fields in complex landscapes, since this is the most cost-effective measure for enhancing RAP. These measures are expected to benefit RAP assemblages and strengthen their role as species pools to spread over the surrounding landscape. As suggested by Solé-Senan et al. (2017), it is expected that these schemes could be more efficient in increasing RAP diversity in the most structurally complex landscapes than in the most simplified.

6 Conclusions

In the last decades, the diversity of arable floras from the Mediterranean farming systems of Spain and Portugal has declined significantly as a result of agriculture intensification and changes in soil management, land use and landscape structure. Some of the arable plants that belong to these segetal communities are representatives from chorological elements with high ecological relevance as: Mediterraneo--Irano-Turanian element, Mediterranean arid element, Ibero-Maghrebian element and Iberian endemism element. The conservation of these species is an urgent necessity because of their intrinsic values as both components of biodiversity and key indicators of traditional and low-intensity agriculture. From a European perspective, Mediterranean agricultural systems still constitute a significant reservoir of diversity of arable flora and every effort must be made to preserve them.

References

AEAC (2020) Asociación Española de Agricultura de Conservación. Suelos Vivos. http://www.agriculturadeconservacion.org/index.php/agricultura-de-conservacion/la-ac-en-espana. With acces on 20 May 2020

Albrecht H, Cambecèdes J, Lang M, Wagner M (2016) Management options for the conservation of rare arable plants in Europe. Bot Lett 163:389–415

Alcántara M (2007) Catálogo de especies amenazadas en Aragón. Flora. Ed. Gobierno de Aragón. Depto de Medio Ambiente, 400 pp

Alignier A, Solé-Senan XO, Robleño I et al (2020) Configurational crop heterogeneity increases within-field plant diversity. J Appl Ecol 57:654–663

Alvaro Fuentes J et al (2007) Agricultura de conservación. Dossier Técnico, n° 24. DARP. Generalitat de Catalunya

Armengot L, José-María L, Blanco-Moreno JM, Bassa M, Chamorro L, Sans FX (2011) A novel index of land use intensity for organic and conventional farming of Mediterranean cereal fields. Agron Sustain Dev 31:699–707

Armengot L, Sans FX, Fischer C, Flohre A, José-María L, Tscharntke T, Thies C (2012) The β-diversity of arable weed communities on organic and conventional cereal farms in two contrasting regions. Appl Veg Sci 15:571–579

Arrúe JL, Cantero-Martinez C, López MV, Moreno F, Murillo JM, Pérez de Ciriza JJ, Sombrero JL, Tenorio JL, Zambrana E (2007) Conservation agriculture research in Spain. In: Knowledge Assessment and Sharing on Sustainable Agriculture (KASSA). Ed. Cirad

Baessler C, Klotz S (2006) Effects of changes in agricultural land-use on landscape structure and arable weed vegetation over the last 50 years. Agric Ecosyst Environ 115:43–50

Bañares A, Blanca G, Güemes J, Moreno JC, Ortiz S (eds) (2010) Atlas y Libro Rojo de la Flora Vascular Amenazada de España. Adenda 2010. Dirección General de Medio Natural y Política Forestal (Ministerio de Medio Ambiente, y Medio Rural y Marino). Sociedad Española de Biología de la Conservación de Plantas. Madrid, 170 pp

Barret SCH (1983) Crop mimicry in weeds. Econ Bot 37:255–282

Bassa M, Chamorro L, José-María L, Blanco-Moreno JM, Sans FX (2012) Factors affecting plant species richness in field boundaries in the Mediterranean region. Biodivers Conserv 21:1101–1114

Benton TG, Vickery JA, Wilson JD (2003) Farmland biodiversity: is habitat heterogeneity the key? Trends Ecol Evol 18:182–188

Bolòs O, Vigo J (1984) Flora dels Països Catalans, vol 1. Ed. Barcino, Barcelona

Braun-Blanquet J, Bolòs O (1957) Les groupements végétaux du bassin moyen de l'Ebre. Anales de la Estación Experimental de Aula Dei 5:1–4, 226 p

Cabezudo B, Talavera S, Blanca López G, Salazar C, Cueto M, Valdés B, Hernández-Bermejo JE, Herrera CM, Rodríguez-Hiraldo MC, Navas D (2005) Lista roja de la flora vascular de Andalucía. Consejería de Medio Ambiente. Junta de Andalucía, 126 p

Cantero-Martínez C, Ángas P, Lampurlanés J (2007) Long term yield and wàter-use efficiency under various tillage Systems in Mediterranean rainfed conditions. Ann Appl Biol 150:293–307

Chamorro J, Masalles RM, Sans FX (2016) Arable weed density in Northeast Spain: does organic farming recover functional biodiversity? Agric Ecosyst Environ 223:1–9

Cirujeda A, Aibar J, Zaragoza C (2011) Remarkable changes of weed species in Spanish cereal fields from 1976 to 2007. Agron Sustain Dev 31:675–688

Cirujeda A, Pardo G, Marí AI, Aibar J, Pallavicini Y, González-Andújar JL, Recasens J, Solé-Senan XO (2019) The structural classification of field boundaries in Mediterranean arable cropping systems allows the prediction of weed abundances in the boundary and in the adjacent crop. Weed Res 59:300–311

Devlaeminck R, Bossuyt B, Hermy M (2005) Seed dispersal from a forest into adjacent cropland. Agric Ecosyst Environ 107:57–64

Egan JF, Graham IM, Mortensen DA (2014) A comparison of the herbicide tolerance of rare and common plants in an agricultural landscape. Environ Toxicol Chem 33:696–702

Fagúndez J (2014) The paradox of arable weeds: diversity, conservation, and ecosystem services of the unwanted. In: Benkeblia N (ed) Agroecology, ecosystems, and sustainability. Publisher CRC Press, pp 139–150

Fahrig L, Baudry J, Brotons L, Burel FG, Crist TO, Fuller RJ, Sirami C, Siriwardena GM, Martin J-L (2011) Functional landscape heterogeneity and animal biodiversity in agricultural landscapes. Ecol Lett 14:101–112

FAO (2004) FAO ethics series 3. The ethics of sustainable agriculture intensificaton. FAO Publishing and Management Services. http://www.fao.org/3/j0902e/j0902e00.htm#Contents. With acces on 20 May 2020

Fernández-Suberviola R (2016) Mapa climático de España y Portugal utilizando la clasificación de Trewartha modificada. https://www.tiempo.com/ram/238092/mapa-climatico-de-espana-y-portugal-utilizando-la-clasificacion-de-trewartha-modificada/. With acces on 20 May 2020

Fried G, Petit S, Dessaint F, Reboud X (2009) Arable weed decline in Northern France: crop edges as refugia of weed conservation? Biol Conserv 142:238–243

Gibson RH, Pearce S, Morris RJ, Symondson WOC, Memmott J (2007) Plant diversity and land use under organic and conventional agriculture: a whole-farm approach. J Appl Ecol 44:792–803

Guillerm JL, Maillet J (1982) Western Mediterranean counties in Europe. In: Holzner W, Numata M, M. (eds) Biology and ecology of weeds. Dr. Junk Publishers, The Hague, pp 227–243

Hald AB (1999) Weed vegetation (wild flora) of long established organic versus conventional cereal fields in Denmark. Ann Appl Biol 134:307–314

Hilbig W (1982) Preservation of agrestal weeds. In: Holzner W, Numata M, M. (eds) Biology and ecology of weeds. Dr. Junk Publishers, The Hague, pp 57–59
Hyvönen T, Salonen J (2002) Weed species diversity and community composition in cropping practices at two intensity levels – a six years experiment. Plant Ecol 159:73–81
Hyvönen T, Ketoja E, Salonen J, Jalli H, Tiainen J (2003) Weed species diversity and community composition in organic and conventional cropping of spring cereals. Agric Ecosyst Environ 97:131–149
INE Portugal (2020) Instituto Nacional de Estatística. Statistics Portugal. https://www.ine.pt/xportal/xmain?xpgid=ine_main&xpid=INE. With acces on 20 May 2020
José-María L, Armengot L, Blanco-Moreno JM, Bassa M, Sans FX (2010) Effects of agricultural intensification on land diversity in Mediterranean dryland cereal fields. J Appl Ecol 47:832–840
Jost L (2007) Partitioning diversity into independent alpha and beta components. Ecology 88:2427–2439
Karlsson LM, Milberg PA (2007) A comparative study of germination ecology of four Papaver taxa. Ann Bot 99:935–946
Karlsson L, Ericsson J, Milberg, PA (2006) Seed dormancy and germination in the summer annual Galeopsis speciosa. Weed Res 46:353–361
Kleijn D, Van der Voort LAC (1997) Conservation headlands for rare arable weeds: the effects of fertilizer application and light penetration on plant growth. Biol Conserv 81:57–67
Kleijn D, Kohler F, Báldi A, Batáry P, Concepción ED, Clough Y, Díaz M, Gabriel D, Holzschuh A, Knop E, Kovács A, Marshall EJP, Tscharntke T, Verhulst J (2012) On the relationship between farmland biodiversity and land-use intensity in Europe. Proc R Soc B Biol Sci 276(1658):903–909
MAPA. Ministerio de Agricultura Pesca y Alimentación (2019) Anuario de estadística 2019. https://www.mapa.gob.es/es/estadistica/temas/publicaciones/anuario-de-estadistica/2019/default.aspx?parte=3&capitulo=07&grupo=1&seccion=1. With acces on 20 May 2020
Marshall EJP (1989) Distribution patterns of plants associated with arable field edges. J Appl Ecol 26:247–257
Milberg P, Hallgren E, Palmer MW (2000) Interannual variation in weed biomass on arable land in Sweden. Weed Res 40:311–321
Moreby SJ, Aebischer NJ, Southway SE, Sotherton NW (1994) A comparison of the flora and arthropod fauna of organically and conventionally grown winter wheat in southern England. Ann Appl Biol 125:13–27
Plieninger T, Hui C, Gaertner M, Huntsinger L (2014) The impact of land abandonment on species richness and abundance in the Mediterranean Basin: a meta-analysis. PLoS One 9(5):art. no. e98355
Ponce C, Bravo C, de León DG, Magaña M, Alonso JC (2011) Effects of organic farming on plant and arthropod communities. A case study in Mediterranean dryland cereal. Agric Ecosyst Environ 141:193–201
Ramôa S, Oliveira e Silva P, Vsconcelos T, Portugal J (2013) Análisis de la vegetación espontánia en cultivos de otoño invierno en Alentejo (Portugal). Congreso 2013 de la Sociedad Española de Malherbología. Valencia. Proceedings: 189–194
Recasens J, Conesa JA (2004) Endangered arable weeds in semiarid Mediterranean climate. In: Conference of AAB-EWRS: weeds and biodiversity. York, UK. Proceedings, p. 11
Ricou C, Schneller C, Amiaud B, Plantureux S, Bockstaller C (2014) A vegetation-based indicator to assess the pollination value of field margin flora. Ecol Indic 45:320–331
Robinson RA (2002) Post-war changes in arable farming and biodiversity in Great Britain. J Appl Ecol 39:157–176
Romero A, Chamorro L, Sans FX (2008) Weed diversity in crop edges and inner fields of organic and conventional dryland winter cereal crops in NE Spain. Agric Ecosyst Environ 124:97–104
Rotchés-Ribalta R, Boutin C, Blanco-Moreno JM, Carpenter D, Sans FX (2015a) Herbicide impact on the growth and reproduction of characteristic and rare arable weeds of winter cereal fields. Ecotoxicology 24:991–1003

Rotchés-Ribalta R, Blanco-Moreno JM, Armengot L, José-María L, Sans FS (2015b) Which conditions determine the presence of rare weeds in arable fields? Agric Ecosyst Environ 203:55–61

Rotchés-Ribalta R, Blanco-Moreno JM, Armengot L, Chamorro L, Sans FX (2015c) Both farming practices and landscape characteristics determine the diversity of characteristic and rare arable weeds in organically managed fields. Appl Veg Sci 18:423–431

Rotchés-Ribalta R, Blanco-Moreno JM, Armengot L, Sans FX (2016) Responses of rare and common segetal species to wheat competition and fertilised type and dose. Weed Res 56:114–123

Royo-Esnal A, Necajeva J, Torra J, Recasens J, Gesch RW (2015a) Emergence of field pennycres (Thlaspi arvense L.): comparison of two accessions and modeling. Ind Crop Prod 66:161–169

Royo-Esnal A, Gesch RW, Forcella F, Torra J, Recasens J, Necajeva J (2015b) The role of light in the emergence of weeds. Using Camelina microcarpa as an example. PLoS One 10(12): e10.1371

Royo-Esnal A, Gesch RW, Necajeva J, Forcella F, Edo-Tena E, Recasens J, Torra J (2019) Germination and emergence of Neslia paniculata (L.) Desv. Ind Crop Prod 129:455–462

Rydberg NT, Milberg P (2000) A survey of weeds in organic farming in Sweden. Biol Agric Hortic 18:175–185

Sáez L, Aymerich P, Blanché C (2010) Llibre vermell de les plantes vasculars endèmiques i amenaçades de Catalunya. Ed. Argania Editio, 811 pp

Sans FX, Berner A, Armengot L, Mäder P (2011) Tillage effects on weed communities in an organic winter wheat-sunflower-spelt cropping sequence. Weed Res 51:413–421

SEMh (2020) Sociedad Española de Malherbología. http://semh.net/grupos-de-trabajo/cprh/#problemas. With acces on 22 May 2020

Sirami C, Gross N, Baillod et al (2019) Increasing crop heterogeneity enhances multitrophic diversity across agricultural regions. Proc Natl Acad Sci U S A 116(33):16442–16447

Sociedade Portuguesa de Botânica (SPBotânica) – Associação Portuguesa de Ciência da Vegetação – SPB & PHYTOS (2020) A Lista Vermelha da Flora Vascular de Portugal Continental. https://listavermelha-flora.pt/projeto/. With acces on 20 May 2020

Solé-Senan XO, Juárez-Escario A, Conesa JA, Torra J, Royo-Esnal A, Recasens J (2014) Plant diversity in Mediterranean cereal fields: unraveling the effect of landscape complexity on rare arable plants. Agric Ecosyst Environ 185:221–230

Solé-Senan XO, Juárez-Escario A, Robleño IA, Conesa JA, Recasens J (2017) Using the response-effect trait framework to disentangle the effects of agricultural intensification on the provision of ecosystem services by Mediterranean arable plants. Agric Ecosyst Environ 247:255–264

Solé-Senan XO, Juárez-Escario A, Conesa JA, Recasens J (2018) Plant species, functional assemblages and partitioning of diversity in a Mediterranean agricultural mosaic landscape. Agric Ecosyst Environ 256:163–172

Storkey J, Westbury DB (2007) Managing arable weeds for biodiversity. Pest Manag Sci 63:517–523

Storkey J, Meyer S, Still KS, Leuschner C (2012) The impact of agricultural intensification and land-use change on the European arable flora. Proc R Soc B 279:1421–1429

Sutcliffe OL, Kay QON (2000) Changes in the arable flora of central Southern England since the 1960. Biol Conserv 93:1–8

Torra J, Royo-Esnal A, Recasens J (2015) Germination ecology of five arable Ranunculaceae species. Weed Res 55:503–513

Torra J, Edo E, Royo-Esnal A, Recasens J (2016) Temperature and light requirements for germination and emergence of three arable Papaveraceae species. Weed Sci 64:248–260

Torra J, Recasens J, Royo-Esnal A (2018) Seedling emergence response of rare arable plants to soil tillage varies by species. PLoS One 13(6):e0199425

Torra J, Forcela F, Recasens J, Royo-Esnal A (2020) Emergence patterns of rare arable plants and conservation implications. Plan Theory 2020(9):309

Tørresen KS, Skuterud R, Tandsæther HJ, Hagemo MB (2003) Long-term experiments with reduced tillage in spring cereals. I. Effects on weed flora, weed seedbank and grain yield. Crop Prot 22:185–200
Tscharntke T, Klein AM, Kruess A, Steffan-Dewenter I, Thies C (2005) Landscape perspectives on agricultural intensification and biodiversity – ecosystem service management. Ecol Lett 8:857–874
Ulber L, Steinman H-H, Klimek S (2010) Using selective herbicides to manage beneficial and rare weed species in winter wheat. J Plant Dis Prot 117:233–239
Van Calster H, Vandenberghe R, Ruysen M, Verheyen K, Hermy M, Decoq G (2008) Unexpected high 20th century floristic losses in a rural landscape in northern France. J Ecol 96:927–936
Van Elsen T (2000) Species diversity as a task for organic agriculture in Europe. Agric Ecosyst Environ 77:101–109
Whittingham MJ (2007) Will agri-environment schemes deliver substantial biodiversity gain, and if not why not? J Appl Ecol 44:1–5
Wilson PJ, Aebischer NJ (1995) The distribution of dicotyledonous arable weeds in relation to distance from the field edge. J Appl Ecol 32:295–310
Wilson PJ, King M (2003) Arable plants – a field guide. English Nature & WildGuides, Old Basing

Chapter 8
The Status of Arable Plant Habitats in Greece – The Cradle of Arable Farming in Europe

Stefan Meyer and Erwin Bergmeier

Abstract Nowhere in Europe had agriculture been practiced earlier than in Greece. In the Thessalian lowlands as well as in other fertile plains of the mainland and the Aegean Islands, Neolithic agriculture developed more than 8000 years ago. This long history of agriculture, in conjunction with an often relatively low level of cultivation intensity in cereal production, is reflected in a high species diversity and abundance of arable plants that includes rarities of the East Mediterranean and even regional endemics. However, since about the 1980s agricultural intensification has increased rapidly in Greece and the cropland area under traditional management has been declining. This development has been associated with a dramatic decline in arable plant populations especially on the plains, the focus areas of agriculture. Unfortunately, the arable flora in the Eastern Mediterranean region is, even more than in Central Europe, a largely neglected issue, a 'Cinderella of nature conservation' and there are far too few in-depth investigations of the local variation and hot spots of phytodiversity. To date there are hardly any conservation measures in traditional low-intensity agro-ecosystems. To safeguard the unique arable flora of Greece we suggest mandatory remuneration of traditional farming methods by implementation of performance-linked agri-environment schemes (AES) at the national and European Union level.

Keywords Neolithic agriculture · Urbanization · Subsistence agriculture · Important arable plant areas · Archaeophytes · Segetal plant diversity

S. Meyer (✉)
Albrecht-von-Haller-Institute for Plant Sciences, Plant Ecology and Ecosystems Research, University of Göttingen, Göttingen, Germany
e-mail: smeyer1@gwdg.de

E. Bergmeier
Albrecht-von-Haller-Institute for Plant Sciences, Vegetation and Phytodiversity Research, University of Göttingen, Göttingen, Germany

C. Hurford et al. (eds.), *The Changing Status of Arable Habitats in Europe*,
https://doi.org/10.1007/978-3-030-59875-4_8

1 Introduction

The Mediterranean is characterised by unique species assemblages of plants and represents a global focus of species conservation (Myers et al. 2000). Especially in the Eastern Mediterranean, the development of agriculture is closely linked with the development of civilization and the European cultural landscape. Nowhere in Europe was agriculture practiced earlier than in Greece. In the fertile mainland plains such as in Thessaly as well as on the Aegean Islands, Neolithic agriculture developed more than 8000 years ago and reached agriculturally less productive regions in the late 6th to 3rd millennium BC (Halstead 2000; von Schnurbein 2014). In early agriculture, winter-annual cereals (wheat, barley) and pulses (lens, vetch, pea) were the main crops. In the classical Greece of the 1st millennium BC, the most famous Mediterranean agricultural products such as olives, cereals and wine already were, as today, widespread. Currently, about 16% of the area of Greece is used as farmland, and grain is grown on slightly more than half (11,890 km^2) of this area (Storkey et al. 2012). Abandoned terraced land shows that, particularly in the south and on the islands, formerly more land used to be cultivated (Fig. 8.1). Nowadays, compared to other European countries, the use of herbicides and fertilizer and the average wheat yield is still relatively low (Storkey et al. 2012). Some regions still contain a considerable proportion of traditional and moderately intense farming sustaining soil fertility and agrobiodiversity. The long history of Greek agriculture in conjunction with a relatively low level of cultivation intensity in today's cereal production is reflected in a relatively high species diversity and abundance of arable plants compared to elsewhere in Europe, including many that are already extinct or extremely rare elsewhere (Bergmeier and Strid 2014; Bergmeier and Meyer 2018).

However, since about the middle of the last century as part of the trend towards global agricultural intensification (mechanization of tillage, introduction of large-scale chemical weed control) the crop area under traditional management has been rapidly declining. This development was associated with a dramatic decline in arable

Fig. 8.1 Examples of traditional arable landscapes in Greece: abandoned terraces near Olymbos on the island of Carpathos (South Aegean), 24 April 2018 (*left*) and the famous fertile plains of Avlona on the island of Carpathos (South Aegean) that are still used for cereal and vegetable production, 25 April 2018 (*right*). The arable terraces on the mountain slopes have been abandoned for decades. ©Stefan Meyer

plant populations, first in Central and Western Europe and later also in the Mediterranean (Storkey et al. 2012; Meyer et al. 2013a; Bergmeier and Strid 2014). As elsewhere, with the onset of mechanization, productive sites have increasingly been intensified in Greece, whilst agronomically marginal sites like remote and hard-to-reach fields that were characterised by subsistence farming were abandoned through emigration of farmers in the 1950s and 1980s or transferred into pastures (Lienau 1976; Bergmeier and Meyer 2018).

As elsewhere in Europe, many arable plant species are archaeophytes in Greece, considered native further east and travelling westward with the expanding prehistoric agriculture. Other species may have been native in the country, in naturally more or less disturbed habitats such as scree and riverbanks, from where they spread into cultivated fields. The distribution of arable plants of traditional agriculture in Greece is by no means random but follows north-south as well as west-east gradients (Bergmeier and Strid 2014), quite like the biogeographic patterns of other flora species. The pool of plant species associated with traditional agriculture differs from that in other European countries and includes endemic species such as Doerfler's tulip (*Tulipa doerfleri*) (Fig. 8.2) on Crete and Spruner's ox tongue (*Anchusa stylosa* subsp. *spruneri*) in Central Greece (Georghiou and Delipetrou 2010; Bergmeier and Strid 2014).

Many species from the Middle East or the East Mediterranean reach Europe only in the arable lands of Greece such as *Bellevalia speciosa*, *Bongardia chrysogonum*, *Euphorbia arguta*, *Euphorbia microsphaera*, *Lactuca aculeata*, *Leontice leontopetalum*, *Lolium subulatum*, *Medicago blancheana*, *Nigella orientalis*, *Silene*

Fig. 8.2 *Tulipa doerfleri*, a tulip of arable fields endemic to Central Crete, plateau of Gious Kambos, 28 April 2012. ©Ute Bergmeier. (Crotty 2008)

Fig. 8.3 Examples of arable plants that are only found in Greece at the edge of their range: The Yellow Nigella (*Nigella orientalis*) (*top left*) was rediscovered on the island of Lesvos (East Aegean) in 2016 after over 100 years, 2 May 2016 ©Stefan Meyer; The Lion's Foot (*Leontice leontopetalum*) (*top right*) can be found in Europe only in Greece. Formerly widespread on fertile soils, this plant decreased on arable land due abandonment or transformation to other habitats (pastures, grasslands). The plant is adapted to traditional arable farming crop rotation cycles with periods of cereal cultivations and 1–2-year fallow periods, 6 May 2015 ©Stefan Meyer. (*Bottom*) – Field in early spring with *Leontice leontopetalum* and *Anemone coronaria*, island of Chios (East Aegean), 17 March 2013. ©Ute Bergmeier. (Crotty 2010)

longipetala, *Tulipa clusiana* and *Vicia sibthorpii* (Fig. 8.3). Some species evolved with agriculture, originating from ancestral forms growing in steppe-like habitats further east. Others are more widespread, though still classed as rare Mediterranean arable species, and are more common in Greece than in the Central and Western Mediterranean (e.g. *Bupleurum odontites*, *Bupleurum subovatum*, *Euphorbia aleppica*, *Hypecoum pendulum*, *Ridolfia segetum*) (Bergmeier and Strid 2014; Dimopoulos et al. 2013). Among the most well-represented plant genera in arable fields in Greece are chamomiles (*Anthemis*, *Cota*), larkspurs (*Consolida*), spurges (*Euphorbia*), vetchlings (*Lathyrus*), poppies (*Papaver*), cornsalads (*Valerianella*) and vetches (*Vicia*). The most abundant wild plants in many species-rich arable fields in the country constitute Asteraceae (*Anthemis*, *Cota*, *Glebionis*), yellow-flowered Brassicaceae (*Hirschfeldia*, *Raphanus*, *Rapistrum*, *Sinapis*) and Fabaceae (*Lathyrus*, *Medicago*, *Vicia*).

The high proportion of wild geophytes in fields particularly in southern mainland Greece and the islands is also remarkable (Strid and Tan 2003). Such bulbous or tuberous plants belong to species that are well adapted for habitats of arable land and tolerate annual tilling by traditional methods. As a consequence of deep-ploughing, artificial irrigation and use of herbicides some species with below-ground storage organs are now threatened with extinction, e.g., *Tulipa undulatifolia*, *Leontice leontopetalum*, *Bongardia chrysogonum*, *Silene longipetala*, *Vicia sibthorpii*, *Allium cyrilli* (Fig. 8.4) and *Bellevalia ciliata*. Others are more resilient and able to survive at field edges, in abandoned fields or locally in semi-natural habitats, e.g., *Bellevalia trifoliata*, *Gladiolus italicus*, *Muscari comosum*, *Ornithogalum narbonense* and *Ornithogalum nutans* (Fig. 8.5). An extreme example of an invasive plant in arable fields and olive plantations is the goat's-foot (*Oxalis pes-caprae*), a South African member of the sorrel family flowering in early spring, which spreads effectively by means of small subterranean propagules. It is now a troublesome weed all over the Mediterranean.

Fig. 8.4 The Cyrilli Garlic (*Allium cyrilli*, and its related species *Allium nigrum*) with is characteristic bulbs is distributed on deep, productive soils in the lower areas of Greece. The plant is especially sensitive to deep ploughing and irrigation of fields and can be used as an indicator for sustainable, biodiversity-friendly arable farming, 8 May 2015. ©Stefan Meyer

Fig. 8.5 Extremely species rich fields with a high share of geophytes such as *Gladiolus italicus*, *Muscari comosum*, *Allium neapolitanum* and *Ornithogalum narbonense* on the island of Astypalaia (Cyclades), 27 April 2017. ©Stefan Meyer

Additional arable plant species, many of which are rare or even nearly extinct in other regions/countries, still occur in significant numbers in Greek fields. However, Bergmeier and Strid (2014) and Bergmeier and Meyer (2018) highlighted significant declines in population numbers of several species due to abandonment of fields and changes in agronomic practice such as:

I. increasing density and therefore competition from cultivated plants,
II. increased use of herbicides and synthetic fertilizers,
III. the use of heavy farm machinery and deep tillage,
IV. the turn in management toward mere fodder production and early grazing,
V. the urbanization in the greater Athens area and in other locations,
VI. the early breaking of stubble, and
VII. the uncontrolled utilization of fields in marginal areas for wind turbines and solar energy.

Populations of some arable plants are declining and have become sporadic or locally rare through the use of foreign seeds that, in contrast to the earlier practice of sowing own seed corn collected from the previous year, prevents the distribution and re-sowing of these obligatory arable plants such as corn cockle (*Agrostemma githago*) or bearded darnel (*Lolium temulentum*). Such species with diaspores similar to those of the crops, harvested with the crop and being returned to the fields with the seeds, had adapted to grow along with the cultivated plants and share their life cycle.

2 Conservation Activities

The arable flora in the Mediterranean region is, even more than in Central Europe, a largely neglected issue, a 'Cinderella of nature conservation' (Meyer et al. 2013b, 2014). EU agricultural policy still favours industrial export-oriented farming over subsistence agriculture, and arable phytodiversity has so far been almost invisible in EU nature conservation policy and NATURA 2000 sites so that practical conservation strategies are now urgently needed. The Greek arable flora is significant from botanical, biogeographical, environmental, genetic and agricultural perspectives. Consequently, there is an urgent need for action to ensure the survival of arable plant communities of sufficient quality and quantity. However, up to now existing scientific studies are limited in number, completeness and geographical scope. They provide a thorough floristic and biogeographical overview but no spatially explicit fine-scale analysis for applied purposes (Rechinger 1951; Turland et al. 2004; Bergmeier 2006; Bergmeier and Strid 2014; Bergmeier and Meyer 2018).

All in all, there are far too few in-depth investigations of the phytodiversity and appropriate conservation measures in traditional low-intensity agro-ecosystems in the Eastern Mediterranean region (Turland et al. 2004; Türe and Böcük 2008; Bergmeier 2012; Bergmeier and Strid 2014) but awareness has increased noticeably in recent years (e.g. Bergmeier and Meyer 2018; Meyer et al. 2018). Bergmeier and Strid (2014) used the Flora Hellenica Database to provide the first survey of wild plant species of traditional agriculture in Greece based on about 19,500 field and literature records. This survey showed that arable plants had largely been neglected even in the current (second edition) of the Red Data Book of Greece (Phitos et al. 2009), and that it is misleading with respect to the present status of the arable flora. For each of the 138 taxa identified by Bergmeier and Strid (2014), life form, rareness (on national scale), local abundance (habitat occupancy), regional restrictedness and the Red List status of the arable plants using IUCN criteria are given. Distribution patterns show a pronounced south–north gradient, and the unique character of the East Aegean arable flora is revealed. Most arable species are annuals (83%). Perennial herbs, mostly bulbous or tuberous plants (geophytes), account for 17%, and a major proportion of the latter must be considered as threatened at different levels.

We suggest conservation measures for the unique arable flora of Greece such as mandatory remuneration of traditional farming methods (abstaining from pesticides and synthetic fertilisers) by implementation of results-based agri-environment schemes (AES) at national and European Union level. To reflect the geographical and ecoregional variation, in-situ measures should focus on species-rich arable fields in all phytogeographical regions of Greece (Fig. 8.6). These in-situ (or on-farm)

Fig. 8.6 Arable fields in a structurally rich biodiverse landscape near Vrondero, Prespa, Northwest Greece, 1000 m, 11 June 2017 (*left*). *Centaurea adamii*, here visited by Marbled White, *Melanargia galathea*, (*right*) is a rare biennial plant of scattered distribution from South and Southeast Europe to Southwest Asia, typically found in these fields, 12 July 2017. ©Erwin Bergmeier

agroecosystem priority areas must be cultivated according to the local needs to sustain arable biodiversity. According to (Bergmeier and Strid 2014), the three most important aspects for such a network of reserves are:

1. Selection of suitable areas and species-rich fields for the conservation of regionally and geographically differentiated biodiversity.
2. Financial aid and long-term perspectives for farmers to operate according to the principles of traditional and ecologically-oriented agriculture.
3. Customized management (without the use of herbicides and synthetic fertilizers, with regular but shallow tillage and low or irregular crop sowing density).
4. Preserving the know-how of a sustainable agriculture, and the genetic diversity of old cultivated varieties.

The most prominent Greek cereal field complexes in traditional agricultural practice can be found on upland valley basins and plains of the Aegean islands and in the lowland and upland plains of the mainland. They are among the most valuable for segetal plant diversity conservation. Surrounded by mountains of metamorphic (e.g., gneiss, marble, quartzite) or igneous rock (e.g. granite, andesite) and filled with debris and colluvia, such plains, often karst-generated endorheic plateaus known as oropédio (Fig. 8.7), are favourable for agriculture. At altitudes between 200 and over 1000 m a.s.l. they may have been cultivated since millennia,

Fig. 8.7 Arable fields separated by stone walls in an oropédio (karst plateau) near Avlona, island of Carpathos (South Aegean), 280 m, 14 April 2017. ©Ute Bergmeier

although many and in particular the highest and remote ones are today abandoned, entirely or in major parts, probably for reasons of climate or shortage of labour (Bergmeier and Meyer 2018).

Since 2016 the authors have been involved in a case study to determine how a differentiated hotspot analysis can lead to effective and goal-oriented nature conservation concepts for arable plant communities in the Aegean archipelago. The aim is to establish a novel approach using the concept developed by Plantlife (UK) in 2005 to assess the conservation value of plant communities on arable land and to identify priority areas (Important arable plant areas – IAPA; Byfield and Wilson 2005). The results and practical recommendations will help to put nature conservation administrations, farmers, and program planners in the agricultural administration in a position to use the often very limited available funding purposefully and efficiently.

References

Bergmeier E (2006) The diversity of segetal weeds in Crete (Greece) at species and community level. Annali di Botanica ns 6:53–64

Bergmeier E (2012) Segetal weed vegetation database of Greece. In: Dengler J, Oldeland J, Jansen F, Chytrý M, Ewald J, Finckh M, Glöckler F, Lopez-Gonzalez G, Peet RK, Schaminée JHJ (eds) Vegetation databases for the 21st century. Biodiversity & Ecology 4:387

Bergmeier E, Meyer S (2018) Segetal plants of winter-annual crop fields in the Aegean islands – viewed in the contexts of landscape and traditional agricultural practice. Berichte der Reinhold-Tüxen-Gesellschaft 30:73–84

Bergmeier E, Strid A (2014) Regional diversity, population trends and threat assessment of the weeds of traditional agriculture in Greece. Bot J Linn Soc 175:607–623

Byfield AJ, Wilson PJ (2005) Important Arable Plant Areas: identifying priority sites for arable plant conservation in the United Kingdom. Plantlife International, Salisbury

Crotty FV, (Photographer) (2008) Diverse mesofauna sample [photograph]. Rothamsted, North Wyke, Devon, UK

Crotty FV, (Photographer) (2010) Mites separated into main groups [photograph]. Rothamsted, North Wyke, Devon, UK

Dimopoulos P, Raus T, Bergmeier E, Constantinidis T, Iatrou G, Kokkini S, Strid A, Tzanoudakis D (2013) Vascular plants of Greece: an annotated checklist. Berlin: Botanischer Garten und Botanisches museum Berlin-Dahlem; Athens: Hellenic botanical society. Englera 31

Georghiou K, Delipetrou P (2010) Patterns and traits of the endemic plants of Greece. Bot J Linn Soc 162(2):130–422

Halstead P (2000) Land use in postglacial Greece: cultural causes and environmental effects. In: Halstead P, Frederick C (eds) Landscape and land use in postglacial Greece. Sheffield Academic Press, Sheffield. (Sheffield Studies in Aegean Archaeology 3:110–128)

Lienau C (1976) Bevölkerungsabwanderung, demographische Struktur und Landwirtschaftsform im W-Peloponnes. Giessener Geographische Schriften 37:1–119

Meyer S, Wesche K, Krause B, Leuschner C (2013a) Dramatic losses of specialist arable plants in Central Germany since the 1950s/60s – a cross-regional analysis. Divers Distrib 19 (9):1175–1187

Meyer S Hilbig W, Steffen K, Schuch S (2013b) Ackerwildkrautschutz. Eine Bibliographie. BfN-Skripten 351: 222 p

Meyer S, van Elsen T, Metzner J, Blümlein B, Kaerlein M, Gottwald F, Wehke S, Dieterich M, Wahmhoff W, Stock R, Leuschner C (2014) Agrobiodiversitätsschutz durch Schutzäcker – 100 Äcker für die Vielfalt. Natur und Landschaft 89 (9/10): 434–441

Meyer S, Bergmeier E, Panitsa M, Sfakianou D, Georgiadis N, Dimitropoulos G (2018) Segetal plant diversity on Lemnos Island (Greece) – status quo and implications for conservation efforts within the TerraLemnia-project. In: Proceedings of the 2nd Mediterranean plant conservation week 'Conservation of Mediterranean plant diversity: complementary approaches and new perspectives', La Valetta, Malta, 12–16 November, 2018. Further information: https://terra-lemnia.net

Myers N, Mittermeier RA, Mittermeier CG, da Fonseca GAB, Kent J (2000) Biodiversity hotspots for conservation priorities. Nature 403:854–858

Phitos D, Constantinidis T, Kamari G (eds) (2009) Vivlío erythrón dhedhoménon ton spánion & apeiloúmenon fitón tis Elládhas [The Red Data Book of rare and threatened plants of Greece]. 2 vols. Hellenic Botanical Society, Patras. (in Greek)

Rechinger KH (1951) Phytogeographia Aegaea. Denkschriften der Österreichischen Akademie der Wissenschaften. Mathematisch-naturwissenschaftliche Klasse 15(2,2):1–208

Storkey J, Meyer S, Still KS, Leuschner C (2012) The impact of agricultural intensification and land-use change on the European arable flora. Proc R Soc B 279:1421–1429

Strid A, Tan K (2003) Bulbous and tuberous plants of Greece. Bocconea 16(2):757–762

Türe C, Böcük H (2008) Investigation of threatened arable weeds and their conservation status in Turkey. Weed Res 48:289–296

Turland N, Phitos D, Kamari G, Bareka P (2004) Weeds of the traditional agriculture of Crete. Willdenowia 34:381–406

von Schnurbein S (ed) (2014) Atlas der Vorgeschichte. Europa von den ersten Menschen bis Christi Geburt. 3. Aufl. p 244. wbg Theiss, Darmstadt, Germany

Part III
Research and Surveillance Projects

Chapter 9
Soil Organisms Within Arable Habitats

Felicity V. Crotty

Abstract Soil is often discussed in terms of its physical or chemical composition, the potential for soil erosion, flooding or poor crop yields, but it is soil biodiversity that is driving the functionality of the soil. Within the soil, reside the most abundant and diverse groups of organisms found on earth, packed within a three-dimensional space. This chapter provides the reader with a greater understanding of soil biodiversity, covering aspects of the microbiota (bacteria and fungi), mesofauna (springtails and mites) and macrofauna (earthworms), as well as how they link together as part of the soil food web. Outlining the main ways agricultural management effects soil biodiversity within arable habitats and how the promotion of greater crop rotation and plant diversity across farms through agri-environment schemes, native arable plants and wildflower strips and hedgerows can protect belowground fauna which will lead to healthier soils and increases in biodiversity as a whole.

Keywords Soil organisms within arable habitats · Soil biodiversity · Mycorrhizae · Tillage systems · Agro-chemicals · Agricultural intensification · Habitat heterogeneity

1 Overview of Soil Biodiversity

Soil is the most biodiverse ecosystem on Earth, it has been referred to as an enigma (Anderson 1975; Nielsen et al. 2010), a black box (Tiedje et al. 1999; Wilkinson 2008) or a jigsaw (Briones 2014). However, the best analogy is that soil "is like a poor man's tropical rainforest" (Giller 1996), as you do not need to go to the tropics to find a habitat as diverse and abundant with life – even though the majority of organisms residing there are barely visible without magnification. In one teaspoon of soil (1 g) there are more bacterial cells than the entire human population on earth, up

F. V. Crotty (✉)
Royal Agricultural University, Cirencester, UK
e-mail: Felicity.Crotty@rau.ac.uk

C. Hurford et al. (eds.), *The Changing Status of Arable Habitats in Europe*,
https://doi.org/10.1007/978-3-030-59875-4_9

Table 9.1 Categories and characteristics of soil organisms

Time	Characteristics	Organisms
Permanent	Entire life cycle in the soil	Acari, Collembola, earthworms
Temporal	Part of life cycle in the soil – larvae that develop within the soil prior to emergence as adults	Insect larvae e.g. Coleoptera larvae, Diptera larvae
Periodical	Frequently enter into the soil – To avoid harsh conditions (sunlight/drying out)	Slugs, ants and larvae
Transitory	An inactive phase in the soil (egg, pupa and hibernation) not an active period	Some insects e.g. butterflies, moths
Accidental	Animals that fall down on to soil or are transported by wind or water runoff	Insect larvae, canopy insects

Adapted from Bot and Benites (2005)

to 10 billion bacterial cells (Raynaud and Nunan 2014), with up to 40,000 species (Tiedje 1995); as well as several hundred metres of fungal hyphae (Turbé et al. 2010). In rich organic soils, there can be representatives of 20 different lineages of Arthropoda, the most diverse phylum of living organisms (Behan-Pelletier 2003). The opacity of the environment, combined with the abundance and diversity of life within the soil, has meant that the impact of agricultural practices has been largely overlooked, due to the resilience and functional redundancy perceived within the ecosystem.

Soil organisms range in scale across several orders of magnitude – from micro to mega (1 μm to 10+ cm). The microflora (bacteria, fungi) are less than 5 μm, and microfauna (protozoa, nematodes and tardigrades) less than 10 μm, whilst the mesofauna (springtails, mites and many other invertebrates) are those organisms with less than 2 mm body width (Swift et al. 1979). Macrofauna (earthworms, ants, termites) are those invertebrates with greater than 2 mm body width, whilst the megafauna (rabbits, moles, rodents) are vertebrates that reside within the soil. Due to the varying lifecycles of insects, it has been estimated that around 90% of all insects spend either all or part of their lives within the soil (Gaugler 1990), however as well as the permanent soil fauna (Acari, Collembola and earthworms), this could also be temporal, periodical or transitory time spent within the soil (Table 9.1).

The abundance and diversity found within the soil food web allows for a greater capacity for functional redundancy within this ecosystem (Nielsen et al. 2011). It is perceived that there is a vast diversity of organisms with similar functions across the soil habitat. This overlap in diversity and function is unusual compared to different environments – like the tropical rainforest. Organisms within a tropical rainforest do not have large overlapping niches, as appears within the soil food web, which is why there is so much diversity within these habitats. One of the reasons for these overlapping niches, is the differences in scale of the organisms; for example, different species of bacteria break down plant material chemically, whilst springtails and earthworms fragment the litter, breaking the plant material's physical structure.

The organisms living within the soil are also functionally diverse, as well as biologically. For example, free living bacterial species have many different

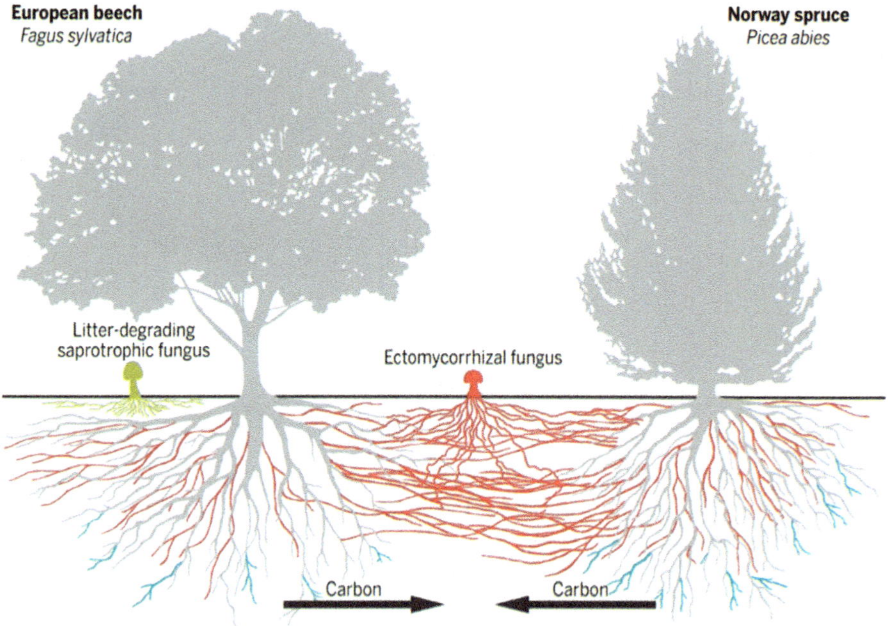

Fig. 9.1 Extent of interconnectedness of mycorrhizal fungal network linking different tree species. Carbon can move from one plant to another through these hyphal networks (van der Heijden 2016)

functions: mineralisers convert organic compounds to their mineral components; oxidisers convert nutrients into less available plant forms; nitrifyers convert ammonium to nitrate, making it more available for plant uptake; whilst denitrifyers convert nitrate to nitrogen (particularly within poorly drained soils), reducing the amount of nitrogen available for plant uptake. Studies have found that bacterial diversity strongly relates to soil pH, this suggests that soil bacteria are driven by the same ecological processes that govern higher organisms (Griffiths et al. 2011). Fungi are the other main group of microorganisms within the soil; these can be either saprotrophic (decomposers) or form symbiotic relationships with plants (mycorrhizal). Mycorrhizal fungi form associations with all gymnosperms and more than 83% of dicotyledonous and 79% of monocotyledonous plants (Kennedy and de Luna 2005). There are two main forms – arbuscular (form arbuscules within the plant root, designed for nutrient and water exchange) and ectomycorrhizal (where the hyphae do not penetrate the root cortex and form fruiting bodies (sporocarps)). Mycorrhizal hyphae are thinner in diameter than the thinnest of plant roots, allowing the mycorrhizal hyphae to penetrate smaller pores within the soil profile to obtain nutrients and water to exchange with the host plant for photosynthate products (Fig. 9.1). They have the potential to enhance plant growth when nutrients are limiting or the plant is growing in a stressed environment (drought or compaction), however the addition of

Fig. 9.2 Diversity of mesofauna extracted from an agricultural grassland soil sample, within this picture at least 12 species of Collembola are visible as well as three species of mite and a staphylinid beetle (Crotty 2008)

fertiliser and other nutrients, reduces the necessity of this symbiotic relationship, reducing its strength.

Nematodes are part of the microfauna and are one of the most abundant animal classes on earth; these microscopic, unsegmented round worms have been calculated to outnumber humans by 57 billion to one (van den Hoogen et al. 2019). Nematodes have been identified to make up five main functional groups, bacterivores, fungivores, herbivores, omnivores and predators. However, research has focused on plant parasitic/herbivorous nematodes e.g. potato cyst nematode, due to the amount of damage they cause (thought to be up to $100 billion USD per year). Within agriculture nematicide application regularly occurs to reduce the economic impact of plant parasitic nematodes, however studies have shown that non-target nematodes are also reduced (Wada and Toyota 2008; Timper et al. 2012; Waldo et al. 2019) this will have a knock-on impact on the functioning of the whole soil food web. Nematode abundance has been found to correlate with soil carbon (van den Hoogen et al. 2019), thereby decreasing the overall abundance of nematodes within the soil profile is likely to reduce the functionality of carbon cycling within the soil ecosystem.

The other most abundant group of organisms within the soil is the mesofauna (Fig. 9.2) – springtails (Collembola), mites (Acari) and other insects with <2 mm body width. Springtails and mites can be found in densities from 100,000 per m^2 up to 1 million individuals per m^2 (Behan-Pelletier 2003). However, because of their small size, the contribution of mesofauna to total soil respiration and biomass is low, typically 1–5% in temperate ecosystems (Hopkin 1997). Springtails play an essential role in the decomposer food web consuming recently fallen plant material and other decaying organisms. Larger species (>3 mm in length), directly comminute dead leaves whilst smaller species (<2 mm) are more likely to be microorganism feeders (Bernard and Felderhoff 2007) controlling bacterial and fungal populations. A few species are herbivorous e.g. *Sminthurus viridis*, although these are not normally a problem unless they are invasive (Crotty et al. 2016a). In general, Collembola are not

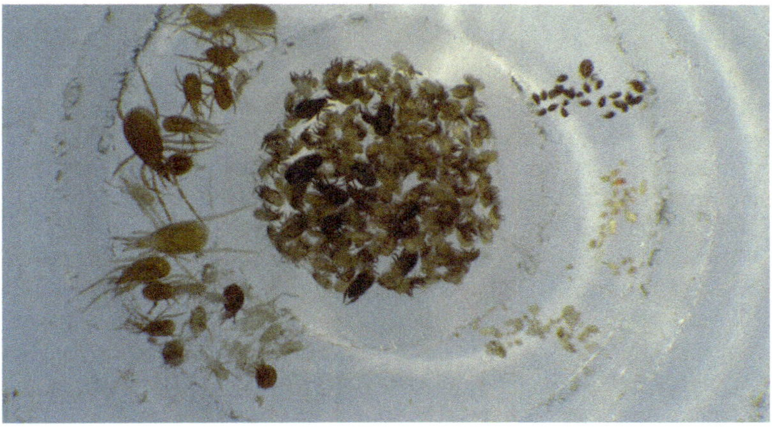

Fig. 9.3 Diversity of mites extracted from an agricultural grassland soil sample, separated into the main groups, left side of picture Mesostigmata, centre and top right Oribatid mites (centre Macropyline Oribatids; top right Brachypyline Oribatids), middle right side of picture Prostigmata, bottom right side Astigmata (Crotty 2010)

actively tunnelling and burrowing soil animals, but still have a large effect on soil structure and composition, for example some soils contain massive quantities of collembolan faecal pellets, which can provide a beneficial release of essential nutrients to crops as they are slowly broken down by microbes (Hopkin 1997).

Mites (Acari) are the other most populous group within the mesofauna (Fig. 9.3). Mites have far greater functional diversity than springtails, with a large number of predatory species, as well as decomposer species. There are three main lineages of mites in soil (Mesostigmata, Prostigmata and Oribatida); Mesostigmata, usually form about 20% of the total mite fauna (Klompen 2016) and are mostly predatory (except the Uropodidae, which also feed on decaying plant material). Prostigmata are the most functionally diverse group, with orders of magnitude size differences between the Heterostigmata and the Trombidiidae, as well as having a range of functions from decomposers to predatory. Prostigmata usually form about 40–50% of the total mite abundance (Klompen 2016). Oribatid mites are usually very common, although only in stable environments, and can be very rare in conventionally tilled agricultural fields. However, Astigmata (part of the Oribatid group) which are often ignored in studies of soil fauna, as they are identified as "dust mites", are often found in disturbed soils such as cultivated fields.

Earthworms are often cited as the most important group of soil organisms, referred to as "ecosystem engineers" as they affect the whole environment and either directly or indirectly have an impact on the other species inhabiting the same space (Jones et al. 1994). This is due to the large amount of soil processed by earthworms, improving soil structure, soil water infiltration rates, and recycling of nutrients. Earthworms can be divided into three main functional groups (epigeic, endogeic and anecic) or two ecotypes (detritvores and geophages). Epigeic species are detritivorous, living near the soil surface consuming recently fallen plant material. Endogeic species are geophagous, burrowing horizontally within the soil profile, consuming and processing the soil. Anecic species are detritivorous, however they transport dead plant material from the soil surface into their permanent vertical burrows, where it decomposes before consumption. Anecic earthworms thus directly contribute to the movement of organic matter within the soil profile and also water infiltration rates. A meta-analysis of studies on the effect of earthworms on crop yields has shown that earthworms can increase crop yields up to 25% when comparing fields with and without earthworms (van Groenigen et al. 2014). Knowing how many earthworms there are within a habitat has been postulated as a way of assessing how healthy a soil is. However, large scale surveying is limited and the analysis by Phillips et al. (2019) shows that earthworm abundance and diversity peaks at higher latitudes (e.g. temperate zones), which is the opposite to aboveground fauna.

2 Managing Arable Habitats to Promote Soil Organisms

Crop establishment method is one of the main factors affecting soil organisms. Mechanical disturbance of the soil habitat has the biggest impact on all soil organisms. Historically tillage was thought to promote bacterial growth (and the rest of the food chain linked to it within the soil food web), whilst damaging hyphal networks, reducing fungal growth (Hedlund et al. 2004). Tillage increases nutrient mobility within the soil, increasing decomposition and mineralisation (Hendrix et al. 1986). However, recent studies are showing an enhancement of the bacterial community when tillage is reduced, with greater bacterial diversity in no-till systems (Schmidt et al. 2018). Dependent on function, tillage can affect the fungal community to a large extent, particularly when combined with crop type. Mycorrhizal fungal species have been shown to be present in greater abundance in no-till systems, compared to pathogenic fungi which have been found to be more abundant where soil is ploughed (Detheridge et al. 2016). No-till systems may also promote the relationship between mycorrhizal fungi and crop plants because plants preferentially follow old root channels as they grow, and therefore are more likely to encounter fungal propagules than in tilled soil (Kabir 2005).

Assessing the effect of tillage on the next level of the soil food web – Nematodes, Collembola and mites, there are varying effects dependent on where the organisms live within the soil pore. For example, Nematodes live within the water filled pore space, and this, along with their small size means they are relatively unperturbed by tillage processes (Timper et al. 2012). There are varying effects on the mesofauna (Collembola and mites) dependent on their lifecycle strategy and functional group. Studies focusing on Collembola diversity and abundance under tillage systems have found contrasting results, with some studies showing increases in abundance and diversity when tillage was reduced (Bedano et al. 2006; Coulibaly et al. 2017), however the review by van Capelle (2012) found individual collembolan density and species diversity decreased when tillage was reduced. Within the Acari, tillage affects different lineages to a varying extent, for example Oribatids can become locally extinct in regularly tilled arable soils, whereas the Heterostigmata (Prostigmata) and Astigmata can be found to be extremely abundant. However, these differences are largely down to lifecycle strategy and functional group as stated previously. Oribatids are relatively long lived and slow to develop – moving less than 8 metres per annum (Lehmitz et al. 2012) thereby less likely to recolonise an area when their food source (decomposing plant material) has been removed, compared to a predatory mite with a short lifecycle and active foraging ecology.

Macrofauna are most likely to be effected by tillage processes, due to their large size and functional ecology. Frequent studies have shown earthworms to be negatively affected by tillage as found in the global meta-analysis by Briones and Schmidt (2017). The impact of tillage on earthworms is fivefold – firstly, through direct damage to the earthworms due to their large size; secondly, through the physical disruption of the soil itself, damaging earthworm burrows necessitating investment of energy and fitness in the rebuilding of burrows; thirdly, through the burial of crop residues, effecting food availability and location within the soil profile; fourthly, through the enhanced risk of predation as the earthworms are moved to the soil surface, and finally, through the change in climatic conditions (water content and temperature) within the soil – increasing the risk of drying out and exposure to direct sunlight. Studies showing a variation in earthworm responses are likely due to different species, timing of tillage operation, soil, crop and climatic variations (Kladivko 2001). Overall the critical review of tillage systems by van Capelle et al. (2012), highlight the necessity to assess the impact of these systems on all soil organisms, as tillage driven impacts may differ dependent on taxa, soil type and soil depth.

Another broad ranging factor affecting soil organisms within arable habitats is the application of pesticides. All pesticides have wider implications than their target pest. Fungicides applied to cereal crops to prevent rust are likely to affect mycorrhiza association within the soil. Nematicides used to reduce potato cyst nematode will also reduce populations of bacterial and fungal feeding nematodes. Insecticides often applied as a foliar spray are used to reduce herbivorous insects, however they are commonly found to reduce the natural predator populations that would normally reduce these insect pests. Insecticide seed dressings or granules will also reduce soil faunal populations, including springtails, mites, beetles, fly larvae, pscopterans, nematodes and earthworms (Pisa et al. 2017). Molluscicides have also been found to detrimentally impact non-target soil fauna for example Collembola behaviour, reproduction and survival were found to be negatively affected by molluscicides (Cardoso et al. 2015). Herbicides have also been found to negatively affect soil fauna – for example glyphosate has been found to decrease root mycorrhization and AMF spore biomass (Zaller et al. 2014); inhibited growth of the soil bacterium *Pseudomonas* species (Aristilde et al. 2017); as well as reducing earthworm activity and reproduction rates (Gaupp-Berghausen et al. 2015) in microcosm studies.

The likelihood of pesticides effecting non-target organisms is dependent on application mode, timing of application and climatic conditions at the time (and after). Seed dressings may have a greater effect than spray applications because they are systemic pesticides that protect the seed in the initial phase and are then incorporated by the growing plant. For example, seed dressings in winter wheat significantly reduced the surface activity of earthworms with no difference whether it was an insecticidal or fungicidal dressing used (Van Hoesel et al. 2017). It is not just above-ground pollinators that are affected by neonicotinoids. Studies have shown residues within soil from neonicotinoid seed dressings have significantly reduced soil fauna populations including Collembola and earthworms (Pisa et al. 2017). If rainfall follows foliar application of pesticides, these chemical compounds will reach the soil faster than expected and the likelihood of detrimental effects will increase. Pesticide wash-off has been found to be greater for high and infrequent rainfall compared to light and frequent rainfall (Aslam et al. 2018), mulch has also been shown to reduce the amount of pesticide infiltrating the soil – a second benefit of mulching as well as reducing soil erosion. However, if there is a sustained period of drought there is also the potential for prolonged detrimental effects as the active compounds may remain on the soil surface for longer than predicted leading to increased contact with soil fauna (Ng et al. 2014).

Application of fertilisers can have an effect on soil faunal populations, this can be both positive (application of a food source for soil organisms to utilise) or negative (soil acidification – changing the pH of the soil to have less hospitable conditions). Mineral (artificial) fertilisers have limited direct effects on the majority of soil fauna, however they have been found to weaken the association of mycorrhizal fungi with crop plants, due to a reduction in the necessity for the symbiotic relationship between mycorrhiza and crops. Organic fertiliser both slurry (dung and urine) and farm yard manure (dung, urine and bedding material), have been shown to increase the abundance of soil fauna, particularly detritivorous organisms like earthworms (Leroy et al. 2008; van Eekeren et al. 2009), Collembola, nematodes and microbes (Ponge et al. 2013). Different organic fertilisers have been shown to have varying effects on the different earthworm functional groups – detritivorous (epigeic and anecic species) and geophagous (endogeic species). Detritivorous earthworms were found in increased abundance when farmyard manure was used as fertiliser (in comparison to slurry), whereas geophagous earthworms were found to show no difference (Onrust and Piersma 2019). However, over-fertilisation with slurry or biodigestate will have detrimental effects on soil biodiversity, even if it does increase the amount of organic matter within the soil, this has been seen on studies in Oribatid mites (Bosch-Serra et al. 2014) and earthworms (Bhogal et al. 2016).

3 Relationship Between Arable Crops and Soil Organisms

Intensification within arable cropping systems has led to a decrease in aboveground biodiversity, with fewer varieties of cultivars used across regions as well as genetic diversity of crops. Since the 1960s, farmers have stopped planting crop varieties of locally adapted populations (landraces) in favour of more widely adapted varieties produced through plant breeding systems (modern varieties) (Millennium Ecosystem Assessment 2005). However, in the last decade there has been increased interest in historic varieties and native relatives, to increase the naturally occurring resistance genes within a crop and could lead to improved synergy with the local soil habitat. Crop rotation has also reduced, with some farmers focusing on simplified rotations, with some fields regularly growing wheat after wheat. Agri-environment schemes have led to increases in the number of crops within a farm rotation, through common agricultural policy payments for a minimum of three crops being grown on farm during the same harvest year. Increasing the amount of plant diversity through crop rotations and the addition of cover crops has been found to significantly increase the amount of carbon and nitrogen stored within the soil, which in turn enhances soil biodiversity (McDaniel et al. 2014). Crop type can influence soil biodiversity, for example a number of studies have found earthworm abundance to be enhanced when clover is used as a forage crop or intercrop (Schmidt et al. 2003; Crotty et al. 2015); it is hypothesised this is due to enhanced quantity, quality and continuity of food supply throughout the year. Crotty et al. (2016b), investigating the legacy effect of forage crops as part of a cereal rotation showed a dramatic decrease in earthworm abundance in all treatments when comparing their 2015 forage results to earthworm abundance after a change to a cereal rotation (Fig. 9.4). This decrease in abundance is likely to be due to the reduction in plant litter available from cereal crops during the growing season, compared to forage crops. The influence of winter-sown in comparison to spring-sown crops is also a factor effecting soil biodiversity abundance. Spring-sown crops, if a cover crop is not incorporated into the rotation, leave the soil bare from September to February. This reduces the amount of food available during the winter for soil organisms. Due to the nature of the soil environment, within temperate regions soil organisms are more likely to be inactive when the soil is dry through the summer, rather than over winter. Some studies have assessed the abundance of soil fauna after winter or spring sown crops, and found reduction in abundances (e.g. Collembola (Frampton and van den Brink 2002)), however this is an area that needs further investigation.

Concern has been growing about the lack of heterogeneity within the agricultural landscape, of particular concern are areas where arable plant species are native, as arable fields represent an important habitat for these species (Albrecht et al. 2016: Storkey, this volume, Chap. 2). Rare arable plant species are adapted to less intensive traditional farming methods, hence the recent declines; reintroduction studies have focused on aboveground interactions with invertebrates, soil, climate and emergence patterns (e.g. Torra et al. 2020), with little consideration for below-ground fauna. However, rare arable plants are sometimes included within wildflower strips to

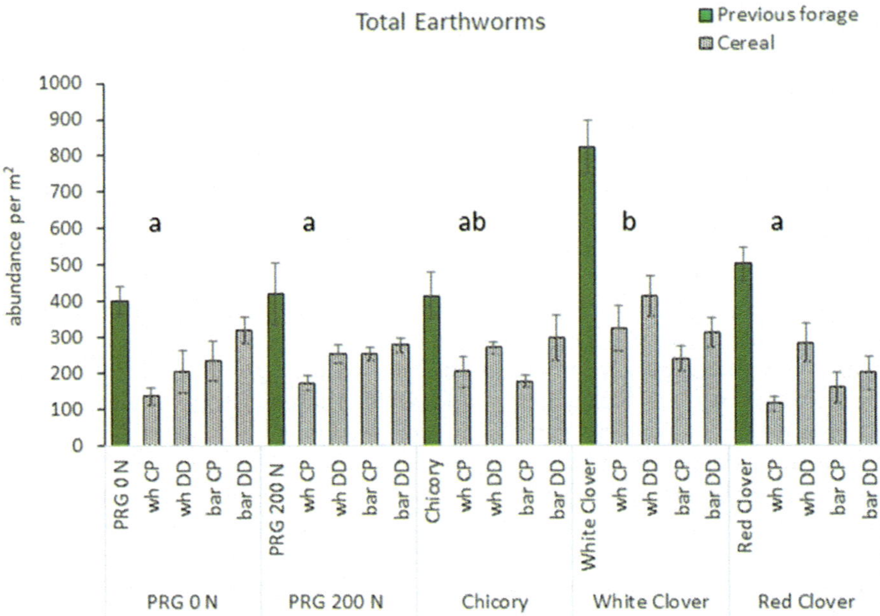

Fig. 9.4 Change in earthworm abundance per m^2 after changing from a 3-year ley to a cereal rotation (Crotty et al. 2016b). Leys were *Lolium perenne* with no nitrogen fertiliser applied for 3 years (PRG 0 N), *L. perenne* with 200 kg N per year applied annually (PRG 200 N), *Cichorium intybus*, *Trifolium repens* and *T. pratense*. Post-forage cereal was planted and abundance measured – year one post forage was spring wheat with plots split and either ploughed (Wh CP) or direct drilled (Wh DD); year two post forage was winter barley with plots remaining split as plough (Bar CP) or direct drilled (Bar DD)

increase biodiversity within agroecosystems. Sown wildflower strips have been found to be very beneficial for insect conservation, both for aboveground insects and soil macrofauna (Haaland et al. 2011). Conservation headlands have also been incorporated into agri-environment schemes, with reductions in pesticides, fertiliser and sowing rates, and uncropped margins and plots to increase arable diversity utilising the seed bank, as there is a reduction in intensive farming methods within these areas, they are also likely to promote below ground biodiversity. Set aside schemes and the incorporation of grass leys promote species rich sowing, and also reduce pesticide use. Arable farmers are utilising grass leys as a way to reduce grass weeds (particularly against black grass *Alopecurus myosuroides*) without continued herbicide spraying, this will lead to an increase in soil organic matter and can also promote soil biodiversity within an intensive arable landscape. An important and often forgotten consideration is the link between soil organisms and plant diversity and abundance, studies have found gastropods and earthworms can disperse endangered herbs within the agri-environment (Türke et al. 2013). There is limited research regarding the link between soil macrofauna and rare arable plants, and could explain reductions in germination rates within intensive arable systems,

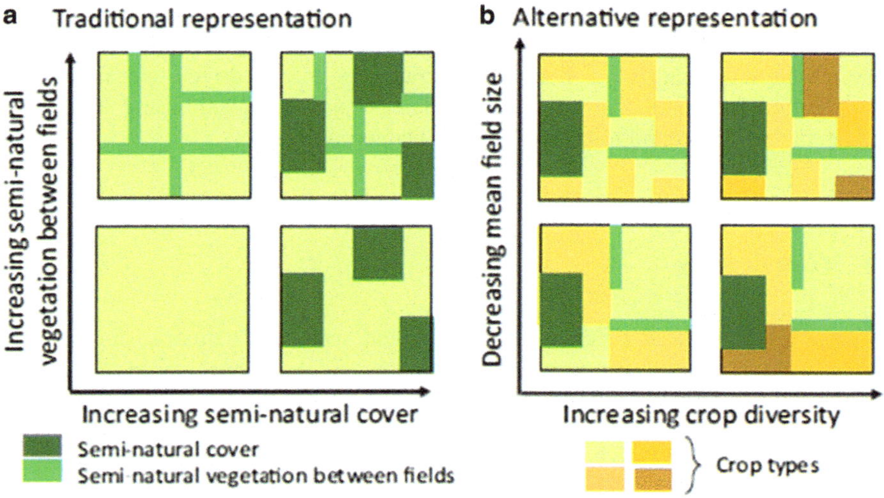

Fig. 9.5 (**a**) Traditional and (**b**) alternative representations of agricultural landscape heterogeneity (Sirami et al. 2019)

warranting further investigation. Tillage system, soil organic matter content, presence of surface residues and seed traits are important factors influencing surface seed predation by invertebrates (Barberi et al. 2010). Increasing biodiversity within agriculture has been the focus of numerous studies, often focussing on the benefits of smaller fields and non-crop elements (margins, hedges and grasslands) within an arable landscape (Šálek et al. 2018). However, increasing crop heterogeneity has also been found to be increasingly beneficial (Fig. 9.5). Viewing and valuing the arable landscape as biodiverse whilst still applying agricultural intensification principles, can show that crop heterogeneity is a major driver of multitrophic diversity in agricultural landscapes (Sirami et al. 2019).

Hedgerows within the arable landscape are also a major source of biodiversity and can reduce the impact of soil erosion, surface runoff and spray drift. Hedgerows often act as corridors throughout the landscape, with many aboveground fauna preferentially following them, whilst foraging, rather than crossing arable fields directly. Hedgerows have been found to host a distinct arbuscular mycorrhizal community, with greater earthworm abundance than arable fields; having less compacted soil with lower soil moisture but higher concentrations of nitrate and phosphate than adjacent fields (Holden et al. 2019). Hedgerows have also been found to elicit increased benefits within arable fields, although these benefits decrease with distance from the hedgerow. For example, the meta-analysis by Laura et al. (2017) found increased crop yields within the arable fields up to 20 times the distance of the height of the hedgerow, as well as higher soil carbon stocks within the soil below the hedgerow and adjacent to the hedgerow. However, the effect of orientation of hedgerow – perpendicular or parallel to tramlines within an arable field, is often not accounted for in research studies. When a field margin is perpendicular to a tramline there will be greater traffic from farmers using this area to turn back onto the cropped field with heavy machinery (sprayers, combine harvester etc.), whereas a margin parallel with the tramlines will be driven over much less. This difference in field movements will lead to soil compaction and impact the potential benefits found to be provided by hedgerows.

4 Conclusion

The soil is the most abundant and diverse habitat, with a large number of organisms of varying sizes residing within the same three-dimensional space. This large biodiversity is directly linked to the functional diversity that also occurs within the soil habitat. Due to the variation in size of these organisms, some organisms with very similar functions can occur but are found in very different space – for example within the micro-aggregate, the pore water or within an earthworm burrow. Resilience within an ecosystem decreases with the increasing number of trophic levels due to the increasing complexity of the ecosystem, therefore it takes longer to reestablish complex food web relationships when they are disturbed (Nielsen 2019). Within the soil food web, the number of trophic levels is difficult to distinguish, with studies showing habitats with similar taxa having different soil food webs, using different basal resources (Crotty et al. 2014). Disturbance selects for shorter food chains; in farmed soils, disturbance can be chemical (pesticides, fertilizers) or physical (cultivation, organic matter incorporation, removal of surface organic layer). The frequency of soil disturbance by physical or chemical agricultural inputs and other disturbances is important to the overall assemblage of soil biota and food chain length (Nielsen 2019). Intensification of arable farming has been occurring since World War II, both in the UK and across Europe, there has been a fourfold increase in crop yields over this time, but a 50% decrease in hedgerows and widespread declines of many groups of organisms (Robinson and Sutherland 2002). Greater understanding of the benefits of soil biodiversity to crop yields and educating farmers to the profitability of maintaining a healthy soil through the incorporation of wider crop rotations, agri-environment schemes, wildflower strips, and hedgerows will lead to increasing soil biodiversity.

References

Albrecht H, Cambecèdes J, Lang M, Wagner M (2016) Management options for the conservation of rare arable plants in Europe. Bot Lett 163:389–415

Anderson JM (1975) The enigma of soil animal species. In: Vanek J (ed) Progress in soil zoology: proceedings of the fifth international colloquium on Progress in soil zoology. W Junk/Prague Academia, The Hague, pp 51–58

Aristilde L, Reed ML, Wilkes RA, Youngster T, Kukurugya MA, Katz V, Sasaki CRS (2017) Glyphosate-induced specific and widespread perturbations in the metabolome of soil pseudomonas species. Front Environ Sci 5

Aslam S, Iqbal A, Lafolie F, Recous S, Benoit P, Garnier P (2018) Mulch of plant residues at the soil surface impact the leaching and persistence of pesticides: a modelling study from soil columns. J Contam Hydrol 214:54–64

Barberi P, Burgio G, Dinelli G, Moonen AC, Otto S, Vazzana C, Zanin G (2010) Functional biodiversity in the agricultural landscape: relationships between weeds and arthropod fauna. Weed Res 50:388–401

Bedano JC, Cantu MP, Doucet ME (2006) Soil springtails (Hexapoda: Collembola), symphylans and pauropods (Arthropoda: Myriapoda) under different management systems in agroecosystems of the subhumid Pampa (Argentina). Eur J Soil Biol 42:107–119

Behan-Pelletier VM (2003) Acari and Collembola biodiversity in Canadian agricultural soils. Can J Soil Sci 83:279–288

Bernard EC, Felderhoff KL (2007) Biodiversity explosion: Collembola (springtails) of Great Smoky Mountains National Park. Southeast Nat 6:175–182

Bhogal A, Taylor M, Nicholson F, Rollett A, Williams JR, Newell-Price P, Chambers B, Litterick A, Whittingham M (2016) Effect of repeated digestate and compost applications on soil and crop quality. DC-Agri WRAP

Bosch-Serra ÀD, Padró R, Boixadera-Bosch RR, Orobitg J, Yagüe MR (2014) Tillage and slurry over-fertilization affect oribatid mite communities in a semiarid Mediterranean environment. Appl Soil Ecol 84:124–139

Bot A, Benites J (2005) The importance of soil organic matter. Key to drought-resistant soil and sustained food production. Food and Agriculture Organisation, Rome

Briones MJI (2014) Soil fauna and soil functions: a jigsaw puzzle. Front Environ Sci 2:1–22

Briones MJI, Schmidt O (2017) Conventional tillage decreases the abundance and biomass of earthworms and alters their community structure in a global meta-analysis. Glob Chang Biol 23:4396–4419

Cardoso DN, Santos MJG, Soares AMVM, Loureiro S (2015) Molluscicide baits impair the life traits of Folsomia candida (Collembola): possible hazard to the population level and soil function. Chemosphere 132:1–7

Coulibaly SFM, Coudrain V, Hedde M, Brunet N, Mary B, Recous S, Chauvat M (2017) Effect of different crop management practices on soil Collembola assemblages: a 4-year follow-up. Appl Soil Ecol 119:354–366

Crotty FV (Photographer) (2008) Diverse mesofauna sample [photograph]. Rothamsted, North Wyke, Devon, UK

Crotty FV (Photographer) (2010) Mites separated into main groups [photograph]. Rothamsted, North Wyke, Devon, UK

Crotty FV, Blackshaw RP, Adl SM, Inger R, Murray PJ (2014) Divergence of feeding channels within the soil food web determined by ecosystem type. Ecol Evol 4:1–13

Crotty FV, Fychan R, Scullion J, Sanderson R, Marley CL (2015) Assessing the impact of agricultural forage crops on soil biodiversity and abundance. Soil Biol Biochem 91:119–126

Crotty FV, Fychan R, Benefer CM, Allen D, Shaw P, Marley CL (2016a) First documented pest outbreak of the herbivorous springtail Sminthurus viridis (Collembola) in Europe. Grass Forage Sci 71:699–704

Crotty FV, Fychan R, Sanderson R, Rhymes JR, Bourdin F, Scullion J, Marley CL (2016b) Understanding the legacy effect of previous forage crop and tillage management on soil biology, after conversion to an arable crop rotation. Soil Biol Biochem 103:241–252

Detheridge AP, Brand G, Fychan R, Crotty FV, Sanderson R, Griffith GW, Marley CL (2016) The legacy effect of cover crops on soil fungal populations in a cereal rotation. Agric Ecosyst Environ 228:49–61

Frampton GK, van den Brink PJ (2002) Influence of cropping on the species composition of epigeic Collembola in arable fields. Pedobiologia 46:328–337

Gaugler R (1990) Entomopathogenic nematodes in biological control. CRC Press, Boca Raton. https://doi.org/10.1201/9781351071741

Gaupp-Berghausen M, Hofer M, Rewald B, Zaller JG (2015) Glyphosate-based herbicides reduce the activity and reproduction of earthworms and lead to increased soil nutrient concentrations. Sci Rep 5:12886

Giller PS (1996) The diversity of soil communities, the 'poor man's tropical rainforest. Biodivers Conserv 5:135–168

Griffiths RI, Thomson BC, James P, Bell T, Bailey M, Whiteley AS (2011) The bacterial biogeography of British soils. Environ Microbiol 13:1642–1654

Haaland C, Naisbit RE, Bersier L-F (2011) Sown wildflower strips for insect conservation: a review. Insect Conserv Divers 4:60–80

Hedlund K, Griffiths B, Christensen S, Scheu S, Setala H, Tscharntke T, Verhoef H (2004) Trophic interactions in changing landscapes: responses of soil food webs. Basic Appl Ecol 5:495–503

Hendrix PF, Parmelee RW, Crossley DA, Coleman DC, Odum EP, Groffman PM (1986) Detritus food webs in conventional and no-tillage agroecosystems. Bioscience 36:374–338

Holden J, Grayson RP, Berdeni D, Bird S, Chapman PJ, Edmondson JL, Firbank LG, Helgason T, Hodson ME, Hunt SFP, Jones DT, Lappage MG, Marshall-Harries E, Nelson M, Prendergast-Miller M, Shaw H, Wade RN, Leake JR (2019) The role of hedgerows in soil functioning within agricultural landscapes. Agric Ecosyst Environ 273:1–12

Hopkin SP (1997) Biology of springtails. Oxford University Press, Oxford

Jones CG, Lawton JH, Shachak M (1994) Organisms as ecosystem engineers. Oikos 69:373–386

Kabir Z (2005) Tillage or no-tillage: impact on mycorrhizae. Can J Plant Sci 85:23–29

Kennedy AC, de Luna LZ (2005) Rhizosphere. In: Hillel D (ed) Encyclopedia of soils in the environment. Elsevier, Oxford, pp 399–406

Kladivko EJ (2001) Tillage systems and soil ecology. Soil Tillage Res 61:61–76

Klompen H (2016) Mites. In: Lal R (ed) Encyclopedia of soil science. Marcel Dekker, New York, pp 1488–1490

Laura VV, Bert R, Steven B, Pieter DF, Victoria N, Paul P, Kris V (2017) Ecosystem service delivery of agri-environment measures: a synthesis for hedgerows and grass strips on arable land. Agric Ecosyst Environ 244:32–51

Lehmitz R, Russell D, Hohberg K, Christian A, Xylander WER (2012) Active dispersal of oribatid mites into young soils. Appl Soil Ecol 55:10–19

Leroy BLM, Schmidt O, Van den Bossche A, Reheul D, Moens M (2008) Earthworm population dynamics as influenced by the quality of exogenous organic matter. Pedobiologia 52:139–150

McDaniel MD, Tiemann LK, Grandy AS (2014) Does agricultural crop diversity enhance soil microbial biomass and organic matter dynamics? A meta-analysis. Ecol Appl 24:560–570

Millennium Ecosystem Assessment (2005) Ecosystems and human well-being: synthesis. Washington, DC

Ng EL, Bandow C, Proença DN, Santos S, Guilherme R, Morais PV, Römbke J, Sousa JP (2014) Does altered rainfall regime change pesticide effects in soil? A terrestrial model ecosystem study from Mediterranean Portugal on the effects of pyrimethanil to soil microbial communities under extremes in rainfall. Appl Soil Ecol 84:245–253

Nielsen UN (2019) Soil fauna assemblages: global to local scales. Cambridge University Press, Cambridge

Nielsen UN, Osler GHR, Campbell CD, Neilson R, Burslem DFRP, van der Wal R (2010) The enigma of soil animal species diversity revisited: the role of small-scale heterogeneity. PLoS One 5:Article No.: e11567

Nielsen UN, Ayres E, Wall DH, Bardgett RD (2011) Soil biodiversity and carbon cycling: a review and synthesis of studies examining diversity–function relationships. Eur J Soil Sci 62:105–116

Onrust J, Piersma T (2019) How dairy farmers manage the interactions between organic fertilizers and earthworm ecotypes and their predators. Agric Ecosyst Environ 273:80–85

Phillips HRP, Guerra CA, Bartz MLC, Briones MJI, Brown G, Crowther TW, Ferlian O, Gongalsky KB, van den Hoogen J, Krebs J, Orgiazzi A, Routh D, Schwarz B, Bach EM, Bennett J,

Brose U, Decaëns T, König-Ries B, Loreau M, Mathieu J, Mulder C, van der Putten WH, Ramirez KS, Rillig MC, Russell D, Rutgers M, Thakur MP, de Vries FT, Wall DH, Wardle DA, Arai M, Ayuke FO, Baker GH, Beauséjour R, Bedano JC, Birkhofer K, Blanchart E, Blossey B, Bolger T, Bradley RL, Callaham MA, Capowiez Y, Caulfield ME, Choi A, Crotty FV, Dávalos A, Cosin DJD, Dominguez A, Duhour AE, van Eekeren N, Emmerling C, Falco LB, Fernández R, Fonte SJ, Fragoso C, Franco ALC, Fugère M, Fusilero AT, Gholami S, Gundale MJ, López MG, Hackenberger DK, Hernández LM, Hishi T, Holdsworth AR, Holmstrup M, Hopfensperger KN, Lwanga EH, Huhta V, Hurisso TT, Iannone BV, Iordache M, Joschko M, Kaneko N, Kanianska R, Keith AM, Kelly CA, Kernecker ML, Klaminder J, Koné AW, Kooch Y, Kukkonen ST, Lalthanzara H, Lammel DR, Lebedev IM, Li Y, Lidon JBJ, Lincoln NK, Loss SR, Marichal R, Matula R, Moos JH, Moreno G, Morón-Ríos A, Muys B, Neirynck J, Norgrove L, Novo M, Nuutinen V, Nuzzo V, Rahman PM, Pansu J, Paudel S, Pérès G, Pérez-Camacho L, Piñeiro R, Ponge J-F, Rashid MI, Rebollo S, Rodeiro-Iglesias J, Rodríguez MÁ, Roth AM, Rousseau GX, Rozen A, Sayad E, van Schaik L, Scharenbroch BC, Schirrmann M, Schmidt O, Schröder B, Seeber J, Shashkov MP, Singh J, Smith SM, Steinwandter M, Talavera JA, Trigo D, Tsukamoto J, de Valença AW, Vanek SJ, Virto I, Wackett AA, Warren MW, Wehr NH, Whalen JK, Wironen MB, Wolters V, Zenkova IV, Zhang W, Cameron EK, Eisenhauer N (2019) Global distribution of earthworm diversity. Science 366:480–485

Pisa L, Goulson D, Yang E-C, Gibbons D, Sánchez-Bayo F, Mitchell E, Aebi A, van der Sluijs J, MacQuarrie CJK, Giorio C, Long EY, McField M, Bijleveld van Lexmond M, Bonmatin J-M (2017) An update of the Worldwide Integrated Assessment (WIA) on systemic insecticides. Part 2: Impacts on organisms and ecosystems. Environ Sci Pollut Res

Ponge J-F, Pérès G, Guernion M, Ruiz-Camacho N, Cortet J, Pernin C, Villenave C, Chaussod R, Martin-Laurent F, Bispo A, Cluzeau D (2013) The impact of agricultural practices on soil biota: a regional study. Soil Biol Biochem 67:271–284

Raynaud X, Nunan N (2014) Spatial ecology of bacteria at the microscale in soil. PLoS One 9: e87217

Robinson RA, Sutherland WJ (2002) Post-war changes in arable farming and biodiversity in Great Britain. J Appl Ecol 39:157–176

Šálek M, Hula V, Kipson M, Daňková R, Niedobová J, Gamero A (2018) Bringing diversity back to agriculture: smaller fields and non-crop elements enhance biodiversity in intensively managed arable farmlands. Ecol Indic 90:65–73

Schmidt O, Clements RO, Donaldson G (2003) Why do cereal-legume intercrops support large earthworm populations? Appl Soil Ecol 22:181–190

Schmidt R, Gravuer K, Bossange AV, Mitchell J, Scow K (2018) Long-term use of cover crops and no-till shift soil microbial community life strategies in agricultural soil. PLoS One 13:e019295

Sirami C, Gross N, Baillod AB, Bertrand C, Carrié R, Hass A, Henckel L, Miguet P, Vuillot C, Alignier A, Girard J, Batáry P, Clough Y, Violle C, Giralt D, Bota G, Badenhausser I, Lefebvre G, Gauffre B, Vialatte A, Calatayud F, Gil-Tena A, Tischendorf L, Mitchell S, Lindsay K, Georges R, Hilaire S, Recasens J, Solé-Senan XO, Robleño I, Bosch J, Barrientos JA, Ricarte A, Marcos-Garcia MÁ, Miñano J, Mathevet R, Gibon A, Baudry J, Balent G, Poulin B, Burel F, Tscharntke T, Bretagnolle V, Siriwardena G, Ouin A, Brotons L, Martin J-L, Fahrig L (2019) Increasing crop heterogeneity enhances multitrophic diversity across agricultural regions. Proc Natl Acad Sci 116:16442

Swift MJ, Heal OW, Anderson JM (1979) Decomposition in terrestrial ecosystems. Blackwell Scientific Publications, Oxford

Tiedje JM (1995) Approaches to the comprehensive evaluation of prokaryote diversity of a habitat. In: Allsopp D, Hawksworth DL, Colwell RR (eds) Microbial diversity and ecosystem functioning. CAB International, Wallingford, pp 73–87

Tiedje JM, Asuming-Brempong S, Nusslein K, Marsh TL, Flynn SJ (1999) Opening the black box of soil microbial diversity. Appl Soil Ecol 13:109–122

Timper P, Davis R, Jagdale G, Herbert J (2012) Resiliency of a nematode community and suppressive service to tillage and nematicide application. Appl Soil Ecol 59:48–59

Torra J, Forcella F, Recasens J, Royo-Esnal A (2020) Emergence patterns of rare arable plants and conservation implications. Plan Theory 9:309

Turbé A, De Toni A, Benito P, Lavelle P, Lavelle P, Ruiz N, Van der Putten WH, Labouze E, Mudgal S (2010) Soil biodiversity: functions, threats and tools for policy makers. In: Bio Intelligence Service, IRD, and NIOO (ed) European Commission (DG Environment). Paris

Türke M, Blattmann T, Knop E, Kindermann A, Prestele J, Marquez L, Eisenhauer N, Fischer C (2013) Weeds and endangered herbs have unforeseen dispersal helpers in the agri-environment: gastropods and earthworms. Renew Agric Food Syst 28:380–383

van Capelle C, Schrader S, Brunotte J (2012) Tillage-induced changes in the functional diversity of soil biota – a review with a focus on German data. Eur J Soil Biol 50:165–181

van den Hoogen J, Geisen S, Routh D, Ferris H, Traunspurger W, Wardle DA, de Goede RGM, Adams BJ, Ahmad W, Andriuzzi WS, Bardgett RD, Bonkowski M, Campos-Herrera R, Cares JE, Caruso T, de Brito CL, Chen X, Costa SR, Creamer R, Mauro da Cunha Castro J, Dam M, Djigal D, Escuer M, Griffiths BS, Gutiérrez C, Hohberg K, Kalinkina D, Kardol P, Kergunteuil A, Korthals G, Krashevska V, Kudrin AA, Li Q, Liang W, Magilton M, Marais M, JAR M, Matveeva E, Mayad EH, Mulder C, Mullin P, Neilson R, TAD N, Nielsen UN, Okada H, JEP R, Pan K, Peneva V, Pellissier L, Carlos Pereira da Silva J, Pitteloud C, Powers TO, Powers K, Quist CW, Rasmann S, Moreno SS, Scheu S, Setälä H, Sushchuk A, Tiunov AV, Trap J, van der Putten W, Vestergård M, Villenave C, Waeyenberge L, Wall DH, Wilschut R, Wright DG, Yang J-I, Crowther TW (2019) Soil nematode abundance and functional group composition at a global scale. Nature 572:194–198

van der Heijden MGA (2016) Underground networking. Science 352:290

van Eekeren N, de Boer H, Bloem J, Schouten T, Rutgers M, de Goede R, Brussaard L (2009) Soil biological quality of grassland fertilized with adjusted cattle manure slurries in comparison with organic and inorganic fertilizers. Biol Fertil Soils 45:595–608

van Groenigen JW, Lubbers IM, Vos HMJ, Brown GG, De Deyn GB, van Groenigen KJ (2014) Earthworms increase plant production: a meta-analysis. Sci Rep 4:636

Van Hoesel W, Tiefenbacher A, König N, Dorn VM, Hagenguth JF, Prah U, Widhalm T, Wiklicky V, Koller R, Bonkowski M, Lagerlöf J, Ratzenböck A, Zaller JG (2017) Single and combined effects of pesticide seed dressings and herbicides on earthworms, soil microorganisms, and litter decomposition. Front Plant Sci 8. https://doi.org/10.3389/fpls.2017.00215

Wada S, Toyota K (2008) Effect of three organophosphorous nematicides on non-target nematodes and soil microbial community. Microbes Environ 23:331–336

Waldo BD, Grabau ZJ, Mengistu TM, Crow WT (2019) Nematicide effects on non-target nematodes in bermudagrass. J Nematol 51:1–12

Wilkinson DM (2008) Testate amoebae and nutrient cycling: peering into the black box of soil ecology. Trends Ecol Evol 23:596–599

Zaller JG, Heigl F, Ruess L, Grabmaier A (2014) Glyphosate herbicide affects belowground interactions between earthworms and symbiotic mycorrhizal fungi in a model ecosystem. Sci Rep 4

Chapter 10
Agricultural Intensification, Sustainable Farming and the Fate of Arable Bryophytes in Switzerland

Irene Bisang and Ariel Bergamini

Abstract Land-use changes including agricultural intensification are major drivers of biodiversity loss. As other countries, Switzerland introduced agri-environment schemes (AES) to counteract declining agro-biodiversity. Among numerous bryophytes of the agricultural landscape, arable bryophytes have adapted to recurrent tillage. They are short-lived above-ground, while their diaspores may build up a persistent diaspore bank in the soil. We monitored the field hornwort and the Carolina hornwort, characteristic species of the Central European arable bryophyte flora over a period of almost three decades, and studied effects of Biodiversity Promoting Areas (BPAS), a key element of Swiss AES, on arable bryophyte diversity. Hornwort occurrences strongly declined, and stubble-fields left unprocessed until late autumn, the favourite habitat of arable specialist bryophytes, nearly disappeared from 1991 to 2018. The loss of suitable habitats follows from decreased cereal cultivation and direct post-harvest management. Different BPA-fallows, designed to promote birds, flowering plants, or ecosystem services, harboured arable bryophyte specialists, but their richness decreased with fallow age. Diaspore banks contained arable bryophyte specialists not occurring above-ground and may serve as a reservoir during unsuitable cultivation. Targeted modifications of current AES, aiming to maintain habitats favourable for arable bryophytes, are necessary to safeguard the typical arable bryophytes in the Swiss arable landscape.

Keywords Agricultural intensification · Arable bryophytes · *Anthoceros agrestis* · Hornworts · Changing agricultural practices · Stubble fields · *Phaeoceros carolinianus*

I. Bisang (✉)
Department of Botany, Swedish Museum of Natural History, Stockholm, Sweden
e-mail: Irene.bisang@nrm.se

A. Bergamini
WSL Swiss Federal Research Institute, Birmensdorf, Switzerland
e-mail: ariel.bergamini@wsl.ch

C. Hurford et al. (eds.), *The Changing Status of Arable Habitats in Europe*,
https://doi.org/10.1007/978-3-030-59875-4_10

1 Setting the Scene

Land-use change, combined with urban development, is considered the major driver for the dramatic loss of biodiversity and ecosystem services in Europe (IPBES 2018). It is reinforced by production-based subsidies that have contributed to unsustainable intensification of agricultural and forestry practices. Bryophytes, small green plants of hidden beauty (Box 10.1), are affected by these processes, as are larger and more eye-catching flowering plants, trees, birds, bees and butterflies (e.g., Donald et al. 2001; Lichtenberg et al. 2017). Of the more than 1800 European bryophytes, ca 20% are directly affected by agriculture and forestry management or by emissions resulting from nutrient loads and pesticide applications. Another third of the European bryophyte flora is impacted by modifications to natural ecosystems, which are often initiated in efforts to improve agricultural production (Hodgetts et al. 2019). In Switzerland, 10–15% of the bryophyte species have been found in habitats that are utilized for agriculture in one way or another, and many of their populations have decreased (Schnyder et al. 2004). Before the profound structural changes in the European landscape took place, agricultural activities had created and maintained a variety of habitats over thousands of years (Price 2000). Complex mosaics of crop fields, cultivated at low intensities, sustained high levels of biodiversity (Tscharntke et al. 2005; Emmerson et al. 2016). They harboured numerous plant species, which have adapted to and require regular disturbance through tillage. Agricultural disturbances create space for the germination, establishment and successful growth of these so-called ‘weeds’ (Nowak et al. 2019). Many other organisms, for example beneficial arthropods, rely directly on segetal native plants. Beyond being directly associated with arable plants, many organisms also depend on extensively managed, diverse arable landscapes to maintain species richness and density and to provide enhanced ecosystem services (Isaacs et al. 2009; Kennedy et al. 2013; Rusch et al. 2016).

Box 10.1: Bryophytes – Diversity, Life History and Eco-Physiological Characteristics

Bryophytes (mosses, liverworts and hornworts) are a diverse group of currently ca 18,000 known species of land plants that share a common life cycle. As in all extant land plants, their life cycle is characterized by the alternation of a gametophytic and a sporophytic generation. In contrast to all other land plants, the haploid gametophyte is the dominant vegetative stage. The green photosynthetic bryophyte gametophytes, which are either leafy or thalloid, are mostly a few centimetres tall. They develop sexual organs. Liquid water or a water film is necessary for the motile spermatozoids to reach the sessile egg cell. If fertilization is successful, a diploid embryo is formed which develops into a sporophyte. The sporophyte remains attached to the gametophyte and depends on it for most of its sustenance. The sporophyte produces, through

(continued)

Box 10.1 (continued)
meiosis, one-celled spores, which are the first stage in the haploid gametophyte generation. Many bryophyte species or populations rarely form sporophytes, relying instead on specialized vegetative diaspores such as gemmae, bulbils or tubers, or even remain non-reproductive during their entire life span. Bryophytes are "poikilohydric" plants, which means that they do not possess, or to a very limited degree, structural or functional mechanisms (e.g., conducting tissues) to regulate cell water content. The environment largely determines the water content of bryophyte tissues. Most species also have the capacity to tolerate dehydration over long periods and to recover quickly from it without physiological damage.

The field hornwort (*Anthoceros agrestis* Paton) and the Carolina hornwort (*Phaeoceros carolinianus* (Michx.) Prosk.) are the two focus species of our monitoring investigation (Fig. 10.2). Hornworts belong to a separate evolutionary lineage; thus, they represent a phylogenetically distinct element of the European arable flora (de Sousa et al. 2019). They comprise 200–250 species globally. Hornwort species diversity peaks in tropical environments, whereas only eight species occur in Europe and three in Switzerland (Swissbryophytes 2004–2019). Two European species that do not occur in Switzerland (Nees' hornwort, *Anthoceros neesii* Prosk. and round notothylas, *Notothylas orbicularis* (Schwein.) Sull.) are endangered in Europe, and Nees' hornwort is one of the few true endemic bryophyte species in Central Europe (Hodgetts et al. 2019).

2 Agricultural Intensification, Biodiversity Decline and Agro-Biodiversity Promotion

About one-quarter (10,450 km^2; excluding summer farm pastures) of the total Swiss area is agricultural land, and ca. 10% (3980 km^2) is arable land including fodder meadows (BfS 2019). Agricultural cultivation is concentrated to the lowlands, the so-called Swiss Plateau. Typically, crop rotation has been applied in the area since the end of the eighteenth century, nowadays alternating between cereal, maize, root crops and grassland used as mown fodder meadows and pastures.

Structural changes, plant breeding success, seed cleaning, and the mechanization and increased application of artificial fertilizers and pesticides during the third "agricultural revolution" (Baumann 2011) are the main drivers of the severe decline of various organism groups in agricultural ecosystems worldwide, among them bryophytes (e.g., Donald et al. 2001; Zechmeister and Moser 2001; Richner et al. 2015; Habel et al. 2019; Hodgetts et al. 2019). Agri-environment schemes (AES) were introduced in many countries in Europe and globally as a response to declining agro-biodiversity, to advocate sustainable farming practices and to compensate farmers for lower incomes when applying ecological management (Warren et al.

2008; AGRI 2018). In Switzerland, cross-compliance measures ("ökologischer Leistungsnachweis" (ÖLN); i.e., proof of ecological performance) have linked the direct trade-and production-independent payments to farmers progressively to the fulfilment of explicit ecological directives since 1993. They were legally formalised in the Agricultural Act in 1998 (Schweizerischer Bundesrat 1998; BAFU and BLW 2008). In order to qualify for these ÖLN payments, a number of requirements concerning crop rotation, pesticide application, soil conservation and fertilizer management must be met at the farm level (Schweizerischer Bundesrat 2020a). In addition, each farm has to manage at least 7% of its agricultural land area as so-called Biodiversity Promoting Area (BPA; formerly called Ecological Focus Areas). Farmers can choose from a catalogue of around 15 different types of BPA including, for example, rotational fallows, flower strips, conservation headlands, extensive pastures and high-stem orchards (BLW 2019). By 2017, 165,000 ha, or 16%, of Swiss agricultural land was managed as BPAs. BPAs are usually optimized to support the diversity of birds, arthropods or flowering plants, or certain ecosystem services such as pollination or natural pest control (e.g., Aviron et al. 2009; Tschumi et al. 2016; Zingg et al. 2019). Switzerland has further set up Environmental Objectives for Agriculture. They are targeted at, inter alia, the conservation of "national priority species", threatened species of different organism groups for which Switzerland has an international responsibility for their long-term survival (UZL: BAFU and BLW 2008, 2016; BAFU 2019).

3 Status of the Arable Bryophyte Flora Under Current Management Regimes

Bryophytes (Box 10.1) of arable fields (Fig. 10.1) are short-lived species that develop to reproductive maturity and form spores, or commonly asexual diaspores, within a short period, sometimes during one growing season. The germination of annual arable bryophyte species may occur from spring onwards in Central Europe, though many germinate only after the crop harvest in late summer. Both sexual and asexual diaspores of arable bryophytes can be long-lived in a persistent diaspore bank (i.e., with diaspores surviving more than 1 year and up to more than 5 years; Bisang 1996; Cailliau and Price 2007; Bisang et al. 2009; Studer 2016). Of the ca. 1100 bryophyte species occurring in Switzerland, approximately 30 species in the agricultural landscape depend on habitats with bare soil, i.e., they are largely confined to arable fields and are considered as 'arable specialists' (BAFU and BLW 2008; Swissbryophytes 2004–2019). About half of them have decreased (Schnyder et al. 2004). Before agriculture spread in Europe, many of these species probably grew in naturally disturbed habitats.

The field hornwort and the Carolina hornwort are two characteristic species of the segetal flora north of the Alps in Switzerland and in neighbouring countries. (Box 10.1; Fig. 10.2) In these regions, they grow on sub-neutral soils and develop

Fig. 10.1 Typical bryophyte vegetation in an arable field with a crystalwort (*Riccia sorocarpa)*, an arable specialist bryophyte, in the centre of the image. (Photo: Ariel Bergamini)

Fig. 10.2 The field hornwort (*Anthoceros agrestis*) in an arable field, with sporophytes that contain mature (black) spores. (Photo Lars Hedenäs)

optimally in stubble fields that are not further cultivated or ploughed until late in the year. Because hornworts, in particular the field hornwort, are sensitive to frost, many populations die at the beginning of the winter in Central Europe, which implies a very short life span above-ground for them. Their spores, however, may remain viable for several years in a diaspore bank (Bisang 1996, 1998, 2004; Köckinger 2017). The Carolina hornwort is classified as endangered in Switzerland. It is protected by Swiss national law and is one of the 17 national priority bryophytes in Switzerland bound to the crop-habitat type (Schnyder et al. 2004; Schweizerischer Bundesrat 2017; BAFU 2019). In the ongoing update of the Swiss bryophyte Red List, also the field hornwort is considered vulnerable (N. Schnyder 2020, pers. comm.). Both hornwort species are targeted in the Swiss Environmental Objectives for Agriculture. Taken together with their phylogenetic distinctiveness (Box 10.1), the two hornworts are excellent model organisms for studying the impact of agricultural management on the characteristic arable bryophyte flora in the Swiss Plateau.

Back in the 1980s, the lead author and collaborators studied the distribution of the field hornwort and the Carolina hornwort in the Swiss Plateau, basing their study on field surveys, Swiss natural history collections, and floristic literature. The field work was conducted before the onset of the Swiss ÖLN. Although not directly proven by the herbarium record, we concluded that the species had been more widespread in the beginning of the twentieth century, but were under-collected, possibly due to their short life span and their commonness in former times (Bisang 1992; Urmi et al. 1993; Hofmann et al. 2007). To form a less collector-biased picture of the impact of arable management on the bryophyte flora, we repeatedly monitored selected fields in annual intervals from 1991–1995, 2005–2007 and 2016–2018. All these fields had known hornwort occurrences in the 1980s. We calculated a hornwort index (HI) for each field and each survey year. The HI quantified the occurrence and population size of hornworts in cultivated fields in the Swiss Plateau, and could be related to different types of crop management and weather conditions (Bisang et al. 2009; Bisang et al. 2019). The situation of hornworts represents the status of other arable specialist bryophytes in these fields. This unique data set spanning almost 30 years, including the implementation and continuous modification of the ÖLN, revealed a significant decrease in HI, with the mean value declining from 0.36 to 0.13 from the first to the third survey period (Fig. 10.3). Changes in management practices between 1991 and 2018 seem mainly responsible for the decrease in HI. Generally, hornworts showed a much higher affinity for stubble fields than for other crops (unpubl. data). However, over the nearly three decades of the survey, stubble fields that are left unmanaged after the harvest during autumn have become rare in the Swiss Plateau. Hornworts and other short-lived arable bryophyte specialists have therefore lost much of their optimal habitat. From 1990 to 2018, the area of cereal cultivation shrank by one third, while the area cultivated as fodder meadows increased by 36% (BfS 2019). Amended directives for soil conservation introduced as part of the ÖLN, which aim to reduce nitrate leaching by increasing the duration of vegetation cover, have obliged farmers, since 2005, to swiftly cultivate fields that are harvested before 1 August with a winter crop or intermediate crop (Decrem et al.

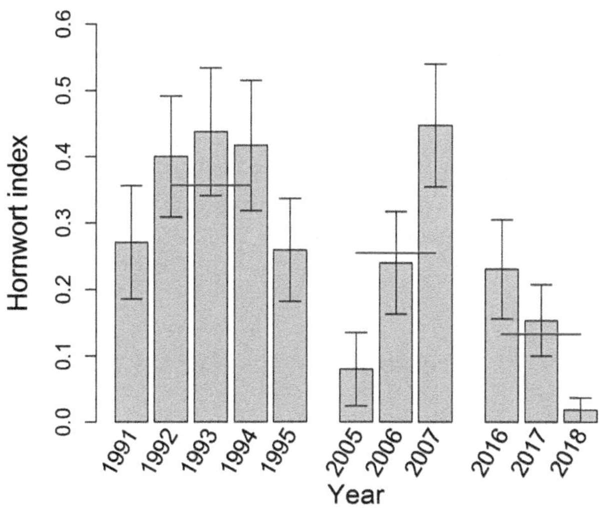

Fig. 10.3 The mean hornwort index (HI), in 28 cultivated fields in the Swiss Plateau, decreased from 1991 to 2018
A HI of 1 indicates that all fields harboured well-developed hornwort populations; a HI of 0 denotes that no hornworts occurred in any of the fields. The mean and SE of the HI per year across study fields (excluding fields that were ploughed at the time of the survey) are presented. Red lines denote the mean HI for each survey period: 1991–1995, 2005–2007 and 2016–2018. Unpubl. data from Bisang et al. (2019)

2007; Schweizerischer Bundesrat 2020a). These so-called catch-crops or cover-crops consist of Turnip rape, grass or clover species, or specific species mixtures. Moreover, breeding efforts have resulted in the planting of early-ripening and high-yield crops, which enables earlier harvests and subsequent post-harvest cultivation. The ÖLN regulations also allow broad-spectrum herbicide application in arable farming under certain conditions. Consequently, hornworts are affected by mechanical management regimes, by competition with crop plants, and by herbicides. As a result, they may not establish above-ground, perish imminently after direct impacts, or are unable to reach sexual maturity before winter.

In our data set, high humidity during summer, which is correlated with high precipitation, favoured hornwort occurrences (e.g., Fig. 10.3: 2007), while high temperature, a large number of hot days (days with temperature > 30 °C) and high radiation in the summer months inhibited their development (unpubl. data). These findings are in line with previous evidence of direct effects of precipitation and temperature conditions on bryophyte growth and productivity (Callaghan et al. 1997; Wiklund and Rydin 2004). However, we found that weather effects on HI were weaker than management effects, and they were partly mediated through effects on field management and crop development (unpubl. data). For example, delayed ploughing in wet summers may have a positive effect on hornwort occurrences. Future climate scenarios predict drier summers with more hot days in

Switzerland, conditions that we have shown to negatively affect hornworts (NCCS 2018). Hence, hornworts and other arable bryophyte species will be under increasing climate stress in the future, even if bryophyte-friendly agri-environment schemes are adopted. On the other hand, the predicted higher winter temperatures may lead to some above-ground populations surviving until spring, as has been observed occasionally (Swissbryophytes 2018; pers. obs.). If mild winters will be beneficial for the long-term survival of hornworts in the intensively cultivated Swiss Plateau, however, remains to be investigated.

4 Do Biodiversity Promoting Areas (BPAs) Benefit Arable Bryophytes?

Numerous European and worldwide studies have evaluated the suitability of different types of agri-environment schemes (AES) for the promotion of biodiversity (e.g., Ansell et al. 2016). These evaluations have revealed limited to moderately positive effects that vary among species and organism groups and depend substantially on the landscape context (Kleijn et al. 2006; Whittingham 2011; Lichtenberg et al. 2017). Evidence is emerging that rare, specialized or endangered species are seldom favoured by such schemes (Aviron et al. 2009; Jeanneret et al. 2010; Aviron et al. 2011).

However, bryophytes have not been considered in the AES evaluations to date. In the Swiss Plateau, we investigated four types of fallows of different ages, all of which are among the BPA options for Swiss farmers: annual flower strips, wildflower strips, rotational fallows and improved field margins (Schweizerischer Bundesrat 2020a). Across 48 sites, we recorded 44 bryophyte species, 13 of which were considered arable specialists and included the field hornwort and the Carolina hornwort (Valentini 2014; Studer 2016). All BPA-types harboured bryophytes (Fig. 10.4a, b). Species richness was higher in the BPA-fallows than in the fields under conventional cultivation (with conventional cultivation or conventional management, we henceforth denote current farming practices that are not subject to specific BPA management) (Fig. 10.4b). Richness of bryophyte specialists decreased with increasing age of the fallows; as vegetation grows denser, litter accumulates, and soil disturbance becomes minimal over time (Fig. 10.5a).

Specialist species richness differed between fallow-types, indicating that they are more sensitive to environmental conditions than generalist species (Fig. 10.4a, b). Total bryophyte cover and the proportion of bare soil at the study sites positively affected species richness in the fallows (Valentini 2014; Studer 2016).

The observed beneficial effects for arable bryophytes in habitats with reduced competition from other plants agrees with the findings from the hornwort monitoring presented above (Bisang et al. 2019). Bryophyte species composition varied little

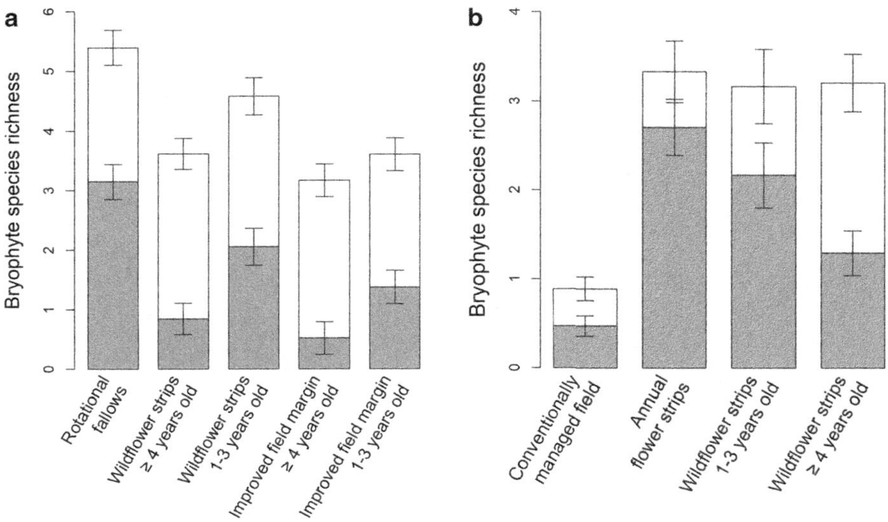

Figs. 10.4 Species richness of arable bryophyte specialists (grey) and of generalists (white) in Biodiversity Promoting Areas (BPAs, different fallow-types) and in conventionally managed arable fields in the Swiss Plateau. Conventional management (in this and following figure texts, and main text body) denotes current farming practices that are not subject to specific BPA management.
The mean and SE of the number of species per study plot are presented
(**a**). Total species richness did not differ between BPA-types, but specialist species richness differed between rotational fallows and old wildflower strips and old improved field margins.
*Six sites for each BPA-type; eight plots per site. Linear mixed effects model with BPA-type as a fixed factor ($F = 4.60$; $P < 0.01$), and region and site as random factors; data from Valentini (*2014*)*
(**b**). Total species richness differed between BPAs and conventional management ($P < 0.001$), but not between BPA-types. Specialist species richness differed between conventional management and annual flower strips and young wildflower strips, but not between conventional management and old wildflower strips.
*Eighteen sites, including one pair each of a BPA and a conventionally managed field; four plus four plots per site. Linear mixed effects model with management type as a fixed factor ($F = 10.65$; $P < 0.001$), and site and management unit (two at each site) as random factors; data from Studer (*2016*)*

between fallow-types, suggesting that abiotic conditions and vegetation structure are more important than BPA-type. Taken together, our findings demonstrate that although arable bryophyte specialists do occur in the existing BPAs to some extent, the most typical or red-listed species are rare even in these fields. For example, the national priority species Carolina hornwort was documented in only one fallow, a wildflower strip (Valentini 2014), or was found only in the diaspore bank of the studied fallows (Studer 2016).

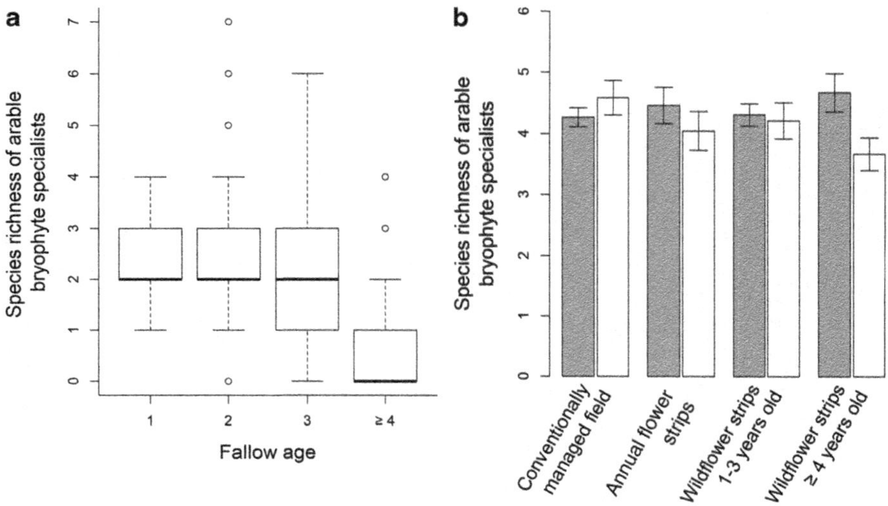

Fig. 10.5 (**a**). Bryophyte specialist species richness in Biodiversity Promoting Areas (BPAs, different fallow-types) in the Swiss Plateau that are ≥4 years old is lower than in younger BPAs *Box-and-whisker diagrams of the number of species per plot in BPAs of different ages. The line denotes the median, the box includes 50% of the values, the whiskers designate the 1st and the 4th quartile; and the circles indicate extreme values. Number of replications: age 1 year, $n = 16$; age 2 year, $n = 40$; age 3 year, $n = 95$; age ≥ 4 year, $n = 87$. Linear mixed effects model with age as a fixed factor ($F = 7.70$; $P < 0.01$), and region and site as random factors; data from Valentini (2014)* (**b**). Species richness in the diaspore bank of arable bryophyte specialists at soil depths of 0–3 cm (grey) and 12–15 cm (white) in Biodiversity Promoting Areas (BPAs, different fallow-types) and in conventionally managed arable fields in the Swiss Plateau
The mean and SE of the number of species per study plot are presented. Eighteen sites including one pair each of a BPA and a conventionally managed field; two depths per plot in four plus four plots per site were sampled. Linear mixed effects model with management type, depth and their interaction as fixed factors ($P > 0.05$), and site, management unit (two at each site) and plot as random factors; data from Studer (2016)

5 Does the Diaspore Bank Serve as a Reservoir for Arable Bryophytes Under Pressure?

Many arable specialists – both flowering plants and bryophytes – can persist for several years in the diaspore bank and re-emerge when above-ground conditions turn favourable again. Hornworts have also been demonstrated to develop gametophytic populations in fields and areas where they have not been observed for several or even for many years (Cailliau and Price 2007; Bisang et al. 2009). In our study, the bryophyte diaspore bank of cultivated fields, including BPAs, contained 14 arable specialists out of 27 species in total. Four out of five species that germinated from soil samples, but which were not recorded at the surface, were typical arable species. These included both the field hornwort and the red-listed Carolina hornwort, and in addition, the red-listed spreading earth-moss (*Physcomitrium patens*; Studer 2016).

Bryophyte diversity in the diaspore bank was higher than that in the surface vegetation under all management regimes. Interestingly, the species diversity of bryophyte specialists in the diaspore bank differed little between BPAs and conventionally cultivated fields and between different soil depths, but there was a tendency of lower diversity at 15 cm depth in wildflower strips that have not been tilled for four or more years (Fig. 10.5b). Considering all species in the diaspore bank, species richness decreased with soil depth in annual flower strips and in old wildflower strips. A persistent diaspore bank is crucial for arable specialists to bridge periods of unsuitable weather and management conditions. Although the germination potential seems to remain under conventional farming, the diaspore bank needs to be replenished often enough by input from reproductive plants. In line with previous investigations, our field study results indicate that diaspore longevity may be limited at greater soil depths (Callaghan et al. 2020). Earthworm activity has been shown to contribute to bryophyte diaspore movement in the soil (van Tooren and During 1988). The observed critical decline of earthworms under agricultural intensification is thus likely to directly affect species composition below- and above-ground, as well as having an indirect effect through a negative impact on soil quality (Blakemore 2018). The potential to establish from the diaspore bank has been shown to decrease with time after disturbance for forest bryophytes (Jonsson 1993). Although segetal plants often build up persistent reservoirs of seeds, spores or vegetative propagules in the soil, longevity varies widely among species. Overall, the seed density of flowering plants in arable soils has declined steeply since 1900 (Sutcliffe and Kay 2000; Dutoit et al. 2003; Tscharntke et al. 2005; Kohler et al. 2011).

6 Safeguarding the Arable Bryophyte Flora in Switzerland – Current State, Recommendations and Perspectives

Awareness of the severe decline and continuous fragility of the segetal flora is increasing in Switzerland and elsewhere. However, the bryophyte fraction of this flora is still barely acknowledged. Like specialist or rare species of other organism groups, bryophyte specialists need tailored modifications of the current ÖLN regulations. The most crucial measure to preserve arable bryophytes in the intensively cultivated Swiss Plateau is to maintain suitable habitats (Figs. 10.6 and 10.7). This implies that fields or set-aside parts of fields with documented occurrences, or at least a high potential for occurrence, of arable specialist bryophytes: (i) are cultivated with a high proportion of cereals during crop rotation; (ii) are not farmed after harvest nor tilled before the end of October, or remain as over-wintering stubbles until spring, to allow the specialist species to complete reproduction; and (iii) are regularly ploughed, to allow diaspores to re-surface and to prevent small arable bryophytes from being suppressed by larger more competitive late-successional and unspecialized bryophytes. We also recommend limiting vehicular traffic during the

Fig. 10.6 A typical arable landscape in the Swiss Plateau in the autumn, with different fields under a regime of crop rotation. (Photo: Irene Bisang)

Fig. 10.7 Low-intensity managed stubble fields are a favourable habitat for arable bryophyte specialists in Central Europe. These are not ploughed, or otherwise cultivated, from the time of harvest until late autumn or the following spring. (Photo: Irene Bisang (Czech Republic))

crop-growing season because heavy machinery impacts sensitive bryophyte populations, causes soil compaction, and negatively affects the water storage capacity of the soil (Keller et al. 2019). Intensive land use, including fertilization practices, alters the biochemistry and quality of the soil, which in turn may affect arable species composition (Thomson et al. 2015; Vašák et al. 2015). Fertilizer and pesticide applications should be regulated, and the application of broad-spectrum herbicides after the harvest must be prohibited. These measures can be implemented by adjusting existing ÖLN-directives. The precise field management should be detailed in user agreements with the farmers, and the farmers' compensation should be comparable to that for other efforts as part of the ÖLN. Among the existing BPA, fallow-types younger than 3 years should be promoted, and the directives for certain BPAs be amended (for example, conservation headlands, annual flower strips, different fallow options; Richner 2006; Fig. 10.8). The actual implementation of action plans for arable weeds and, in particular, for inconspicuous bryophytes often conflicts with farmers' demands for profitability, and diversity-promoting measures are met with severe reluctance. Professional advisory services provided by specialists and the incorporation of attractive segetal flowering plants in the measures are therefore likely to improve the motivation by farmers and the conservation success

Fig. 10.8 Rotational fallows, a Biodiversity Promoting Area (BPA) option in Switzerland, harbour specialist arable bryophyte in the first year. As the fallow vegetation increases in density, typically, the more competitive late-successional bryophyte generalists and flowering plants suppress the specialist arable bryophytes. (Photo Ariel Bergamini)

(Birrer et al. 2018). Conservation of the arable flora is starting to gain some attention in Switzerland in projects related to the biodiversity obligations of the Swiss cantons, and at the policy level (Schweizerischer-Bundesrat 2020b). Round notothylas, another arable hornwort that is not known to occur in Switzerland but is endangered in Europe (Box 10.1), is subject to specific action plans in the German federal state of Hessen (Drehwald 2018).

In Germany, late tillage and over-wintering stubble fields are widely practiced as part of AES to promote the segetal flora. Furthermore, one hundred arable field sanctuaries have been committed to low-intensity agricultural practices in a network across Germany, to ensure conservation of the endangered arable flora (100 Fields for Diversity: Meyer et al. 2014; Gottwald and Stein-Bachinger 2016). While bryophytes are not explicitly targeted by these actions, they undoubtedly profit from such management regimes, as do other organisms that prefer extensively managed, open and regularly disturbed habitats. These include hares inhabiting cultivated fields, insects of various groups, breeding farmland birds, and migrating amphibians and reptiles (Perkins et al. 2011; Tschumi et al. 2016; Meichtry-Stier et al. 2018). In other European countries with still relatively large, but also rapidly decreasing, areas of extensive agricultural cultivation, arable plant diversity conservation has not received political attention (e.g., Czech Republic, pers. communication Jan Kučera; Sutcliffe et al. 2015; Fig. 10.7).

Finally, it is becoming increasingly clear that successful biodiversity promotion measures need to take into account not only local (e.g., soil conditions) but also regional factors such as the structure and complexity of the surrounding landscape, including its topography, the regional diversity of farming regimes, and habitat connectivity. Agro-biodiversity also strongly depends on the general cultivation intensity in the production area (Kleijn et al. 2009; Whittingham 2011; Batáry et al. 2015; Lichtenberg et al. 2017). Segregated, size-limited and possibly sub-optimally placed AES-areas cannot compensate for the simultaneous loss of biodiversity due to the continuous intensification of conventional production (e.g., Tscharntke et al. 2012; Kremen and Merenlender 2018). When adopting or developing AES for bryophytes, we need to consider these components and criteria to conserve the specialist species in their main habitats in the intensively cultivated land.

Acknowledgements We are especially grateful to Luc Lienhard (Natur&Geschichte, Biel, Switzerland), who has contributed significantly to hornwort monitoring since the eighties in a dedicated, competent and diligent way. His collaboration in various aspects of the study is warmly acknowledged. We thank Lisa Studer and Maja Valentini for allowing us to use the data from their master's theses; Lars Hedenäs (Swedish Museum of Natural History, Stockholm, Sweden), Ron Porley (Aljezur, Portugal), Norbert Schnyder (swissbryophytes.ch) and Clive Hurford (Serapias ecological monitoring consultancy) for valuable comments on earlier versions of the manuscript, Melissa Dawes for improving the English; and Lars Hedenäs for consenting to the publication of one of his photos.

References

AGRI (2018) CAP explained. Direct payments for farmers 2015–2020. European Commission: Directorate General for Agriculture and Rural Development AGRI, Brussels. https://doi.org/10.2762/572019

Ansell D, Freudenberger D, Munro N, Gibbons P (2016) The cost-effectiveness of agri-environment schemes for biodiversity conservation: a quantitative review. Agric Ecosyst Environ 225:184–191. https://doi.org/10.1016/j.agee.2016.04.008

Aviron S, Nitsch H, Jeanneret P, Buholzer S, Luka H, Pfiffner L, Pozzi S, Schüpbach B, Walter T, Herzog F (2009) Ecological cross compliance promotes farmland biodiversity in Switzerland. Front Ecol Environ 7. https://doi.org/10.1890/070197

Aviron S, Herzog F, Klaus I, Schüpbach B, Jeanneret P (2011) Effects of wildflower strip quality, quantity, and connectivity on butterfly diversity in a Swiss arable landscape. Restor Ecol 19 (4):500–508. https://doi.org/10.1111/j.1526-100X.2010.00649.x

BAFU (2019) Liste der National Prioritären Arten und Lebensräume. BAFU, Bern

BAFU, BLW (2008) Umweltziele Landwirtschaft. BAFU & BLW, Bern

BAFU, BLW (2016) Umweltziele Landwirtschaft. Statusbericht 2016. BAFU & BLW, Bern

Batáry P, Dicks LV, Kleijn D, Sutherland WJ (2015) The role of agri-environment schemes in conservation and environmental management. Conserv Biol 29(4):1006–1016. https://doi.org/10.1111/cobi.12536

Baumann W (2011) Agrarrevolution. Version 23. 03. 2011. In: Historisches Lexikon der Schweiz (HLS). HLS [CC BY-SA]. https://hls-dhs-dss.ch/de/articles/013827/2011-03-23/. Accessed 15 Jan 2020

BfS (2019) Landwirtschaftliche Strukturerhebung 2018. Federal Statistical Office (BfS). https://www.bfs.admin.ch/bfs/de/home/aktuell/neue-veroeffentlichungen.gnpdetail.2019-0444.html. Accessed 20 Jan 2020

Birrer S, Doppler V, Uebersax A, Rometsch S (2018) Erhaltung und Förderung gefährdeter Schweizer Ackerbegleitflora. Schlussbericht 2017. Projektteam Ressourcenprojekt Ackerbegleitflora, unpublished report

Bisang I (1992) Hornworts in Switzerland – endangered? Biol Conserv 59:145–149. https://doi.org/10.1016/0006-3207(92)90574-7

Bisang I (1996) Quantitative analysis of the diaspore banks of bryophytes and ferns in cultivated fields in Switzerland. Lindbergia 21:9–20

Bisang I (1998) The occurrence of hornwort populations (Anthocerotales, Anthocerotopsida) in the Swiss Plateau: the role of management, weather conditions and soil characteristics. Lindbergia 23:94–104

Bisang I (2004) Population development, demographic structure, and life cycle aspects of two hornworts in Switzerland. Lindbergia 28:105–112

Bisang I, Bergamini A, Lienhard L (2009) Environmental-friendly farming in Switzerland is not hornwort-friendly. Biol Conserv 142:2104–2113. https://doi.org/10.1016/j.biocon.2009.04.006.

Bisang I, Lienhard L, Bergamini A (2019) Entwicklung von Ackermoospopulation und ihren Lebensräumen im Schweizer Mittelland während der letzten 28 Jahren. Schlussbericht an das Bundesamt für Umwelt (BAFU), unpublished report

Blakemore RJ (2018) Critical decline of earthworms from organic origins under intensive, humic SOM-depleting agriculture. Soil Syst 2(2):1–28. https://doi.org/10.3390/soilsystems2020033

BLW (2019) Biodiversitätsbeiträge. Bundesamt für Landwirtschaft (BLW). https://www.blw.admin.ch/blw/de/home/instrumente/direktzahlungen/biodiversitaetsbeitraege.html. Accessed 10 Nov 2019

Cailliau A, Price MJ (2007) Hornworts in the agricultural fields of Geneva: new findings, the soil diaspore bank and ex situ soil cultures. Candollea 62(2):165–172

Callaghan TV, Carlsson BÅ, Sonesson M, Temesváry A (1997) Between-year variation in climate related growth of circum-arctic populations of the moss *Hylocomium splendens*. Funct Ecol 11:157–165. https://doi.org/10.1046/j.1365-2435.1997.00072.x

Callaghan DA, During HJ, Forrest LL, Wilkinson K (2020) Neglected and at risk: bryophyte diaspore banks of coastal dune systems. J Bryol 42(3):223–234. https://doi.org/10.1080/03736687.2020.1743561

de Sousa F, Foster PG, Donoghue PCJ, Schneider H, Cox CJ (2019) Nuclear protein phylogenies support the monophyly of the three bryophyte groups (Bryophyta Schimp.). New Phytol 222 (1):565–575. https://doi.org/10.1111/nph.15587

Decrem M, Spiess E, Richner W, Herzog F (2007) Impact of Swiss agricultural policies on nitrate leaching from arable land. Agron Sustain Dev 27(3):243–253. https://doi.org/10.1051/agro:2007012

Donald PF, Green RE, Heath MF (2001) Agricultural intensification and the collapse of Europes' farmland bird populations. Proc R Soc B 268:25–29. https://doi.org/10.1098/rspb.2000.1325

Drehwald U (2018) Bundesmonitoring 2017 zur Verbreitung des Kugel-Hornmooses (*Notothylas orbicularis*) (Art des Anhangs II der FFH-Richtlinie) in Hessen. Hessisches Landesamt für Naturschutz, Umwelt und Geologie, Giessen, Hessen

Dutoit T, Gerbaud É, Buisson É, Roche P (2003) Dynamique d'une communauté d'adventices dans un champ de céréales créé après le labour d'une prairie semi-naturelle: rôles de la banque de graines permanente. Écoscience 10(2):225–235. https://doi.org/10.1080/11956860.2003.11682770

Emmerson M, Morales MB, Oñate JJ, Batáry P, Berendse F, Liira J, Aavik T, Guerrero I, Bommarco R, Eggers S, Pärt T, Tscharntke T, Weisser W, Clement L, Bengtsson J (2016) How agricultural intensification affects biodiversity and ecosystem services. In: Dumbrell AJ, Kordas RL, Woodward G (eds) Large-scale ecology: model systems to global perspectives, Advances in ecological research, vol 55. Academic, Amsterdam, pp 43–97. https://doi.org/10.1016/bs.aecr.2016.08.005

Gottwald F, Stein-Bachinger K (2016) Landwirtschaft für Artenvielfalt. Ein Naturschutzmodul für ökologisch bewirtschaftete Betriebe. WWF Deutschland. https://www.landwirtschaft-artenvielfalt.de

Habel JC, Ulrich W, Biburger N, Seibold S, Schmitt T (2019) Agricultural intensification drives butterfly decline. Insect Conserv Divers 12(4):289–295. https://doi.org/10.1111/icad.12343

Hodgetts N et al (2019) A miniature world in decline: European Red List of Mosses, Liverworts and Hornworts. International Union for Conservation of Nature (IUCN). https://doi.org/10.2305/IUCN.CH.2019.ERL.2.en

Hofmann H, Urmi E, Bisang I, Müller N, Küchler M, Schnyder N, Schubiger C (2007) Retrospective assessment of frequency changes in Swiss bryophytes over the last two centuries. Lindbergia 32:18–32

IPBES (2018) The IPBES regional assessment report on biodiversity and ecosystem services for Europe and Central Asia. Secretariat of the Intergovernmental Science-Policy Platform on Biodiversity and Ecosystem Services, Bonn

Isaacs R, Tuell J, Fiedler A, Gardiner M, Landis D (2009) Maximizing arthropod-mediated ecosystem services in agricultural landscapes: the role of native plants. Front Ecol Environ 7 (4):196–203. https://doi.org/10.1890/080035

Jeanneret P, Walter T, Birrer S, Herzog F (2010) Evaluating the performance of the Swiss agri-environmental measures for biodiversity: methods, results and questions. Asp Appl Biol 100:35–42

Jonsson BG (1993) The bryophyte diasporebank and its role after small-scale disturbance in a boreal forest. J Veg Sci 4(6):819–826. https://doi.org/10.2307/3235620

Keller T, Sandin M, Colombi T, Horn R, Or D (2019) Historical increase in agricultural machinery weights enhanced soil stress levels and adversely affected soil functioning. Soil Tillage Res 194:104293. https://doi.org/10.1016/j.still.2019.104293

Kennedy CM et al (2013) A global quantitative synthesis of local and landscape effects on wild bee pollinators in agroecosystems. Ecol Lett 16(5):584–599. https://doi.org/10.1111/ele.12082

Kleijn D, Baquero RA, Clough Y, Diaz M, De Esteban J, Fernández F, Gabriel D, Herzog F, Holzschuh A, Jöhl R, Knop E, Kruess A, Marshall JP, Steffan-Dewenter I, Tscharntke T, Verhulst J, West TM, Yela JL (2006) Mixed biodiversity benefits of agri-environment schemes in five European countries. Ecol Lett 9:243–254. https://doi.org/10.1111/j.1461-0248.2005.00869.x

Kleijn D, Kohler F, Báldi A, Batáry P, Concepción ED, Clough Y, Díaz M, Gabriel D, Holzschuh A, Knop E, Kovács A, Marshall EJP, Tscharntke T, Verhulst J (2009) On the relationship between farmland biodiversity and land-use intensity in Europe. Proc R Soc B 276(1658):903–909. https://doi.org/10.1098/rspb.2008.1509

Köckinger H (2017) Die Horn- und Lebermoose Österreichs. Cat Florae Austriae Biosyst Ecol Ser 32:1–382

Kohler F, Vandenberghe C, Imstepf R, Gillet F (2011) Restoration of threatened arable weed communities in abandoned mountainous crop fields. Restor Ecol 19:62–69. https://doi.org/10.1111/j.1526-100X.2009.00641.x

Kremen C, Merenlender AM (2018) Landscapes that work for biodiversity and people. Science 362 (6412):eaau6020. https://doi.org/10.1126/science.aau6020

Lichtenberg EM et al (2017) A global synthesis of the effects of diversified farming systems on arthropod diversity within fields and across agricultural landscapes. Glob Chang Biol:4946–4957. https://doi.org/10.1111/gcb.13714

Meichtry-Stier KS, Duplain J, Lanz M, Lugrin B, Birrer S (2018) The importance of size, location, and vegetation composition of perennial fallows for farmland birds. Ecol Evol 8 (18):9270–9281. https://doi.org/10.1002/ece3.4420

Meyer S, van Elsen T, Blümlein B, Kaerlein M, Metzner J, Gottwald F, Wehke S, Dietrich M, Wahmhoff W, Stock R, Leuschner C (2014) Agrobiodiversitätsschutz durch Schutzäcker. Das Projekt „100 Äcker für die Vielfalt". Nat Landschaft 89(9/10):434–441

NCCS (2018) CH2018 – climate scenarios for Switzerland, technical report. National Centre for Climate Services, Zürich

Nowak S, Waindzoch K, Świerszcz S, Niemczyk M, Spałek K, Nowak A (2019) Crop density rather than ruderal plants explains the response of ancient segetal weeds. Biologia 74 (4):351–359. https://doi.org/10.2478/s11756-018-00178-8

Perkins AJ, Maggs HE, Watson A, Wilson JD (2011) Adaptive management and targeting of agri-environment schemes does benefit biodiversity: a case study of the corn bunting *Emberiza calandra*. J Appl Ecol 48(3):514–522. https://doi.org/10.1111/j.1365-2664.2011.01958.x

Price DT (2000) Europe's first farmers. Cambridge University Press, Cambridge

Richner N (2006) Die Schweizer Segetalflora. Untersuchungen zur Erhaltung, Förderung und zum Schutz der Segetalarten in der Schweiz. Master thesis, Department of Environmental System Sciences, ETH, Zürich

Richner N, Holderegger R, Linder HP, Walter T (2015) Reviewing change in the arable flora of Europe: a meta-analysis. Weed Res 55:1–13. https://doi.org/10.1111/wre.12123

Rusch A, Chaplin-Kramer R, Gardiner MM, Hawro V, Holland J, Landis D, Thies C, Tscharntke T, Weisser WW, Winqvist C, Woltz M, Bommarco R (2016) Agricultural landscape simplification reduces natural pest control: a quantitative synthesis. Agric Ecosyst Environ 221:198–204. https://doi.org/10.1016/j.agee.2016.01.039

Schnyder N, Bergamini A, Hofmann H, Müller N, Schubiger-Bossard C, Urmi E (2004) Rote Liste der gefährdeten Arten der Schweiz. BUWAL, Bern

Schweizerischer Bundesrat (1998) Bundesgesetz über die Landwirtschaft 910.1 (Landwirtschaftsgesetz, LwG) vom 29. April 1988. Schweizerische Eidgenossenschaft, Bern

Schweizerischer Bundesrat (2017) Liste der geschützten Pflanzen: Anhang 2, Art. 20 Abs. 1 NHV. Stand 1. Juni 2017. In: Natur und Heimatschutzverordnung (NHV). 451.1. Schweizerischer Bundesrat, Bern

Schweizerischer Bundesrat (2020a) Verordnung über die Direktzahlungen an die Landwirtschaft (Direktzahlungsverordnung, DZV) vom 23. Oktober 2013. 910.13. Stand 1. Januar 2020. Schweizerischer Bundesrat, Bern

Schweizerischer Bundesrat (2020b) Botschaft zur Weiterentwicklung der Agrarpolitik ab 2022 (AP22+). Schweizerische Eidgenossenschaft, Bern

Studer L (2016) Bryophyte diaspore banks in ecological focus areas in arable farming in Swiss lowlands. Master thesis, Department of Environmental Systems Science, ETH, Zürich

Sutcliffe OL, Kay QON (2000) Changes in the arable flora of central southern England since the 1960s. Biol Conserv 93(1):1–8. https://doi.org/10.1016/S0006-3207(99)00119-6

Sutcliffe LME et al (2015) Harnessing the biodiversity value of Central and Eastern European farmland. Divers Distrib 21(6):722–730. https://doi.org/10.1111/ddi.12288

Swissbryophytes (2004–2019) Moosflora der Schweiz. Universität Zürich. http://www.swissbryophytes.ch. Accessed at 18 Oct 2019

Swissbryophytes (2018) Nationales Daten- und Informationszentrum der Schweizer Moose. Universität Zürich. http://www.swissbryophytes.ch. Accessed at 18 Oct 2019

Thomson BC, Tisserant E, Plassart P, Uroz S, Griffiths RI, Hannula SE, Buée M, Mougel C, Ranjard L, Van Veen JA, Martin F, Bailey MJ, Lemanceau P (2015) Soil conditions and land use intensification effects on soil microbial communities across a range of European field sites. Soil Biol Biochem 88:403–413. https://doi.org/10.1016/j.soilbio.2015.06.012

Tscharntke T, Klein AM, Kruess A, Steffan-Dewenter I, Thies C (2005) Landscape perspectives on agricultural intensification and biodiversity – ecosystem service management. Ecol Lett 8 (8):857–874. https://doi.org/10.1111/j.1461-0248.2005.00782.x

Tscharntke T, Clough Y, Wanger TC, Jackson L, Motzke I, Perfecto I, Vandermeer J, Whitbread A (2012) Global food security, biodiversity conservation and the future of agricultural intensification. Biol Conserv 151(1):53–59. https://doi.org/10.1016/j.biocon.2012.01.068

Tschumi M, Albrecht M, Bärtschi C, Collatz J, Entling MH, Jacot K (2016) Perennial, species-rich wildflower strips enhance pest control and crop yield. Agric Ecosyst Environ 220:97–103. https://doi.org/10.1016/j.agee.2016.01.001

Urmi E, Schubiger C, Bisang I (1993) Veränderungen in der Moosflora der Schweiz. Diss Bot 196:261–279

Valentini M (2014) Bryophytes of arable fields. Comparison of three ecological compensation measures in the Swiss lowlands. Master thesis, Department of Environmental System Sciences, ETH, Zürich

van Tooren BF, During HJ (1988) Viable plant diaspores in the guts of earthworms. Acta Bot Neerl 37:181–185. https://doi.org/10.1111/j.1438-8677.1988.tb02126.x

Vašák F, Černý J, Buráňová Š, Kulhánek M, Balík J (2015) Soil pH changes in long-term field experiments with different fertilizing systems. Soil Water Res 10:19–23. https://doi.org/10.17221/7/2014-SWR

Warren J, Lawson C, Blecher K (2008) The agri-environment. Cambridge University Press, Cambridge

Whittingham MJ (2011) The future of agri-environment schemes: biodiversity gains and ecosystem service delivery? J Appl Ecol 48(3):509–513. https://doi.org/10.1111/j.1365-2664.2011.01987.x

Wiklund K, Rydin H (2004) Colony expansion of *Neckera pennata*: modelled growth rate and effect of microhabitat, competition, and precipitation. Bryologist 107:293–301. https://doi.org/10.1639/0007-2745(2004)107[0293:CEONPM]2.0.CO;2

Zechmeister HG, Moser D (2001) The influence of agricultural land-use intensity on bryophyte species richness. Biodivers Conserv 10(10):1609–1625. https://doi.org/10.1023/a:1012008828522

Zingg S, Ritschard E, Arlettaz R, Humbert J-Y (2019) Increasing the proportion and quality of land under agri-environment schemes promotes birds and butterflies at the landscape scale. Biol Conserv 231:39–48. https://doi.org/10.1016/j.biocon.2018.12.022

Chapter 11
Invertebrate Trends in an Arable Environment: Long-Term Changes from the Sussex Study in Southern England

J. A. Ewald, N. J. Aebischer, and N. W. Sotherton

Abstract The Game & Wildlife Conservation Trust's Sussex Study is one of the world's longest running studies of the cereal ecosystem. We report here on 50 years of changes in invertebrate abundance, farming management and weather on the study area and review relevant research contributions arising from this work.

The abundance of six groups of chick-food invertebrates all showed annual variations. Symphyta, Lepidoptera & Neuroptera increased by 26% over the 50 years, Chrysomelidae & Curculionidae and Non-aphid Hemiptera remained stable, Araneae & Opiliones, Aphididae and Carabidae & Elateridae declined by 32%, 57% and 76% respectively. Of three chick-food indices examined, the grey partridge chick-food index declined by 18%, the yellowhammer index by 57% and the corn bunting chick-food index increased by 44%.

Overall invertebrate abundance declined by 17%, though this was influenced by the very numerous Collembola (−1% change in abundance), Thysanoptera (+29%) and Aphididae (see above). Discounting these large groups, overall abundance of the remaining invertebrates declined by 34%.

We review the results of previous research on the drivers of these changes, considering changes in cropping, use of pesticides and changes in weather.

Keywords Sussex study · Long-term monitoring · Invertebrate abundance · Agricultural intensification · Farmland birds · Climate change

1 Introduction

Reports of declines in the abundance and biomass of insects have received much attention, both in the scientific community and more broadly (Dirzo et al. 2014; Hallmann et al. 2017; van Klink et al. 2020; Wepprich et al. 2019). A review of

J. A. Ewald (✉) · N. J. Aebischer · N. W. Sotherton
Game & Wildlife Conservation Trust, Hampshire, UK
e-mail: jewald@gwct.org.uk; naebischer@gwct.org.uk; nsotherton@gwct.org.uk

C. Hurford et al. (eds.), *The Changing Status of Arable Habitats in Europe*,
https://doi.org/10.1007/978-3-030-59875-4_11

reports of insect declines highlighted four main drivers: agricultural intensification, climate change, changes in land use, and light and environmental pollution (Sánchez-Bayo and Wyckhuys 2019). Analysis of long-term studies of arable systems, where data have been collected at the same sites in the same way over many years, would help to clarify the relationship between agricultural land use and changes in insect abundance (Leather 2018; Macgregor et al. 2019; Wauchope et al. 2019).

One of the most well-researched examples of the effects of agricultural intensification on farmland wildlife and the arable ecosystem comes from southern England, on the Sussex Downs. For over 50 years the Game & Wildlife Conservation Trust (GWCT) has monitored the status of grey partridge *Perdix perdix* and its associated farmland ecosystem on the Sussex study area (Aebischer and Ewald 2012; Campbell et al. 1997; Potts 1986, 2012; Southwood and Cross 2002; Sotherton 1991; Sotherton et al. 2014; Sotherton et al., this volume – Chap. 18). In one of the most comprehensive and longest-running studies of its kind, invertebrates in cereal crops have been sampled since 1970, using standardised methods, in order to better understand the food resources available to chicks of grey partridges and other farmland birds. Information is also recorded on land use, agricultural management and weather (Potts and Vickerman 1974; Ewald et al. 2015). The results of this monitoring have fed into associated research that investigates ways to counteract some of the negative effects of changes in farming methods that have been identified, allowing both farming and conservation to take place with measures to address these negative effects (Sotherton 1991; Thomas et al. 1992; Sotherton et al., this volume – Chap. 18). The monitoring has also provided information on the diversity and trends in abundance of invertebrates of cereal ecosystems (Aebischer 1991; Potts 1986, 2012; Potts and Vickerman 1974). Here we report on trends in the abundance of invertebrates important in the diet of farmland birds, as well as overall trends in invertebrate abundance.

2 Setting the Scene: The Sussex Study Area

The GWCT has collected data on invertebrate abundance in cereal crops, as well as information on cropping, on a core area of 32 km^2 of farmland on the Sussex Downs in southern England from 1970 to the present day (Potts 1986, 2012; Aebischer 1991; Ewald and Aebischer 1999, 2000). The study area is situated across rolling downland between the rivers Adur and Arun (Fig. 11.1); the dominant soils are chalk rendzinas with abundant flint, isolated caps of clay on higher parts and post-glacial deposits along the lower parts of a series of 'dry valleys'. There is no natural supply of surface water. The majority (95%) of the core study area is made up of six large farms that, in 1970, ranged in size from 245 to 900 ha (average 505 ha), while the remainder is made up of six smaller farms or parts of farms. In its interaction with the farmers on the study area the GWCT seeks to monitor their farming management and its effects on the cereal ecosystem. The GWCT does not set out to change the farm

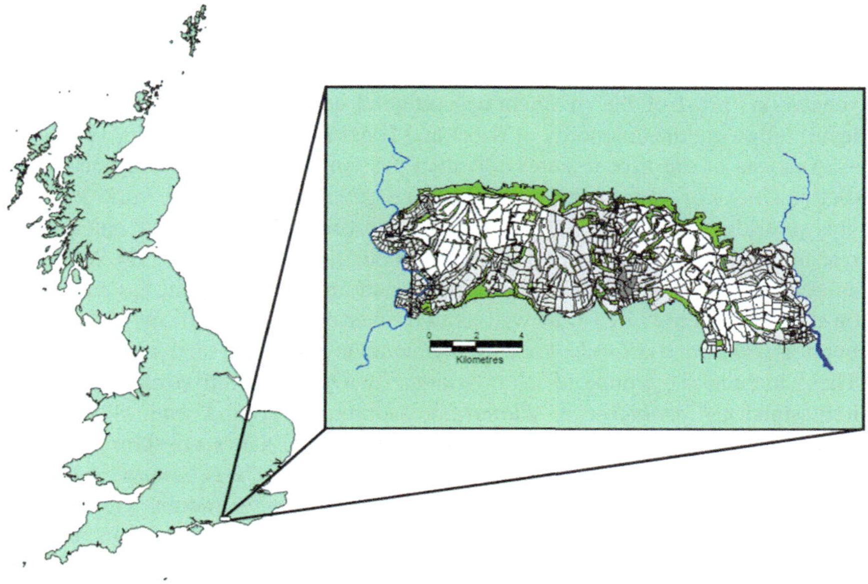

Fig. 11.1 Location of the Sussex Study Area

management; it provides advice if asked, and information from the monitoring to the farmers and land managers on the study area. They are free to make their own agronomic and other decisions, with the intention that their management reflects what would be the case across most mixed arable farmland in the UK.

The landscape of the Sussex Study area is predominantly agricultural, with 57% of the area as arable fields, 28% as grass fields, 3% in chalk downland, 8% as woodland, scrub and rough grassland, 4% as roads and tracks, and < 1% as hedges and buildings.

3 Invertebrate Data Collection, Identification and Analysis

The original impetus of the Sussex Study was to investigate declines in the abundance of grey partridges, so invertebrate monitoring was designed to provide information on resources available to grey partridges on the study area. Hence invertebrates are sampled in the third week of June, corresponding to the time of peak grey partridge chick hatch (Potts 1980) and the time when grey partridge chicks are most reliant on invertebrates for food (Southwood and Cross 2002). Approximately 100 cereal fields are sampled each year, with one invertebrate sample per field, using a D-Vac suction trap (Dietrick 1961). Each sample consists of five sub-samples, each of 0.092 m^2, taken at 5-m intervals along a diagonal transect into the field, starting at the field edge. A D-vac samples invertebrates mainly in the vegetation layer and primarily collects small invertebrates (< 4 mm; Doxon et al. 2011; McCravy 2018). Invertebrates are euthanised on collection, the samples initially deep frozen then stored in alcohol after removal of unwanted soil and vegetation. Most of the invertebrates sampled are identified to order and family level, following the taxonomy of Kloet and Hincks (1964, 1972, 1976, 1977, 1978).

Analysis of the invertebrates collected has concentrated on those that form the diet of chicks of grey partridges and other farmland birds. Early work defined the grey partridge chick-food index (CFI) that explained the variation in chick survival rate on a farm-by-farm basis (Potts 1986; Potts and Aebischer 1991). Most of the subsequent analysis of invertebrate change and farm management has concentrated on the components of CFI, as well as two other chick-food indices, those for corn bunting (*Emberiza calandra*) and yellowhammer (*Emberiza citrinella*, Table 11.1). These include six groups of taxa: *Araneae & Opiliones* (all sizes of spiders and harvestman); *Carabidae & Elateridae* (adults of ground and click beetles); *Symphyta, Lepidoptera & Neuroptera* (adults, larvae and shed skins of sawflies, butterflies, moths and lacewings); *Chrysomelidae & Curculionidae* (adults and larvae of leaf beetles and weevils); Non-aphid *Hemiptera* (adults and nymphs of plant bugs/hoppers, excluding aphids) and *Aphididae* (adults & nymphs of aphids), as well as the three fore-mentioned chick-food indices (Ewald et al. 2016). Additionally, we considered overall invertebrate abundance; because it is heavily influenced by three numerically very important taxa, *Collembola* (springtails), *Thysanoptera* (thrips) and *Aphididae*, we also considered invertebrate abundance without these three taxa.

To give an idea of general trends in invertebrate abundance in cereal fields in Sussex over the last 50 years, we calculated the average annual number per sample of

Table 11.1 Composition and calculations of published chick-food indices, detailing the life-cycle stages included in each index, which have been used in the analysis of invertebrates collected from crops during the Sussex Study

Index (Abbreviation)	Calculation	Reference
Grey partridge chick food index (CFI)	0.000368*Aphididae (all) +0.00614* [non-Aphididae Homoptera (adults & larvae) & Heteroptera (adults & larvae)] + 0.1199* [Neuroptera (all) & Lepidoptera (adults & larvae) & Symphyta (adults & larvae)] + 0.1411* [Carabidae (adults) & Elateridae (adults)]- + 0.0832*[Chrysomelidae (adults & larvae) & Curculionidae (adults & larvae)]	Potts and Aebischer (1991)
Corn bunting four-food index (4FI)	Opiliones (all sizes) + Orthoptera (all sizes) + [Neuroptera (all) & Lepidoptera (adults & larvae)] + Symphyta (adults & larvae)	Brickle et al. (2000)
Yellowhammer index (YHI)	Araneae (all sizes) + Homoptera (adults & larvae) + Heteroptera (adults & larvae) + Lepidoptera (larvae) + Coleoptera (adults & larvae) + Tipulidae (adults, larvae & skins)	Stoate et al. (1998) and Moreby and Stoate (2001)

invertebrates in each taxonomic group and chick-food index. We revealed the underlying long-term trend in the annual means by using least-squares regression to fit a generalised additive model (GAM, Hastie and Tibshirani 1990; Fewster et al. 2000) of year with five degrees of freedom (one for every decade of data available: Aebischer 2019). The change between the beginning and end of each 50-year time series was calculated as the percentage change between the GAM values for the first and last years. Model fitting was carried out in Genstat Release 17.1 for Windows.

4 Results

4.1 Trends in Chick Food Abundance

Plant bugs/hoppers (Non-aphid *Hemiptera*) varied from year-to-year, with a 6% decline in abundance over the 50 years. Leaf beetles & weevils (*Chrysomelidae & Curculionidae*) declined through the mid-1980s and then increased through the 2010s, with subsequent declines; there was no change in abundance overall (Fig. 11.2). Ground & click beetles (*Carabidae & Elateridae*) after showing an increase in the 1980s, declined subsequently, with a 76% decline in abundance over 50 years. Aphids (*Aphididae*) and yellowhammer index (YHI) declined after the aphid outbreaks seen in the mid-1970s, with both declining by 57% overall (Figs. 11.2 and 11.3). Spiders & harvestmen (*Araneae & Opiliones*), caterpillars and lacewings (*Symphyta, Lepidoptera & Neuroptera*, Fig. 11.2) and corn bunting four-food index (4FI, Fig. 11.3), after initial declines, increased from the mid-1990s. Overall spiders & harvestmen declined by 32%, caterpillars and lacewings increased by 26% and the corn bunting index increased by 44%. Grey partridge chick-food index (CFI, Fig. 11.3) declined in the early part of the study, before showing moderate increases since the turn of the century, with a 18% decline over 50 years (Fig. 11.3).

4.2 Trends in Invertebrate Numbers Overall

Overall invertebrate abundance showed large yearly variation, with a 17% decline (Fig. 11.4). Three taxa (*Collembola, Aphididae* and *Thysanoptera*) that are common in the Sussex invertebrate samples may influence the overall numbers. We have seen that aphid numbers have declined, after outbreaks in the mid-1970s (Fig. 11.2); the abundance of springtails (*Collembola*) showed large yearly variation, with indications of a decrease until the mid-1990s, followed by a slight increase thereafter with

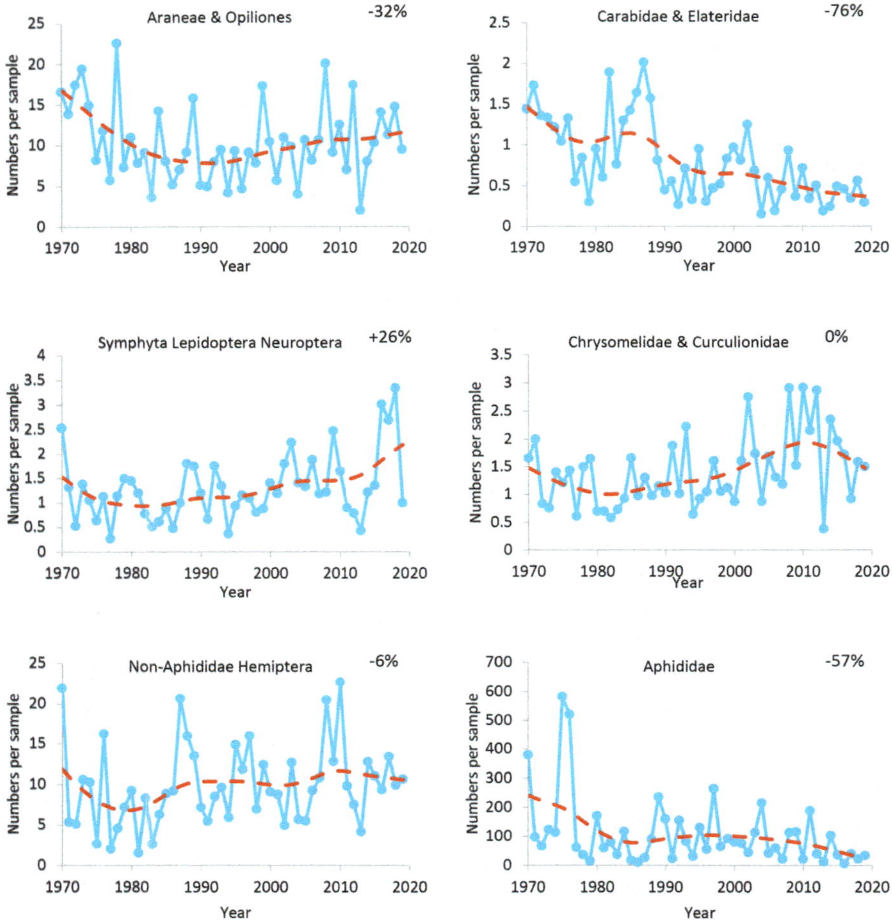

Fig. 11.2 Average annual invertebrate densities (numbers per sample) in the Sussex Study from 1970 to 2019 for six important groups of chick-food insects: Araneae & Opiliones, Carabidae & Elateridae, Symphyta, Lepidoptera & Neuroptera, Chrysomelidae & Curculionidae, Non-Aphid Hemiptera and Aphididae. The long-term trend (dashed line) was obtained by fitting a generalised additive model to year with 5 degrees of freedom

abundance only slightly down (−1%) from that at the beginning of the Sussex Study. The abundance of thrips (*Thysanoptera*) also showed large annual variation, with abundance increasing 29% overall (Fig. 11.4). Invertebrate abundance, after excluding these three taxa, declined until 1990, showed a shallow increase until the mid-2000s, then declined subsequently, with abundance declining by 34%.

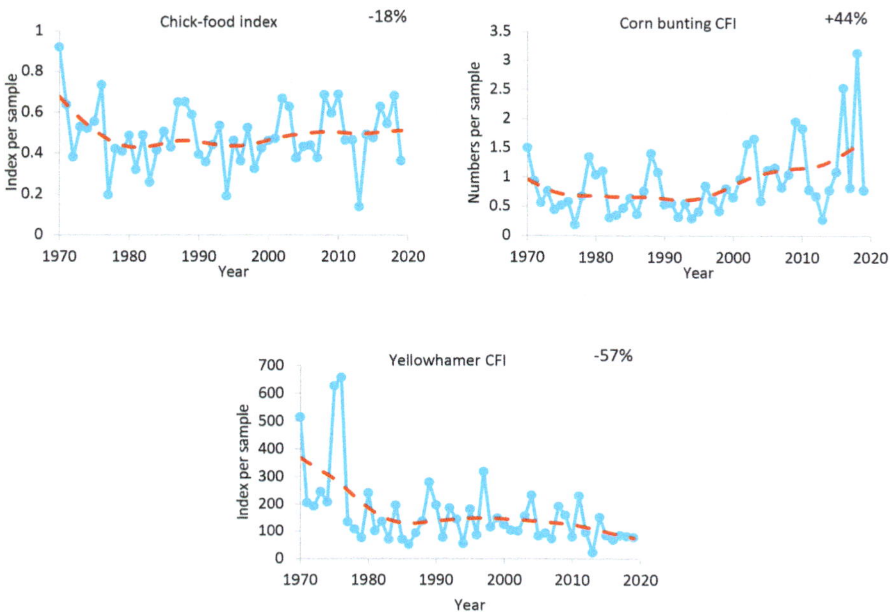

Fig. 11.3 Average annual abundance in the Sussex Study of each of the chick-food indices (blue line) from 1970 to 2019. The long-term trend (dashed line) was obtained by fitting a generalised additive model to year with 5 degrees of freedom

4.3 *Changes in the Agricultural Landscape*

Field size increased in the first half of the first 50 years of the Sussex study (Table 11.2a), with the size of arable fields averaging 12.5 ha in the 1970s, increasing to 15.6 ha in the 1990s. It declined in the last two decades, reflecting measures brought in through the uptake of the agri-environmental measures designed to fund farmers to undertake work to improve the farmed environment and its biodiversity (Ewald et al. 2012; Potts 2012). Fields averaged 10.6 ha in the 2010s. Cropping information is collected annually on a field-by field basis. It is classified into six types: rotational or ley grassland (grass crops established through undersowing the previous year's cereal crop, and kept as grassland for at most 5 years), non-rotational grassland (grass crops established through direct drilling or fields in grass for over 5 years), set-aside (land compulsorily taken out of production under the MacSharry reforms of the European Economic Community's Common Agricultural Policy from 1992 to 2007), spring cereals (spring barley and spring wheat), winter cereals (winter barley, winter oats or winter wheat, including bearded wheat) and break crops (fodder rape, kale, linseed, maize, oilseed rape, peas and beans).

In the 1970s, the move from undersowing to direct drilling was already underway and has continued, with rotational grass making up only 4% of the cropped area in

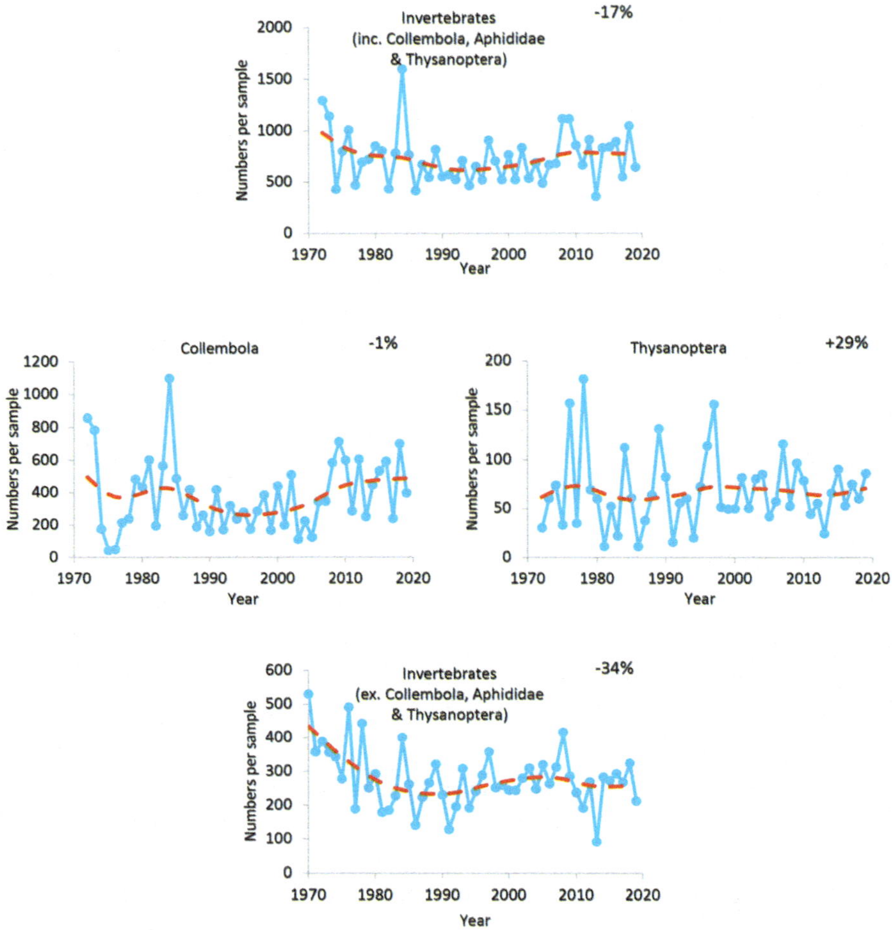

Fig. 11.4 Average annual invertebrate densities (numbers per sample) in the Sussex Study from 1970 to 2019 for all invertebrates combined, two numerically important taxa (Thysanoptera and Collembola), and all invertebrates excluding Thysanoptera, Collembola and the third numerically most important taxon, Aphididae (cf. Fig. 1). The long-term trend (dashed line) was obtained by fitting a generalised additive model to year with 5 degrees of freedom

the last three decades (Table 11.2a). The area of grass crops generally (90% now direct-drilled) increased from a third of cropped area in the 1970s to nearly 40% in the last decade. Cereal crops (both spring and winter) covered 60% on average of the study area in the first 20 years, with that now declining to just over 40% of the area. In the 1970s 55% of cereals were spring sown, though from the 1980s until the 2000s the situation was reversed, with 65% of cereals being winter-sown; in the last decade, cereals were evenly split between spring- and winter-sown. In the 1990s and 2000s, set-aside occupied 5–6% of cropped area. The area sown to break crops

Table 11.2 Changes through time in the (a) size of arable fields and cropping composition, (b) pesticide use, and (c) average daily mean temperature and average monthly total rainfall across April–June, on the Sussex Study area over the past five decades (decadal means ± standard errors)

(a)

Decade	Field size (ha)	Cropping composition (%)					
		Rotational grass	Non-rotational grass	Set-aside	Spring cereals	Winter cereals	Break crops
1970s (N = 10)	12.5 ± 0.1	13.8 ± 0.7	20.2 ± 1.8	–	32.9 ± 2.3	27.4 ± 1.4	5.8 ± 0.5
1980s (N = 10)	14.5 ± 0.3	8.4 ± 1.2	25.8 ± 0.6	–	16.2 ± 1.1	45.0 ± 1.5	4.6 ± 0.4
1990s (N = 10)	15.6 ± 0.1	4.3 ± 0.4	31.4 ± 0.6	4.9 ± 1.2	7.6 ± 0.9	39.6 ± 1.3	12.1 ± 1.1
2000s (N = 10)	13.8 ± 0.4	3.6 ± 0.3	33.2 ± 0.6	6.5 ± 1.2	15.8 ± 2.6	29.6 ± 2.1	11.3 ± 0.7
2010s (N = 10)	10.6 ± 0.1	4.5 ± 0.2	35.2 ± 0.5	–	20.9 ± 1.0	21.9 ± 1.0	17.5 ± 0.7

(b)

Decade	Herbicide spray area (%)	Fungicide spray area (%)	Insecticide spray area (%)
1970s (N = 10)	98.5 ± 4.5	23.3 ± 6.3	10.9 ± 2.9
1980s (N = 10)	145.3 ± 8.7	128.7 ± 15.5	41.7 ± 15.1
1990s (N = 10)	193.6 ± 16.6	173.3 ± 9.7	114.5 ± 9.2
2000s (N = 10)	270.9 ± 11	219.1 ± 8.3	99.0 ± 8.2
2010s (N = 3)	284.6 ± 7.2	209.5 ± 19.8	100.3 ± 6.9

(c)

Decade	Daily mean temperature (°C)	Monthly total rainfall (mm)
1970s (N = 10)	11.0 ± 0.2	53.6 ± 6.7
1980s (N = 10)	11.3 ± 0.2	54.8 ± 6.1
1990s (N = 10)	11.7 ± 0.2	56.8 ± 6.5
2000s (N = 10)	12.3 ± 0.1	58.1 ± 5.9
2010s (N = 9)	12.2 ± 0.3	58.5 ± 9.7

has increased from 6% to 18% over five decades. The current situation is of an even split between spring-sown, winter-sown and break crops, each at 20% of area, with the remaining two-fifths in grass.

Field-by-field information on pesticide use was collected from the farmers and agronomists working on the Sussex Study area (Ewald and Aebischer 2000; Ewald et al. 2016). This included details about the application of herbicides (foliar and residual), fungicides (foliar and seed treatment) and insecticides (foliar and seed treatment), as well as the timing of applications. Detailed information on the active ingredients used on different crops can be found in Ewald and Aebischer (2000), updated in Ewald et al. (2016). Pesticide applications were divided into those applied in the autumn/winter (post-harvest from the previous year until the end of February), and those applied in the spring/summer (beginning of March to June).

We measured intensity of pesticide use as percentage spray area (Ewald et al. 2016), which takes into account the number of times a field is treated with a pesticide (i.e. if a field is treated twice then its spray area would be twice the area of the field). Percentage spray areas of herbicide, fungicide and insecticide changed markedly over the period of study (Table 11.2b). Herbicide use was already widespread in the 1970s, and the area sprayed has increased throughout the study period, with each field treated receiving three applications on average. Fungicide use increased over the first four decades, then levelled off. Insecticide use remained low until the 1990s, when it appeared to peak, although it fluctuated around 100% thereafter. Information on pesticide use is published by crop type in a series of reports by the Pesticide Usage Survey Group (Davis et al. 1990, 1991, 1993; Garthwaite and Thomas 1999, 2003; Garthwaite et al. 1995, 2003, 2005, 2006, 2010, 2011, 2013; Sly 1977, 1986; Steed et al. 1979; Thomas et al. 1997). A comparison between Sussex and the national reports (adjusting for crop type so as to compare like with like) showed that herbicide and fungicide intensity of use was lower in Sussex than nationally, while insecticide intensity of use was similar to national figures. This implies that Sussex results on the effects of herbicide and fungicide are likely to underestimate national effects, but those on insecticide effects are likely to reflect the national picture.

4.4 Changes in Weather

Monthly data on daily mean temperature and total rainfall were obtained from the UKCP09 5×5-km gridded climate dataset provided by the Meteorological Office (Hollis et al. 2019; Perry and Hollis 2005). We chose the months of April, May, June as they represented the time most likely to influence invertebrate development before sampling (Aebischer 1990; Ewald et al. 2015). For each year from 1970 to 2018, we averaged each weather variable across the eight grid squares that contained at least part of the Sussex study area, then averaged the result over the 3 months to reflect long-term changes in temperature and precipitation. The decadal monthly spring mean temperature increased from 11.2 °C in the 1970s to 12.2 °C in the 2010s

(Table 11.2c). There was a small apparent increase in total precipitation (Ewald et al. 2015).

5 Discussion

The long-term changes in invertebrate abundance reported here appear to be less negative than those recorded for the biomass of aerial invertebrates in one of the most well-publicised recent reports of insect decline (Hallmann et al. 2017). Although there was an early decline in abundance in Sussex, this stabilised after the 1990s, with some increases in the mid-2000s, in contrast to a decline in biomass of 70% in Hallmann et al. (2017). The long-term trend in aphids from the Rothamsted Insect Survey (RIS) showed that, although total aphid annual totals fluctuated widely, overall numbers appeared to be in a steady state, with a non-significant decline of −7.6% over 48 years (Bell et al. 2020), in contrast to the long-term declines identified here. The RIS reports abundance from samples collected across the whole year, while Sussex samples are from the third week of June. The differences between the two schemes may be due to this difference in sampling.

5.1 Review of Results from Sussex Study Invertebrate Monitoring

Early work on the Sussex Study highlighted the effect that changes in farming and management of crops were having on invertebrate numbers and on partridges. For instance, there was a move away from undersowing as a means of establishing ley crops, which negatively affected insect abundance. The widespread use of herbicides resulted in less weedy fields and a negative effect on invertebrates in those fields (Potts and Vickerman 1974). This was taken further with the identification of invertebrate taxa that were important in the diet of grey partridge chicks (Potts 1986). Potts and Aebischer (1991) explained variation in grey partridge chick survival rate at a farm level in Sussex on the basis of five invertebrate groups that led to the development of the grey partridge chick-food index (CFI), used here and elsewhere to monitor food resources for grey partridge chicks (Smith et al. 2020).

The importance of the cereal ecosystem was recognised by Potts (1991), who identified the major impacts of intensification of modern cereal growing and the associated environmental damage. He put forward a possible solution – 'extensification', where inputs are reduced for environmental reasons. Aebischer (1991) summarised the changes in cropping and increases in field size on the Sussex study up until 1989, highlighting a decline in the abundance of invertebrates excluding Collembola and in the abundance of aphids, which are reflected in the analysis reported here. We found that the decline in invertebrate abundance had reversed slightly through the early 2000s, whereas the decline in aphid abundance was ongoing.

5.2 Effects of Changes in Pesticide Use on Invertebrates

Aebischer (1990) used 20 years of data on the abundance of sawflies in Sussex to examine the effect of pesticides on this non-target invertebrate taxon using time-series modelling. Annual sawfly densities were found to be related, with a 1-year lag, to the proportion of cereal fields that were undersown and to summer rainfall and temperature. In 1989, 7 km^2 of cereal crops on the study area were treated with dimethoate, an organophosphate insecticide, in the summer; sawfly density in the area treated with dimethoate was less than one-tenth of would have been expected without this treatment. Modelling suggested that, if management returned to that immediately before the large insecticide application, it would take 7 years for sawfly abundance to return to its equilibrium level.

Building on this research, a series of reports looked at the effect of the detailed field-by-field pesticide application on the abundance of invertebrates making up the grey partridge CFI and indices for other farmland bird chick-food indices (Ewald and Aebischer 1999; GCT 2007; Ewald et al. 2016). There were some negative effects on chick-food groups related to those herbicides used before cultivation and establishment of a crop and where multi-site/non-penetrative foliar fungicides were used (Ewald et al. 2016). The main finding of this work reinforced that of previous assessments of the Sussex Study dataset: foliar insecticide use, adjusted for the use of other types of pesticides, is associated with significantly lower abundances of all groups of chick-food invertebrates. Additionally, the use of foliar insecticide is associated with a carry-over effect in the year following an application, with the abundance of seven of the nine chick-food invertebrate groups examined significantly lower. Regarding seed treatment, aphid abundance was negatively affected by neonicotinoid seed treatments, with seed treatments as a group negatively impacting the abundance of four other chick-food invertebrate groups. The overarching negative effect of foliar insecticide applications remained when controlling for seed treatments.

5.3 Effects of Changes in Weather on Invertebrates

Ewald et al. (2015) investigated the sensitivity and resilience of invertebrates in cereal fields to extreme weather events and examined the effect of long-term changes in temperature, rainfall and pesticide use on invertebrate abundance. The abundances of the most common 26 taxa were considered. Effects of extreme weather events on eleven of those taxa were identified, but the results were short-lived, with abundance returning to the long-term trend within a year of the extreme weather event. Some long-term trends in invertebrate abundance correlated with temperature and rainfall, indicating that climate change may affect them. Pesticide use was however more important in explaining the trends, suggesting that reduced pesticide use would mitigate the effects of climate change.

Acknowledgements Our thanks to the farmers, land managers, agronomists and gamekeepers on the Sussex Study area, who allowed us access and answered our questions over the last 50 years. Dick Potts started the Sussex Study in 1968; the longevity and continued relevance of the results from the Sussex Study owe much to his farsightedness.

References

Aebischer NJ (1990) Assessing pesticide effects on non-target invertebrates using long-term monitoring and time-series modelling. Funct Ecol 4:369–373

Aebischer NJ (1991) Twenty years of monitoring invertebrates and weeds in cereal fields in Sussex. In: Firbank LG, Carter N, Darbyshire JF, Potts GR (eds) The ecology of temperate cereal fields. Blackwell Scientific Publications, Oxford, pp 305–331

Aebischer NJ (2019) Fifty-year trends in UK hunting bags of birds and mammals, and calibrated estimation of national bag size, using GWCT's National Gamebag Census. Eur J Wildl Res 65 (64):1–13

Aebischer NJ, Ewald JA (2012) The grey partridge in the UK: population status, research, policy and prospects. Anim Biodivers Conserv 35:353–362

Bell JR, Blumgart D, Shortall CR (2020) Are insects declining and at what rate? An analysis of standardised, systematic catches of aphid and moth abundances across Great Britain. Insect Conserv Divers 13:115–126. https://doi.org/10.1111/icad.12412

Brickle NW, Harper DGC, Aebischer NJ, Cockayne SH (2000) Effects of agricultural intensification on the breeding success of corn buntings *Miliaria calandra*. J Appl Ecol 37:742–755

Campbell LH, Avery MI, Donald P, Evans AD, Green RE, Wilson JD (1997) A review of the indirect effects of pesticides on birds. JNCC Report 227, ISSN 0963 8091

Davis RP, Garthwaite DG, Thomas MR (1990) Pesticide usage survey report 78 – arable farm crops, 1988. MAFF, London

Davis RP, Garthwaite DG, Thomas MR (1991) Pesticide usage survey report 85 – arable farm crops in England & Wales, 1990. MAFF, London

Davis RP, Thomas MR, Garthwaite DG, Bowen HM (1993) Pesticide usage survey report 108 – arable farm crops in Great Britain, 1992. MAFF, London

Dietrick EJ (1961) An improved backpack motorised fan for suction sampling of insects. J Econ Entomol 54:394–395

Dirzo R, Young HS, Galetti M, Ceballos G, Isaac NJB, Collen B (2014) Defaunation in the Anthropocene. Science 345(6195):401–406. https://doi.org/10.1126/science.1251817

Doxon ED, Davis CA, Fuhlendorf SD (2011) Comparison of two methods for sampling invertebrates: vacuum and sweep-net sampling. J Field Ornithol 82:60–67

Ewald JA, Aebischer NJ (1999) Pesticide use, avian food resources and bird densities in Sussex, JNCC report 296. Joint Nature Conservation Committee, Peterborough

Ewald JA, Aebischer NJ (2000) Trends in pesticide use and efficacy during 26 years of changing agriculture in southern England. Environ Monit Assess 64:493–529

Ewald JA, Potts GR, Aebischer NJ (2012) Restoration of a wild grey partridge shoot: a major development in the Sussex study, UK. Anim Biodivers Conserv 35:363–369

Ewald JA, Wheatley CJ, Aebischer NJ, Moreby SJ, Duffield SJ, Crick HQP, Morecroft MB (2015) Influences of extreme weather, climate and pesticide use on invertebrates in cereal fields over 42 years. Glob Chang Biol 21:3931–3950
Ewald JA, Wheatley CJ, Aebischer NJ, Duffield SJ, Heaver D (2016) Investigation of the impact of changes in pesticide use on invertebrate populations, Natural England commissioned report NECR182. Natural England, York
Fewster RM, Buckland ST, Siriwardena GM, Baillie SR, Wilson JD (2000) Analysis of population trends for farmland birds using generalized additive models. Ecology 81:1970–1984
Game Conservancy Trust (GCT) (2007) Development of indicator species to measure pesticide impact on farmland wildlife. Defra, London
Garthwaite DG, Thomas MR (1999) Pesticide usage survey report 159 – arable farm crops in Great Britain, 1998. MAFF, London
Garthwaite DG, Thomas MR (2003) Pesticide usage survey report 171 – arable farm crops in Great Britain, 2000. DEFRA, London
Garthwaite DG, Thomas MR, Hart M (1995) Pesticide usage survey report 127 – arable farm crops in Great Britain, 1994. MAFF, London
Garthwaite DG, Thomas MR, Anderson H, Stoddart H (2003) Pesticide usage survey report 187 – arable farm crops in Great Britain, 2002. DEFRA, London
Garthwaite DG, Thomas MR, Anderson H, Stoddart H (2005) Pesticide usage survey report 202 – arable farm crops in Great Britain, 2004. DEFRA, London
Garthwaite DG, Thomas MR, Heywood E, Battersby A (2006) Pesticide usage survey report 213: arable crops in Great Britain 2006 (including aerial applications 2003–2006). Defra, London
Garthwaite DG, Thomas MR, Parrish G, Smith L, Barker I (2010) Pesticide usage survey report 224: arable crops in Great Britain 2008 (including aerial applications 2007–2008). Defra, London
Garthwaite DG, Barker I, Parrish G, Smith L, Chippindale C, Pietravalle S (2011) Pesticide usage survey report 235: arable crops in Great Britain 2010 (including aerial applications 2010). Defra, London
Garthwaite DG, Hudson S, Barker I, Parrish G, Smith L, Pietravalle S (2013) Pesticide usage survey report 250: arable crops in Great Britain 2012 (including aerial applications 2012). Defra, London
Hallmann CA, Sorg M, Jongejans E, Siepel H, Hofland N, Schwan H, Stenman W, Müller A, Sumser H, Hörren T, Gourson D, de Kroon H (2017) More than 75 percent decline over 27 years in total flying insect biomass in protected areas. PLoS One 12(10):e0185809. https://doi.org/10.1371/journal.pone.0185809
Hastie TJ, Tibshirani RJ (1990) Generalized additive models. Chapman & Hall/CRC, New York/Boca Raton
Hollis D, McCarthy M, Kendon M, Legg T, Simpson I (2019) HadUK-grid—A new UK dataset of gridded climate observations. Geosci Data J 6(2):151–159
Kloet GS, Hincks WD (1964) A checklist of British insects. Small-orders and Hemiptera. Royal Entomological Society of London, London
Kloet GS, Hincks WD (1972) A checklist of British insects. Lepidoptera. Royal Entomological Society of London, London
Kloet GS, Hincks WD (1976) A check list of British insects: Diptera and Siphonaptera. Royal Entomological Society of London, London
Kloet GS, Hincks WD (1977) A check list of British insects. Coleoptera and Strepsiptera. Royal Entomological Society of London, London
Kloet GS, Hincks WD (1978) A check list of British insects. Part 4: hymenoptera. Royal Entomological Society of London, London
Leather SR (2018) "Ecological Armageddon" – more evidence for the drastic decline in insect numbers. Ann Appl Biol 172(1):1–3

Macgregor CJ, Williams JH, Bell JR, Thomas CD (2019) Moth biomass increases and decreases over 50 years in Britain. Nat Ecol Evol 3(12):1645–1649. https://doi.org/10.1038/s41559-019-1028-6

McCravy KW (2018) A review of sampling and monitoring methods for beneficial arthropods in agroecosystems. Insects 9(4):170. https://doi.org/10.3390/insects9040170

Moreby SJ, Stoate C (2001) Relative abundance of invertebrate taxa in the nestling diet of three farmland passerine species, Dunnock *Prunella modularis*, Whitethroat *Sylvia communis* and Yellowhammer *Emberiza citrinella* in Leicestershire, England. Agric Ecosyst Environ 86:125–134

Perry M, Hollis D (2005) The generation of monthly gridded datasets for a range of climatic variables over the UK. Int J Climatol 25(8):1041–1054

Potts GR (1980) The effects of modern agriculture, nest predation and game management on the population ecology of partridges, (*Perdix perdix* and *Alectoris rufa*). Adv in Ecol Res 11:1–79

Potts GR (1986) The partridge: pesticides, predation and conservation. Collins, London

Potts GR (1991) The environmental and ecological importance of cereal fields. In: Firbank LG, Carter N, Darbyshire JF, Potts GR (eds) The ecology of temperate cereal fields. Blackwell Scientific Publications, Oxford, pp 3–21

Potts GR (2012) Partridges. Countryside barometer, New naturalist library book 121. Collins, London

Potts GR, Aebischer NJ (1991) Modelling the population dynamics of the Grey Partridge: conservation and management. In: Perrins CM, Lebreton J-D, Hirons GJM (eds) Bird population studies: their relevance to conservation and management. Oxford University Press, Oxford, pp 373–390

Potts GR, Vickerman GP (1974) Studies on the cereal ecosystem. In: Macfadyen A (ed) Advances in ecological research, vol 8. Academic Press (London) Ltd, London, pp 107–197

Sánchez-Bayo F, Wyckhuys KA (2019) Worldwide decline of the entomofauna: a review of its drivers. Biol Conserv 232:8–27

Sly JMA (1977) Pesticide usage survey report 8-review of usage of pesticides in agriculture and horticulture in England & Wales, 1965–1974. HMSO, London

Sly JMA (1986) Pesticide usage survey report 41-review of usage of pesticides in agriculture, horticulture and animal husbandry in England & Wales, 1980–1983. HMSO, London

Smith BM, Aebischer NJ, Ewald JA, Moreby SJ, Potter C (2020) The potential of arable weeds to reverse invertebrate declines and associated ecosystem services in cereal crops. Front Sustain Food Syst 3. https://doi.org/10.3389/fsufs.2019.00118

Sotherton NW (1991) Conservation Headlands: a practical combination of intensive cereal farming and conservation. In: Firbank LG, Carter N, Darbyshire JF, Potts GR (eds) Ecology of temperate cereal fields. British Ecological Society Symposium, Blackwell Scientific Publications, Oxford, pp 373–397

Sotherton NW, Aebischer NJ, Ewald JA (2014) Research into action: grey partridge conservation as a case study. J Appl Ecol 51:1–5

Southwood TRE, Cross DJ (2002) Food requirements of grey partridge *Perdix perdix* chicks. Wildl Biol 8:175–183

Steed JM, Sly JMA, Tucker GG, Cutler JR (1979) Pesticide usage survey report 18 – arable farm crops, 1977. HMSO, London

Stoate C, Moreby SJ, Szczur J (1998) Breeding ecology of farmland Yellowhammers *Emberiza citrinella*. Bird Study 45:109–121

Thomas MB, Wratten SD, Sotherton NW (1992) Creation of 'island' habitats in farmland to manipulate populations of beneficial arthropods: predator densities and species composition. J Appl Ecol 29:524–531

Thomas MR, Garthwaite DG, Banham AR (1997) Pesticide usage survey report 141 – arable farm crops in Great Britain, 1996. MAFF, London van Klink R, Bowler DE, Gongalsky KB, Swengel AB, Gentile A, Chase JM (2020) Meta-analysis reveals declines in terrestrial but increases in freshwater insect abundances. Science 368 (6489):417–420. https://doi.org/10.1126/science.aax9931

Wauchope HS, Amano T, Sutherland WJ, Johnston A (2019) When can we trust population trends? A method for quantifying the effects of sampling interval and duration. Methods Ecol Evol 10 (12):2067–2078. https://doi.org/10.1111/2041-210X.13302

Wepprich T, Adrion JR, Ries L, Wiedmann J, Haddad NM (2019) Butterfly abundance declines over 20 years of systematic monitoring in Ohio, USA. PLoS One 14(7):1–21. https://doi.org/10.1371/journal.pone.0216270

Chapter 12
Ex Situ Conservation and Reintroduction of Vulnerable Arable Plants in Skåne, Sweden

Gabrielle Rosquist and Jakob Sandberg

Abstract Although nature conservation efforts in Skåne, in southern Sweden, have increased in recent decades, the populations of many plant species continue to decline and are at risk of becoming regionally extinct. Ex situ conservation and subsequent reintroductions may be a critical aid to the survival of these species.

In 2015, the regional authority formed a collaboration with two botanical gardens, a botanical society and a university in an effort to save the most threatened vascular plant species in Skåne. The objectives were to aid the conservation of threatened species through seed sampling, cultivation, amplifying, reintroductions and raising awareness of the regional flora. By 2020, about 30 species were included in the ex situ conservation programme and the number increases annually. To conserve possible local adaptations, we store plant material from two discrete populations of each species in isolation from each other at the two botanical gardens.

To date, we have introduced eight ex situ-grown species to new sites, with some showing positive responses. Depending on the species and the site management, subsequent reintroductions may be needed. Long-term population monitoring is needed to evaluate the success of reintroductions and assess their contribution towards conservation status. Nature conservationists and landowners must cooperate closely in carrying out in situ and ex situ conservation.

Keywords Arable weeds · *Arnoseris minima* · Ex situ conservation · Reintroductions · *Ranunculus arvensis* · Skåne · Seed transfer zones · Vulnerable arable plants

---

G. Rosquist (✉)
County Administrative Board Skåne, Malmö, Sweden
e-mail: gabrielle.rosquist@lansstyrelsen.se

J. Sandberg
Fredriksdal Museums and Gardens, Helsingborg, Sweden
e-mail: jakob.sandberg@helsingborg.se

C. Hurford et al. (eds.), *The Changing Status of Arable Habitats in Europe*,
https://doi.org/10.1007/978-3-030-59875-4_12

1 Flora in the Arable Landscape of Southern Sweden

The arable landscape in southern Sweden has been shaped by humans for thousands of years and can almost exclusively be considered a cultural landscape (Emanuelsson et al. 2002; Olsson et al. 2003; Welinder 2011). Farming, grazing, mowing and the use of timber and wood have had a strong impact on the biodiversity of the arable fields, grasslands, wetlands and forests (Emanuelsson et al. 2002; Emanuelsson 2009). The landscape gradually developed into a mosaic with a high variation of habitats and prerequisites for an overall high biodiversity. The human impact on the arable habitats accelerated during the twentieth century as a consequence of the new technologies introduced into farming practices and the introduction of mineral fertilizers and chemical pesticides (Emanuelsson et al. 2002; Emanuelsson 2009; Dänhardt et al. 2013). As smaller entities turned into larger agricultural units, farming became more specialised and the land use more intense as farms increasingly focused on either cultivation or animal husbandry, with a high proportion of the remaining grazing land converted from natural pastures to leys. This resulted in a landscape divided primarily into either arable fields or forests (Emanuelsson 2009; Dänhardt et al. 2013).

In the search for more land to cultivate, ditching and lowering of the water surface of lakes resulted in a general draining of the landscape (Emanuelsson et al. 2002; Emanuelsson 2009). These changes, and the speed at which they occurred, had a large negative impact on the biodiversity (Emanuelsson et al. 2002, Emanuelsson 2009). Today, changes in land use, together with the impact of climate warming, explain many of the recent changes in the regional flora of Skåne, the southernmost county of Sweden (Tyler et al. 2017); these changes are consistent with the IPBES global assessment 2019 (IPBES 2019).

About half of the nationally red-listed species in Sweden can be found in the arable landscape and as many as one third are dependent on this landscape to survive, especially beetles, butterflies and vascular plants (Sandström et al. 2015). Several plant species are now so rare that their long-term survival is uncertain and approximately 25% of the vascular plants in the country are threatened (Eide 2014; Sandström et al. 2015; Eide et al. 2020). During the last decades, nature conservation has accelerated through the protection and management of species and their habitats, but still many plant populations decrease or approach regional extinction. The in situ conservation management of the arable landscape for plant species is not enough and has very limited chances of success. Therefore, ex situ species conservation may be a crucial complement for long-term survival for some of the most threatened vascular plants in the landscape (Guerrant et al. 2004; Havens et al. 2006; Cochrane et al. 2007).

2 Ex Situ Conservation – A Strategy for Threatened Plants in Skåne

In 2015, the regional authority (the County Administrative Board of Skåne), two regional botanical gardens (Fredriksdal Museums and Gardens at the Municipality of Helsingborg and the Botanical Garden at Lund University), Lund University and regional botanists (the Lund Botanical Society) formed a collaboration to change the negative trends for the most threatened vascular plants in Skåne. The purpose of this collaboration is the long-term conservation of the most threatened plants and, in the long run, to stop the loss of plant diversity in the region. The objectives are ex situ conservation of plant species through seed sampling, storage of seeds or cultivation (i.e. gene bank), amplifying of seeds or plug seedlings for possible subsequent reintroductions, as well as informing and raising public awareness of the natural flora in the region (Rosquist et al. 2018). The ex situ conservation measures are expected to bridge the gaps in time and/or space for threatened plant species with increasingly fragmented distributions in Skåne. The plant dispersal will be enhanced in the landscape, and a backup population will be held in cultivation until new and former suitable habitats are created or restored.

During ex situ conservation and subsequent reintroductions certain aspects need to be considered to avoid negative consequences in the plant material, such as changes in life history traits during cultivation and founder effects or outbreeding depressions during sampling and reintroductions. Removing wild plants from their natural environments can result in changes in life history traits in the cultivated material, with potentially large negative effects on the success of subsequent reintroductions (Albrecht et al. 2018). Several studies have, for example, documented a high risk of losing seed dormancy during ex situ cultivation, especially in short-lived species (Ensslin et al. 2017). To avoid changes in the introduced plant material as far as possible, sowing seeds from the first generation achieved in cultivation is to be preferred and new seeds for cultivation need to be collected frequently from natural populations. The success of the reintroductions may also be affected by the size of the source population and the number of individuals included in the sample. Larger source populations tend to have higher gene diversity, and the numbers of individual plants per population from which seeds are sampled need to be as high as possible to transfer a larger amount of gene diversity ex situ (Hufford and Mazer 2003; Mijnsbrugge et al. 2010; Ensslin et al. 2015; Godfroid et al. 2016). However, to preserve as much gene diversity ex situ as possible, a balance is needed between sampling from a large population with an expected high gene diversity or from a small population with possible local adaptations. Preserving material from the smaller population ex situ, may prevent local adaptations from going extinct if the population disappears and this material may, in turn, contribute to the overall gene diversity of the species on a larger geographic scale.

The knowledge of the species population sizes is obtained by the Flora Guardian in Skåne, run by the Swedish Botanical Society. This activity provides data on the status of the source populations through yearly monitoring of the most threatened plant species and the results can thereafter be used to direct the sampling for ex situ cultivation. For example, the yearly monitoring of the last Swedish population of the annual species *Minuartia viscosa* (Schreb.) Schinz & Thell. (sticky sandwort) showed that the number of observed flowering plants was still high, around 500, and sampling was possible without affecting the source population. The population size may also be too small to allow seed sampling, as was the case with *Pulmonaria angustifolia* L. (narrow-leaved lungwort), where sampling was postponed because only 10 flowering plants remained. Recent knowledge of conservation status of each population is also required when applying for necessary permits and dispensation from the responsible authorities.

Since the start of the regional ex situ conservation project in 2015, a total of 30 plant species have been sampled within the project, of which many are weeds from arable fields (Table 12.1). The best possible knowledge about the cultivation of the threatened plant species are used by horticulturists and botanists at the botanical gardens.

When new knowledge about cultivation of the threatened species becomes available e.g., through scientific research, from farmers or from developing new techniques or methods in the botanical gardens, this is carefully documented.

When restoring population fitness or reintroducing rare plant species in situ, local adaptations need to be considered (Hufford and Mazer 2003; Godfroid et al. 2016). Therefore, seeds sampled from different populations have been separated in the ex situ cultivation, which has been possible thanks to the two separate botanical gardens taking part in the project. When reintroducing plant material into restored habitats, the concept of seed transfer zones may preferably be used to optimise the population viability and avoid impact in terms of outbreeding depression in remaining natural populations (Hufford and Mazer 2003). In Sweden, we currently have no defined seed transfer zones and in the case of threatened species we have therefore tried to establish these in suitable habitats as close to the collection sites as possible. To avoid mixing plant material from different populations in the project, the seeds from cultivation have only been reintroduced into sites with no recent observations of the plant species, as illustrated in the three examples of reintroductions described below.

To avoid making unnecessary mistakes during sampling, cultivation and reintroduction of plant material, we need recent knowledge about the species history, biology and ecology before sampling from a population (Hufford and Mazer 2003; Godfroid et al. 2016). In Sweden, such information is to some extent available in information sheets that have been produced during the compilation of the national red list (www.artdatabanken.se) or conservation action programs for particular endangered species (www.naturvardsverket.se). We also need knowledge about the species' natural habitats prior to the reintroductions in situ, to ensure the habitats are suitable for successful reintroduction and the subsequent regrowth of the species. The reintroductions must also follow the guidelines from the IUCN (IUCN/SSC 2014).

The reintroductions have to be carefully documented and transparent to the society. Therefore, all sampling and reintroductions are thoroughly recorded in the citizen science database, the Swedish Species Observation System called Artportalen (www.artportalen.se) to be visible to land owners, agencies,

Table 12.1 Arable plant species included in the ex situ conservation project, their threat category according to the Swedish national red list, whether the species is an arable weed and at which botanical garden the species is in cultivation

Name	Threat category[a]	Arable weed	Ex situ conservation at[b]
Anthemis cotula	EN	x	Fredriksdal
Anthericum liliago	EN		Lund
Arnoseris minima	EN	x	Fredriksdal
Astragalus arenarius	EN		Lund
Cerastium brachypetalum	VU		Fredriksdal, Lund
Chenopodium murale	CR		Fredriksdal
Chenopodium vulvaria	EN		Fredriksdal, Lund
Dianthus superbus	EN		Fredriksdal
Euphrasia officinalis subsp. pratensis	EN		Fredriksdal
Genista pilosa	NT		Lund
Gentiana pneumonanthe	EN		Fredriksdal, Lund
Gentianella campestris ssp. baltica	EN		Fredriksdal
Hypericum humifusum	EN	x	Fredriksdal, Lund
Hypochaeris glabra	VU	x	Fredriksdal
Jacobaea erucifolia	EN		Fredriksdal, Lund
Kickxia elatine	EN	x	Fredriksdal, Lund
Lathyrus sphaericus	EN		Fredriksdal
Medicago minima	VU		Fredriksdal, Lund
Melilotus dentatus	EN		Fredriksdal, Lund
Minuartia viscosa	CR		Fredriksdal, Lund
Misopates orontium	NT	x	Fredriksdal, Lund
Neslia paniculata	EN	x	Fredriksdal, Lund
Ranunculus arvensis	NT	x	Fredriksdal, Lund
Rosa tomentosa	CR		Fredriksdal
Sherardia arvensis	VU	x	Fredriksdal, Lund
Silaum silaus	CR		Fredriksdal
Tephroseris integrifolia	EN		Fredriksdal
Tephroseris palustris	CR		Fredriksdal
Valerianella dentata	VU	x	Fredriksdal, Lund
Veronica praecox	EN		Fredriksdal, Lund

[a]The threat categories in the national red list (SLU Artdatabanken 2020); *CR* critical endangered, *EN* endangered, *VU* vulnerable, *NT* near threatened
[b]Fredriksdal and Lund refers to Fredriksdal Museums and Gardens in the Municipality of Helsingborg and the Botanical Garden at Lund University, respectively

municipalities, NGOs and others interested in conservation activities. It is also important to be aware that the population development after reintroduction takes time and that long-term monitoring of the new population will be necessary (Albrecht et al. 2018). The future monitoring of the reintroductions will be included in the Flora Guardian programme carried out by the Swedish Botanical Society.

Fig. 12.1 Transplanted *Silaum silaus* in a meadow at Fredriksdal Museums and Gardens. (Photo by Jakob Sandberg)

Informing and raising awareness are important for increasing the knowledge of individual species and of biodiversity in general, as well as for understanding the need for species-specific management in situ. At the two public botanical gardens in the project, there are excellent opportunities to inform the public about the threatened species in the region and to raise awareness of the need for conservation actions in their natural habitats. Threatened plants are put into a relevant context of sustainable land use and food production that are illustrated by hay-meadows and pastures grazed by native breeds of cattle. In the meadows at Fredriksdal Museums and Gardens, *Silaum silaus* (L.) Schinz & Thell (pepper saxifrage) and *Dianthus superbus* L. (fringed pink) have been transplanted (Fig. 12.1), while *Sherardia arvensis* L. (field madder) and *Ranunculus arvensis* L. (corn buttercup) have been established in an arable field. The hundreds of thousands of visitors walking around in the historic quarters every year at Fredriksdal will also get information about *Chenopodium vulvaria* L. (stinking goosefoot) and *Arnoseris minima* (L.) Schweigg. & Körte (lamb's succory), that have been introduced in the historical urban environments (Fig. 12.2).

3 Examples of Reintroductions of Arable Weeds

Some of the most threatened plant species in southern Sweden are directly connected to the cultivation of the landscape; the arable weeds. These species have spread in the open landscape for as long as fields have been farmed (Kloth 2007; Olsson et al. 2003). Before the nineteenth century, there was little control of arable weeds and insufficient tillage even favoured weeds with a perennial habit. During the nineteenth and twentieth centuries weed-control measures became more effective, a change associated with increased mechanisation, the introduction of mineral fertilizers, herbicide applications, new varieties of cereals and effective seed cleaning, and an overall intensification of farming. The historical habitats for arable weeds

Fig. 12.2 *Arnoseris minima* growing in the historical urban environments at Fredriksdal Museums and Gardens. (Photo by Jakob Sandberg)

declined, both in area and abundance, and the remaining populations of the threatened weeds became small and isolated. Consequently, the Swedish Environmental Protection Agency developed a nationwide program for the active conservation of threatened weed species, i.e. those adapted to the traditionally farmed landscape (Kloth 2007). According to the guidelines in this programmme, measures should be taken to save remaining threatened arable weed species combining ex situ methods with subsequent reintroductions. Organic farms and protected areas provide suitable environments where farming methods can be adapted for threatened arable weeds.

There are not many localities that still practise pre-industrial agriculture and only a few nature or cultural reserves have been established with the purpose of protecting threatened weed species. In Skåne, there is only one nature reserve established exclusively for such plant species, Dalby Västermark.

The management of this reserve was developed to favour the natural populations of rare arable weed species, such as *Euphorbia exigua* L. (dwarf spurge), *Kickxia elatine* (L.) Dumort. (sharp-leaved fluellen), *Ranunculus arvensis* L. (corn buttercup) and *Sherardia arvensis* L. (field madder). Dalby Västermark and other remaining sites with arable weeds may serve as source populations for future transplantations of these species.

The reintroductions of arable weed species that are described below were made at sites differing both in soil type and the type of ownership (private or state-owned). The reintroduced weed species are all annuals that germinate mainly in spring (Table 12.2) and involve seeds from ex situ plants grown at the botanical garden of Fredriksdal Museums and Gardens.

3.1 Example of Reintroduction on Fertile Soils

Reintroductions of arable weed species were implemented at Rinnebäcks gård, an organic farm on fertile and productive clay soils at the village Värpinge near the city

Table 12.2 Red-listed arable weed species included in the reintroduction conservation project in southern Sweden. All of the listed species are included in a national conservation action plan for threatened weed species (Kloth 2007)

Name	Threat category[a]	Sampled (year)	Donor localities	Reintroduction localities
Anthemis cotula	EN	2011	Ven (Landskrona)	Värpinge 2019
Arnoseris minima	EN	2014	Ekesvång, Genarp (Lund)	Maglehem 2018
Hypochaeris glabra	VU	2015	Revingekvarn (Lund)	Hagestad 2017
Kickxia elatine	EN	2014	Dalby Västermark (Lund)	Värpinge 2019
Misopates orontium	NT	1995	Lerhamn (Höganäs)	Maglehem 2017; Hagestad 2017
Ranunculus arvensis	NT	2011	Dalby Västermark (Lund)	Värpinge 2015, 2019
Sherardia arvensis	VU	2011	Dalby Västermark (Lund)	Värpinge 2015, 2019; Maglehem 2017
Valerianella dentata	VU	2016	Mölle kapell (Höganäs)	Maglehem 2017

[a]The threat categories in the Swedish national red list (SLU Artdatabanken 2020); *EN* endangered, *VU* vulnerable, *NT* near threatened

of Lund in the south-western part of Skåne. The farm has a long tradition of cultivating crops, but to survive in modern agriculture the owner decided to combine economic and ecologically sustainable management. The cultivation includes historical crops grown under conditions chosen to enhance the biodiversity on the property; thus, the attempt to establish threatened arable weeds at this site was a natural step.

In the autumn of 2015, *Ranunculus arvensis* (L.) (corn buttercup) and *Sherardia arvensis* (L.) (field madder) from Dalby Västermark (situated 10 km away) were reintroduced at the farm. About 7000 and approximately 20,000 seeds respectively were sown within five plots in each of two fields, covering a total surface of 900 m^2. The seeds were sown at the edges of the fields to avoid trampling the crop when monitoring the outcome. Prior to the reintroductions of the arable weeds, the field were sown with spelt wheat (*Triticum spelta*), autumn rye (*Secale cereale*) and about ten old varieties of autumn wheat (*Triticum sp.*).

We monitored the outcome of the reintroduction in late spring 2016, when 174 plants of *R. arvensis* were counted (Fig. 12.4). The following year (2017), the number of *R. arvensis* plants remained high (130 plants) despite the frequent harrowing needed to combat thistles (a common problem in organic farming systems) in that year. The successful establishment had thereby doubled the number of localities of the species in Skåne from one to two. However, no *S. arvensis* plants were visible.

In late spring 2019, four of the five plots there held 254 *R. arvensis* plants. Considering that the population size remained large after four years without any

Fig. 12.3 Sowing the arable weeds *Ranunculus arvensis* and *Sherardia arvensis* in fields at an organic farm in Värpinge. (Photo by Jakob Sandberg)

further management changes, apart from being an organic farm, we can regard this a successful conservation measure. During the visit in spring 2019, we made a complementary sowing of approximately 6200 seeds of *R.arvensis* and approximately 22,000 seeds of *Sherardia arvensis* in four of the plots where the previous seeding was successful (Fig. 12.3). In one of the plots, three other species of threatened arable weeds were introduced with 1300 seeds of *Kickxia elatine* (L.) Dumort. (sharp-leaved fluellen), 21,000 seeds of *Misopates orontium* (L.) Raf. (weasel's-snout) and 2000 seeds of *Anthemis cotula* L. (stinking chamomile).

In the future, the development of the populations of threatened arable weeds at Värpinge will be monitored and, if necessary, further complementary sowing will be made, including both the already established *R. arvensis*, as well as the other introduced rare arable weed species until the populations are sustainable. This will improve both the regional conservation status for arable weeds and the overall biodiversity in the arable landscape in the western part of Lund.

3.2 Examples of Reintroductions on Sandy Soils

During autumn 2017, in the nature reserve Maglehem in the eastern part of Skåne, we introduced a conservation measure to preserve and enhance the unique flora and fauna in a sandy landscape after restoring formerly abandoned fields afforested by pine. The historical management of tillage creating bare sandy soil, with nutrient poor fields left fallow for several years after cultivation, favoured the diverse flora

Fig. 12.4 *Ranunculus arvensis* flowering at an organic farm in Värpinge. (Photo by Gabrielle Rosquist)

Fig. 12.5 The traditional way of sowing an old variety of rye in Maglehem nature reserve in south-east Skåne. (Photo by Gabrielle Rosquist)

and fauna of this area. One objective of the nature reserve is to adapt the management of the crop field for arable weeds. To mimic the historical farming practice, we used horses and old agricultural equipment to prepare the ground for sowing and then seeded it with grains of old and local rye varieties (Fig. 12.5). These measures

Fig. 12.6 *Misopates orontium* flowering at Hagestad nature reserve in the south-east of Skåne. (Photo by Gabrielle Rosquist)

were carried out through a collaboration between the Swedish National Heritage Board, the County Administrative Board of Skåne, Fredriksdal Museums and Gardens and "Allkorn", a Swedish association for conserving and developing historical crops.

On completing the crop sowing, the edges of the field were sown with the seeds of four rare weed species not recently observed in the vicinity of Maglehem: *Misopates orontium* (L.) Raf. (weasel's-snout), *Sherardia arvensis* (field madder), *Valerianella dentata* (L.) Pollich (narrow-fruited cornsalad) and *Arnoseris minima* (lamb's succory), with the last one added in spring 2018. All of these species are annuals that germinate in spring or autumn and occur occasionally in sandy habitats in the arable landscape.

In spring 2018, several young plants of *S. arvensis* were observed in the field. Unfortunately, both spring and summer in 2018 were extremely dry and the seedlings that appeared withered early in the growing season. However, we expected a dormant seed bank to remain, and after waiting a year for more suitable weather conditions, we found 50 plants of *A. minima* in the spring of 2019. We then made further sowings of *S. arvensis* and *M. orontium* in different parts of the same field. The reintroduction of threatened arable weeds will continue until the new populations become sustainable in the area.

The coastal dunes of Hagestad nature reserve (situated in the south-eastern part of Skåne) was the location of the third introduction. Here we ploughed an open area in the wooded dunes in early spring 2017 where we subsequently seeded the arable weeds *Hypochaeris glabra* L. (smooth cat's-ear) and *Misopates orontium* (Fig. 12.6) together with buckwheat (*Fagopyrum esculentum* Moench), a historically common crop in this region.

In summer the same year 2017, a total of about 500 specimens of *M. orontium* and about 60 large specimens of *H. glabra* were observed flowering in the ploughed area. The following year was very dry and only 30 plants of *M. orontium* and no *H. glabra* were found, while in 2019, no plants of *M. orontium* and only four plants of *H. glabra* were observed. However, as indicated above for the Maglehem site, some seeds may remain, and subsequent reintroductions will be carried out to augment the new populations.

4 Summary

Humans played a key role in providing the conditions to support a diverse flora in the arable landscape but over the past century the human impact has become increasingly negative. When conservation management in situ is not enough to preserve the most threatened plant species, ex situ conservation and subsequent reintroductions may be the only option. In situ and ex situ conservation requires close cooperation between nature conservationists and landowners and relies on knowledge both about the species and about the impact of cultivation on the reintroduced material.

In the southernmost county of Sweden, we have established a collaboration between nature conservation bodies and botanical gardens to facilitate ex situ conservation and the subsequent reintroduction of several threatened plant species. In this chapter we have provided the early results from three reintroductions of threatened weed species in arable fields. These results show a positive response for some of the introduced species but also indicate the need for subsequent reintroductions. Long-term monitoring of population development will be essential to determine the sustainability of the populations and to evaluate the success of the reintroductions.

References

Albrecht MA, Osazuwa-Peters OL, Maschinski J, Bell TB, Bowles ML, Brumback WE, Duquesnel J, Kunz M, Lange J, McCue KA, McEachern AK, Murray S, Olwell P, Pavlovic NB, Peterson CL, Possley J, Randall JL, Wright SJ (2018) Effects of life history and reproduction on recruitment time lags in reintroductions of rare plants. Cons Biol Vol 0:1–11

Cochrane JA, Crawford AD, Monks LT (2007) The significance of ex situ seed conservation to reintroduction of threatened plants. Aust J Bot 55:356–361

Dänhardt J, Hedlund K, Birkhofer K, Bracht Jörgensen H, Brady M, Brönmark C, Lindström S, Nilsson L, Olsson O, Rundlöf M, Stjernman M, Smith H (2013) Ekosystemtjänster i det skånska jordbrukslandskapet. (CEC Syntes; Vol. CEC Syntes Nr 01). Centrum för miljö- och klimatforskning, Lunds universitet

Eide W (red) (2014) Arter och naturtyper i habitatdirektivet – bevarandestatus i Sverige 2013. ArtDatabanken SLU, Uppsala

Eide W (red) (2020) Tillstånd och trender för arter och deras livsmiljöer – rödlistade arter i Sverige 2020. SLU Artdatabanken rapporterar 24. SLU Artdatabanken, Uppsala

Emanuelsson U (2009) The rural landscape of Europe. How man has shaped European nature. Formas. Fälth & Hässler, Värnamo

Emanuelsson U, Bergendorff C, Billqvist M, Carlsson B, Lewan N (2002) Det skånska kulturlandskapet. Årsbok för Naturskyddsföreningen i Skåne 2001. BTJ Tryck AB, Lund

Ensslin A, Tschöpe O, Burkart M, Joshi J (2015) Fitness decline and adaptation to novel environments in ex situ plant collections: Current knowledge and future perspectives. Biol Conserv 192:394–401

Ensslin A, van de Vyver A, Vanderborrght T, Godefroid S (2017) Ex situ cultivation entails high risk of seed dormancy loss on short-lived wild plants species. J Appl Ecol 55:1145–1154

Godfroid S, le Pajolec S, van Rossum F (2016) Pre-translocation consideration in rare plant reintroduction: implications for designing protocols. Plant Ecol 217:169–182

Guerrant EO, Havens K, Maunder M (2004) Ex situ plant conservation: supporting species survival in the wild. Island Press, Washington, DC

Havens K, Vitt P, Maunder M, Guerrant EO, Dixon K (2006) Ex situ plant conservation and beyond. Bioscience 56(6):525–531

Hufford KM, Mazer SJ (2003) Plant ecotypes: genetic differentiation in the age of ecological restoration. Trends Ecol Evol 18(3):147–156

IPBES (2019) In: Brondizio ES, Settele J, Díaz S, Ngo HT (eds) Global assessment report on biodiversity and ecosystem services of the intergovernmental science- policy platform on biodiversity and ecosystem services. IPBES Secretariat, Bonn

IUCN/SSC (2014) Guidelines on the use of ex situ management for species conservation. Version 2.0. IUCN Species Survival Commission, Gland

Kloth J-H (2007) Åtgärdsprogram för bevarande av hotade åkergräs. Rapport 5659, Naturvårdsverket

Mijnsbrugge KV, Bischoff A, Smith B (2010) A question of origin: where and how to collect seed for ecological restoration. Basic and Applied Ecology 11:300–311

Olsson K-A, Gustafsson M, Johansson H, Snogerup S, Tyler T (eds) (2003) Floran i Skåne – Vegetation och utflyktsmål. [In Swedish]. Lunds Botaniska Förening. Grahns Tryckeri, Lund

Rosquist G, Eriksen B, Persson H, Sandberg J, Svensson Å, Wigermo C (2018) Bevarande av akut hotade kärlväxter. Sven Bot Tidskr 112(5):316–325

Sandström J, Bjelke U, Carlberg T, Sundberg S (2015) Tillstånd och trender för arter och deras livsmiljöer, rödlistade arter i Sverige 2015. ArtDatabanken Rapporterar 17. ArtDatabanken, SLU. Uppsala

SLU Artdatabanken (2020) Rödlistade arter i Sverige 2020. SLU, Uppsala

Tyler T, Herbertsson L, Olsson PA, Fröberg L, Olsson KA, Svensson Å, Olsson O (2017) Climate warming and land-use changes drive broad-scale floristic changes in Southern Sweden. Glob Chang Biol 2018:1–15

Welinder S (2011) Early farming households, 3900–800 BC. In: Myrdal J, Morell M (eds) The agrarian history of Sweden: from 4000 BC to AD 2000. Nordic Academic Press, Lund, pp 18–45

Chapter 13
Wild Pollinators in Arable Habitats: Trends, Threats and Opportunities

Jonathan Storkey, Mark J. F. Brown, Claire Carvell, Lynn V. Dicks, and Deepa Senapathi

Abstract The dramatic declines in pollinator (social and solitary bees, wasps, flies, beetles, butterflies, and moths) abundance and diversity observed in Europe and globally in recent decades have generated widespread scientific and societal concern. It is particularly important to conserve healthy populations of pollinators in arable landscapes because of the 'ecosystem service' they provide to mass flowering crops such as field beans. Changes in land use prior to the 1970s led to declines in more specialist bee species that have now become rare or extinct in arable landscapes, being replaced by a more generalist, resilient pollinator community. Responses of pollinators to agricultural intensification since the 1970s are subtle and need to be interpreted in the context of other drivers including climate change and pathogen load. However, it is incontrovertible that agricultural practices such as the use of pesticides and loss of semi-natural habitat can have a detrimental effect on pollinator communities and the ecosystem service they provide. It is vital, therefore, that arable landscapes are managed in a way that provides sufficient resource and nesting habitat and either avoids or mitigates exposure to pesticides; the efficacy of specific measures to achieve this are reviewed.

J. Storkey (✉)
Sustainable Agriculture Sciences, Rothamsted Research, Harpenden, Hertfordshire, UK
e-mail: jonathan.storkey@rothamsted.ac.uk

M. J. F. Brown
School of Biological Sciences, Royal Holloway, University of London, Egham, Surrey, UK
e-mail: Mark.Brown@rhul.ac.uk

C. Carvell
UK Centre for Ecology & Hydrology, Wallingford, UK
e-mail: ccar@ceh.ac.uk

L. V. Dicks
Department of Zoology, University of Cambridge, Cambridge, UK
e-mail: lynn.dicks@zoo.cam.ac.uk

D. Senapathi
School of Agriculture, Policy and Development, University of Reading, Reading, UK
e-mail: g.d.senapathi@reading.ac.uk

C. Hurford et al. (eds.), *The Changing Status of Arable Habitats in Europe*,
https://doi.org/10.1007/978-3-030-59875-4_13

Keywords Bees · Moths · Pesticides · Agricultural intensification · Habitat management · Ecosystem services · Agri-environment schemes

1 Introduction

Invertebrate pollinators (social and solitary bees, wasps, flies, beetles, butterflies, and moths) are an integral component of ecological networks in arable habitats (Pocock et al. 2012). However, unlike birds (Siriwardena et al. 1998) and weeds (Richner et al. 2015), where specific communities have been defined that are strongly associated with arable habitats, there is not a typically 'arable' pollinator community. Rather, arable fields largely represent a 'client' of pollinator communities supported by semi-natural habitats in the surrounding landscape that provide most of the foraging resources and nesting sites (Osborne et al. 2008). Although arable weeds have a role to play in providing resources, particularly for wild bees (Bretagnolle and Gaba 2015), and mass flowering crops, such as oilseed rape, can bolster pollinator populations (Riedinger et al. 2015), attention has tended to focus on the negative impact of the expansion of cultivated land on pollinators at the national scale through habitat loss and agricultural intensification (Potts et al. 2010). It is particularly important to monitor and understand any changes in the status of pollinators in arable habitats as they provide an indispensable 'ecosystem service' for crops (Kremen et al. 2002; Vanbergen et al. 2013). The cultivation of certain arable crops, such as field beans, requires insect pollination and yields are related to the level of pollination (Pywell et al. 2015). This chapter reviews recent changes in the abundance and diversity of pollinators in arable habitats, the drivers of change and the efficacy of potential mitigation measures. Though primarily written from a UK perspective, the trends and principles discussed are common to intensively managed agricultural landscapes elsewhere in Europe (Biesmeijer et al. 2006; Potts et al. 2010). The emphasis throughout is on wild pollinators as opposed to managed honeybees.

2 Historical Trends in Pollinators and the Contribution of Agricultural Change

Time series of pollinator surveys specific to arable habitats are rare; however, impacts of changes in land use and agricultural practices on pollinators can be inferred from national surveys. The UK has a long history of biological recording and monitoring by volunteers (now termed 'citizen science'). Records of bee and hoverfly species occurrence submitted by volunteer recorders, collated and verified by the Bees, Wasps and Ants Recording Society (BWARS) and the Hoverfly Recording Scheme (HRS), provide valuable information about long-term changes in species' distributions and status. Extracting this information and using it to analyse trends over time, however, has proved difficult, because records are not collected in a standardised way and are subject to many forms of bias (both in time and space). However, recent analytical developments aimed at accounting for these biases while estimating robust trends have shown great promise (Isaac et al. 2014; van Strien et al. 2013). Specifically, occupancy-detection models have made it possible to estimate reliable trends for individual species at the 1 km square scale.

Trends in occupancy of 213 wild bee and hoverfly species between 1980 and 2010 in the UK were combined into a composite indicator of average relative change in species' distributions, to produce the first Indicator of the Status of Pollinating Insects published alongside other UK Biodiversity Indicators in 2015. New Indicators were published in 2017 and 2018, with the latter, and most recent (at the time of writing) reporting changes in distribution of 351 species of wild pollinator across the UK (representing approximately 70% of all species) between 1980 and 2016 (Powney et al. 2017, 2018). Between 1980 and 2016, 14% of the 351 species analysed became more widespread (5% showed a strong increase at above a threshold of +2.81% per annum) and 34% of species became less widespread (13% showed a strong decrease at below a threshold of −2.73% per annum, equating to a decrease in occupancy of −50% over 25 years). When combined into a composite average trend across all species, occupancy declined by 22% between 1980 and 2016 and the pollinator indicator was therefore assessed as declining over this period. In the shorter term between 2011 and 2016, the indicator showed a minor increase of 2%, however given the uncertainty, the short-term trend was assessed as 'stable'. This is perhaps not surprising given the variation in individual species trends and the short time frame of this period, but is consistent with previous findings that declines in species richness of wild bees and hoverflies observed prior to 1990 may have slowed down in recent decades (Carvalheiro et al. 2013). At the European level, a review of the status of European bee species according to IUCN regional Red List assessment guidelines was published in 2014 (Nieto et al. 2014). This indicates that 9.2% of bee species are considered threatened with extinction and a further 5.2% are Near Threatened. However, for 1101 (57%) of the 1965 species assessed, there was not enough information to evaluate their risk of extinction and these species were classified as Data Deficient (Nieto et al. 2014).

Trends in the abundance of butterflies and moths may provide an indicator of the level of threat to flower-visiting insects generally. While our understanding of the functional role of Lepidoptera as pollinators (in particular for crops) remains limited, the major life history differences between butterflies and moths and other pollinator taxa, may make them a weak proxy for bees and hoverflies. Nevertheless, long-term standardised monitoring of individual butterfly and moth species has allowed repeated updates of their status and trends since around 1976 (Fox et al. 2015). For butterflies, both habitat specialists and wider countryside species (in total 76% of all UK species) have decreased significantly in abundance or occurrence (or both) since 1976. However, as with some bee species, the short-term picture between 2005 and 2014 suggests that declines of several threatened butterfly species have been halted and a range of other species have become more abundant and widespread

Fig. 13.1 A Rothamsted light trap, part of a network for catching night flying insects that has been used to monitor populations of moths since 1964. (Photo Rothamsted Research Ltd.)

(Fox et al. 2015). At the European level, 9% of butterfly species are considered threatened with extinction according to IUCN regional Red List guidelines (a similar figure to that for bees). However, since butterflies are extensively recorded and monitored across Europe, only 1% of species were classified as Data Deficient.

The oldest agricultural research station in the world, Rothamsted Research in the UK, has been monitoring moth populations in the UK since 1964 as part of the Rothamsted Insect Survey (RIS) (Storkey et al. 2016). Moths are monitored using a network of 78 light traps across multiple habitats. Each trap uses a 200-W tungsten bulb that produces a wide wavelength spectrum (400–700 nm) that attracts various night flying insects (Fig. 13.1). Analyses of these data for macro moths have confirmed the negative national trends observed in bees and butterflies in abundance and occupancy since the 1980s. The total annual catch of macro moths across the light trap network was observed to decrease by 35% over a 35-year period between 1968 and 2003 with the largest declines in southern regions. Two thirds of the

337 individual moth species declined with a median 10-year population change of −12% (Conrad et al. 2006).

These alarming and consistent national trends in the populations of a suite of pollinator taxa have rightly focussed attention on the possible drivers of the declines. Given that the intensification of crop production in the 1970s, 1980s and 1990s since the green revolution has been strongly implicated in the declines of farmland birds (Chamberlain et al. 2000; Donald et al. 2001) and arable weed diversity (Storkey et al. 2012), it is not surprising that attention has again focussed on changes in arable habitats as a likely culprit for recent pollinator declines (Goulson et al. 2015). However, a multitude of additional factors have been identified as possible causes of national pollinator declines, including climate change, habitat fragmentation, increased pathogen load (Vanbergen et al. 2013) and, for moths, light pollution (Conrad et al. 2006). This means that establishing causal relationships between changes in arable habitat management and pollinators is not straightforward. Indeed, four recent studies from the UK have challenged the assumption that the intensification of agriculture since the 1970s has been a major driver of recent declines in pollinators.

In response to the widespread attention given to papers that predict a catastrophic decline of insect biomass (Sanchez-Bayo and Wyckhuys 2019) – so called 'insectagedden' – Macgregor et al. (2019) conducted a new analysis of RIS moth data between 1967 and 2017. Contrary to the hypothesis that intensification of crop production over the last four decades has been responsible for moth declines, the authors observed an increase in total biomass in traps located in arable habitats prior to 1982 and no significant decline in biomass since. This led the authors to conclude that agricultural intensification has not been the most important driver of site-level insect biomass change in Britain (Macgregor et al. 2019). A second study over a similar time period and using data on individual moth species abundance from the RIS (Bell et al. 2020) also confirmed that counts had remained stable in farmland habitats since the 1960s despite the fact that significant declines were observed in coastal, urban and woodland habitats. Similar trends have been observed for particular groups of bees and hoverflies; a recent analysis using occupancy models observed a 55% decline in species associated with uplands but a 12% increase in bee species known to be dominant crop pollinators between 1980 and 2013 (Powney et al. 2019). Finally, Redhead et al. (2018) published an analysis of the robustness of plant-pollinator networks in the context of alternative extinction scenarios. Landscapes dominated by agriculture were found to be more robust than other land uses to extinctions which was related to the dominance of generalist species.

All of these studies conclude that contemporary arable habitats are now dominated by common generalist pollinator taxa that are able to persist in highly disturbed, fragmented environments and that large-scale changes prior to 1970 may have already filtered out rarer, more specialist species (Ollerton et al. 2014). A study using distributional records of bumble bees from the UK and Ireland (Casey et al. 2015) also found that most of the rarer bumble bee species showed marked range contractions between 1960–1980, with their ranges remaining more stable between 1981–2012. The more common species' ranges contracted, but less severely than the rare species, between 1960 and 1980, and they appear to have expanded in range post-1980. These findings have implications for the conservation of arable habitats for pollinators. While they may be an argument for managing habitats for the conservation of specialised pollinators to support populations of rare arable weeds (Gibson et al. 2006), the focus should mainly be on supporting sufficient pollinator abundance to deliver the required level of ecosystem services. In this regard, recent evidence has shown that maintaining functional diversity of pollinators, as well as abundance is important (Woodcock et al. 2019). The following sections discuss the threats to this ecosystem service and opportunities in arable habitats to enhance pollinator diversity and abundance.

3 Threats to Pollinators in Arable Habitats

Conventional agricultural intensification is defined as the industrial management of large-scale monocultures with high levels of agrichemical inputs, tillage, grazing, mowing and mechanization (Kovacs-Hostyanszki et al. 2017). Such farm management practices, by reducing both foraging (flowering plants including weeds) and nesting (undisturbed soil or vegetation) resources for pollinators, can decrease pollinator diversity and influence the effectiveness of pollination services. Monocultures of mass flowering crops (e.g. oilseed rape) do provide nectar and pollen sources that support wild bee populations (Holzschuh et al. 2016; Riedinger et al. 2015). However, this transient pulse of food does not compensate for the losses of other food sources like wildflowers and weeds that sustain most pollinators throughout their life cycle (Baude et al. 2016; Potts et al. 2016). Moreover, large-scale cultivation of oilseed rape may even disrupt pollination services in the wider landscape (Holzschuh et al. 2016). Evidence from across Europe shows the proportion, contiguity and management intensity of agricultural land in landscapes lowers pollinator abundance and diversity (Marini et al. 2014; Scheper et al. 2015; Steckel et al. 2014). Although the magnitude varies with the type of agriculture or crop, and the management regime (Le Feon et al. 2013; Marini et al. 2014). There is evidence that increasing the diversity of crop types on the farm, maintaining ecological infrastructure or adopting ‘ecologically-intensive’ management practices commensurate with maintaining agricultural production alongside biodiversity conservation and managing ecosystem services, can ameliorate many of the adverse impacts of conventional intensive management (Kovacs-Hostyanszki et al. 2017; Potts et al. 2016).

Bee functional diversity can be lower in agricultural landscapes compared to natural habitats. For example, in a recent American study, Forrest et al. (2015) found that bee assemblages in farmland had low functional diversity compared to those in nearby natural communities. De Palma et al. (2015) collated data for 257 bee species from multiple studies across Europe and found that smaller-bodied species and those with shorter flight seasons were less likely to be present in areas of intensive agriculture. Such filtering and removal of species and their traits from agricultural pollinator communities presents a risk to the sustained pollination function as the environment changes. There is some evidence, however, to suggest that within intensive farmed areas, existing linear landscape features can act as functional biological corridors facilitating pollinator movements (van Geert et al. 2010) and restoring flowering hedgerow can promote bee populations and diversity (M'Gonigle et al. 2015). Evidence is also growing that by increasing wildflower patch sizes and providing high-value floral resources in the local agricultural landscape can lead to increases in wild bee density (Bennett and Isaacs 2014) and promotion of bumble bee colony reproduction and population persistence (Carvell et al. 2017; Carvell et al. 2015).

Chemical pesticides are used broadly across agricultural, rural, suburban, and urban landscapes to control pests and weeds. However, they may have a range of unintended direct and indirect effects on both wild and managed pollinators (Vanbergen et al. 2014). For example, a recent meta-analysis confirmed that insecticides had generally negative effects on learning and memory in bees when applied at field-realistic dosages (Siviter et al. 2018). Considerable attention has been focused recently on the potential impacts of a class of insecticides called neonicotinoids, but other pesticides, herbicides, and fungicides also have the potential to threaten wild and managed pollinators. Here we focus largely on neonicotinoids, as this is where most progress has been made in terms of new knowledge and understanding since Vanbergen et al. (2014).

Neonicotinoids are the most widely used class of insecticides worldwide (Simon-Delso et al. 2015). Despite their approval by numerous regulatory authorities, many scientific studies have demonstrated significant negative impacts on bumble bees and honeybees, reviewed in Godfray et al. (2015). In 2013, three neonicotinoids – clothianidin, imidacloprid, and thiamethoxam – were placed under a limited moratorium within the EU. Following this, the UK Expert Committee on Pesticides reviewed available evidence in 2017, concluded it supported the 2013 restrictions and noted concerns about usage outside of completely enclosed systems. In 2018, the European Commission completely prohibited the outdoor use of these compounds, following an EFSA risk assessment on the threat to bees from these three neonicotinoids, which concluded that risks to bee health from use of these compounds were high outside of permanent greenhouse structures. This decision has been supported by recent farm-scale experiments in Europe that have demonstrated significant negative impacts of neonicotinoids on bumble bee and solitary bee survival and performance at actual field exposure (Rundlof et al. 2015; Woodcock et al. 2017). Analysis relating 18 years of UK national wild bee distribution data for 62 species to amounts of neonicotinoid use in oilseed rape also revealed significant correlations with solitary bee population declines (Woodcock et al. 2016). Laboratory and semi-field experiments have produced an array of results including: impacts on colony founding and population persistence (shown through modelling) in bumble bees (Baron et al. 2017a, b); evidence that bumble bees actively prefer neonicotinoid contaminated food (Kessler et al. 2015); adverse impacts on foraging behaviour and effectiveness but equivocal effects on bumble bee colony health

(Stanley and Raine 2017); and even possible direct impacts of neonicotinoid exposure on pollination services (Stanley et al. 2015).

4 Opportunities to Enhance Pollinator Populations in Arable Habitats

There is good evidence to show that, in arable habitats, populations of wild pollinators can be supported by providing sown nectar flower mixtures (Fig. 13.2), bare ground (as nesting sites), set-aside areas, nest boxes for cavity-nesting solitary bees (not for bumble bees), and uncropped naturally regenerated field margins. This evidence has been reviewed by Garibaldi et al. (2014). For the UK, 17 specific agri-environmental management options either known to, or widely considered to support pollinators were identified by Dicks et al. (2015), with an overview of the evidence supporting their benefits to pollinators and the resources they provide. Some of the options (fallow land, field margins, buffer strips and forest edges) were included as Ecological Focus Areas (EFAs), obligatory on farms with >15 ha of arable land under the Common Agricultural Policy 2014–2020. However, of the EFA options known to have benefits for pollinators, only fallow land was widely adopted by European farmers (Pe'er et al. 2017), and management to ensure benefit to pollinators was not clearly prescribed. There are clear opportunities to improve habitat quality for pollinators through pollinator-friendly management of habitat areas in farmland (Cole et al. 2020). For example, fallows managed to promote mid-successional plant communities that support diverse life forms with overlapping flowering periods and a range of flower morphologies have the greatest potential to support a diverse pollinator community (Robleno et al. 2018). Improving management of single habitats is unlikely to ensure that all pollinator resource requirements are met at landscape scale. A variety of interconnected, well-managed habitats is needed, which complement each other in the resources they offer, to provide diverse floral resources, nesting sites and larval food habitats for multiple taxa (Cole et al. 2020).

Organic farms generally have higher pollinator densities and numbers of wild pollinator species than non-organic farms, although this is not always true. Studies have found more flower resources within fields on organic farms, than on non-organic farms (Gabriel and Tscharntke 2007; Hardman et al. 2016). This difference did not scale to farm level in the UK study, comparing organic farms to non-organic farms with agri-environment schemes, due to high flower densities in non-crop habitats on farms with Entry Level Stewardship and Conservation Grade schemes (Hardman et al. 2016). The recent success of agri-environment schemes in replacing flowering resources lost to arable habitats through land use change and intensification was shown by a national vegetation survey done as part of the UK Countryside Survey undertaken in 1990, 2000 and 2007. These data show that plant species richness on arable and horticulture land increased by more than 30%

Fig. 13.2 Pollinators in arable landscapes rely on supplementary sources of forage to support the pollination service they provide to crops. This can be in the form of sown wildflower strips: a mix of crimson clover and cornflower (top left), mustard and phacelia (top right), *Bombus terrestris* feeding on thistle (middle left), *Bombus pascuorum* feeding from a cornflower (middle right), *Bombus terrestris* and *Apis mellifera* feeding from Phacelia flowers (bottom). (Photos Hannah McGrath)

between 2000 and 2007, partly due to an increase in sown wildflower field margins which are used by wild bees (Norton et al. 2012). This change has been associated with a slow-down in the loss of bee species-richness at a range of spatial scales after 1990 (1990–2009), compared to rates of loss in the previous 80 years (1920–1989) (Carvalheiro et al. 2013).

Scheper et al. (2013) analysed 71 studies across Europe quantifying the efficacy of agri-environment schemes for enhancing pollinator diversity and abundance. Each study compared numbers of individuals or species, relative to a conventionally managed control site. The meta-analysis identified ecological contrasts in floral resources as the main factor explaining variation in the response of pollinators but that this response is moderated by landscape context and farmland type, with more positive responses in croplands (vs. grasslands) located in simple (vs. cleared or complex) landscapes. These results represent insect densities, or numbers of insects using the created habitats as a resource. They cannot be directly interpreted as long-term benefits to pollinator populations, or reversal of pollinator declines. For this, measures of trends in numbers over time (over at least 4 years) or measures of reproductive success or survival are required. Empirical evidence (Blaauw and Isaacs 2014) and bumble bee population modelling (Iles et al. 2018) both indicate that 4 years of monitoring is a minimum time period for wild bee populations to respond to changes in agricultural land management.

However, in recent years, evidence has emerged that enhancing flower-rich habitats in agricultural landscapes can lead to population-level benefits for wild bumble bees, in terms of enhanced reproductive success, bumble bee colony survival or population density. Carvell et al. (2017) show that for three common bumble bee species (*Bombus terrestris*, *B. lapidarius* and *B. pascuorum*) the survival of family lineages from the summer worker to the spring queen stage in the following year increases significantly with the proportion of high-value foraging habitat, including spring floral resources, within 250–1000 m of the natal colony. In a previous study, Carvell et al. (2015) showed that numbers of bumble bee reproductives (males or males and queens) tend to increase as flowers are added to a landscape. In a study of 19 farms in Hampshire and West Sussex (Southern England), Wood et al. (2015) showed that farms with Higher Level Stewardship agreements had higher floral resources, and this was associated with higher bumble bee nest densities for four common bumble bee species, compared to farms with Entry Level Stewardship agreements.

5 Conclusions

The intensification of agriculture since the 1970s has led to dramatic declines in populations of farmland birds and arable plants that are highly dependent on the frequently disturbed, open habitats that are characteristic of arable fields. This has led to the design of in-field conservation measures based on the ecological requirements of vulnerable species. In contrast, the balance of evidence reviewed in this chapter is that changes in crop management over the last four decades has not been a major contributor to the increased rarity of individual pollinator species. Rather, arable habitats are now dominated by common generalist species after previous changes in agricultural practice eliminated rarer and more specialist species. These remaining generalist species have persisted because they are able to exploit a range of habitats and flowering resources in disturbed, fragmented landscapes. However, sustainable arable production relies on the pollination services delivered by these taxa and appropriate management needs to be in place to maintain sufficient diversity and abundance. Again, in contrast to some farmland birds and weeds that require in-field management to provide the appropriate conditions and resources for their populations to recover (Butler et al. 2009), resources for pollinators can largely be provided in un-cropped habitats. This essentially means establishing a spatio-temporal mosaic of semi-natural habitats that provide pollen and nectar resources and nesting sites (Holland et al. 2015; Cole et al. 2020). A future goal of conserving arable habitats should also be to support the recovery of rarer and more specialised species that have been driven extinct in agricultural landscapes.

Acknowledgments The authors thank The Department for Environment, Food and Rural Affairs (Defra), UK, for permission to use extracts from the Defra publication "Management and drivers of change of pollinating insects and pollination services – BE0160."

References

Baron GL, Jansen VAA, Brown MJF, Raine NE (2017a) Pesticide reduces bumblebee colony initiation and increases probability of population extinction. Nat Ecol Evol 1(9):1308–1316

Baron GL, Raine NE, Brown MJF (2017b) General and species-specific impacts of a neonicotinoid insecticide on the ovary development and feeding of wild bumblebee queens. Proc R Soc B Biol Sci 284(1854):8

Baude M, Kunin WE, Boatman ND, Conyers S, Davies N, Gillespie MAK, Morton RD, Smart SM, Memmott J (2016) Historical nectar assessment reveals the fall and rise of floral resources in Britain. Nature 530(7588):85–88

Bell JR, Blumgart D, Shortall CR (2020) Are insects declining and at what rate? An analysis of standardised, systematic catches of aphid and moth abundances across Great Britain. Insect Conserv Divers 13(2):115–126

Bennett AB, Isaacs R (2014) Landscape composition influences pollinators and pollination services in perennial biofuel plantings. Agric Ecosyst Environ 193:1–8

Biesmeijer JC, Roberts SPM, Reemer M, Ohlemuller R, Edwards M, Peeters T, Schaffers AP, Potts SG, Kleukers R, Thomas CD, Settele J, Kunin WE (2006) Parallel declines in pollinators and insect-pollinated plants in Britain and the Netherlands. Science 313(5785):351–354

Blaauw BR, Isaacs R (2014) Flower plantings increase wild bee abundance and the pollination services provided to a pollination-dependent crop. J Appl Ecol 51(4):890–898

Bretagnolle V, Gaba S (2015) Weeds for bees? A review. Agron Sustain Dev 35(3):891–909

Butler SJ, Brooks D, Feber RE, Storkey J, Vickery JA, Norris K (2009) A cross-taxonomic index for quantifying the health of farmland biodiversity. J Appl Ecol 46(6):1154–1162

Carvalheiro LG, Kunin WE, Keil P, Aguirre-Gutierrez J, Ellis WN, Fox R, Groom Q, Hennekens S, Van Landuyt W, Maes D, Van de Meutter F, Michez D, Rasmont P, Ode B, Potts SG, Reemer M, Roberts SPM, Schaminee J, Wallis de Vries MF, Biesmeijer JC (2013) Species richness declines and biotic homogenisation have slowed down for NW-European pollinators and plants. Ecol Lett 16(7):870–878

Carvell C, Bourke AFG, Osborne JL, Heard MS (2015) Effects of an agri-environment scheme on bumblebee reproduction at local and landscape scales. Basic Appl Ecol 16(6):519–530

Carvell C, Bourke AFG, Dreier S, Freeman SN, Hulmes S, Jordan WC, Redhead JW, Sumner S, Wang J, Heard MS (2017) Bumblebee family lineage survival is enhanced in high-quality landscapes. Nature 543(7646):547–549

Casey LM, Rebelo H, Rotheray E, Goulson D (2015) Evidence for habitat and climatic specializations driving the long-term distribution trends of UK and Irish bumblebees. Divers Distrib 21 (8):864–875

Chamberlain DE, Fuller RJ, Bunce RGH, Duckworth JC, Shrubb M (2000) Changes in the abundance of farmland birds in relation to the timing of agricultural intensification in England and Wales. J Appl Ecol 37(5):771–788

Cole LJ, Kleijn D, Dicks LV et al (2020) A critical analysis of the potential for EU Common Agricultural Policy measures to support wild pollinators on farmland. J Appl Ecol 57:681–694

Conrad KF, Warren MS, Fox R, Parsons MS, Woiwod IP (2006) Rapid declines of common, widespread British moths provide evidence of an insect biodiversity crisis. Biol Conserv 132 (3):279–291

De Palma A, Kuhlmann M, Roberts SPM, Potts SG, Borger L, Hudson LN, Lysenko I, Newbold T, Purvis A (2015) Ecological traits affect the sensitivity of bees to land-use pressures in European agricultural landscapes. J Appl Ecol 52(6):1567–1577

Dicks LV, Baude M, Roberts SPM, Phillips J, Green M, Carvell C (2015) How much flower-rich habitat is enough for wild pollinators? Answering a key policy question with incomplete knowledge. Ecol Entomol 40:22–35

Donald PF, Green RE, Heath MF (2001) Agricultural intensification and the collapse of Europe's farmland bird populations. Proc R Soc London Ser B Biol Sci 268(1462):25–29

Forrest JRK, Thorp RW, Kremen C, Williams NM (2015) Contrasting patterns in species and functional-trait diversity of bees in an agricultural landscape. J Appl Ecol 52(3):706–715

Fox R, Brereton TM, Asher J, August TA, Botham MS, Bourn NAD (2015) The state of the UK's butterflies 2015. Butterfly Conservation and the Centre for Ecology & Hydrology, Wareham

Gabriel D, Tscharntke T (2007) Insect pollinated plants benefit from organic farming. Agric Ecosyst Environ 118(1–4):43–48

Garibaldi LA, Carvalheiro LG, Leonhardt SD, Aizen MA, Blaauw BR, Isaacs R, Kuhlmann M, Kleijn D, Klein AM, Kremen C, Morandin L, Scheper J, Winfree R (2014) From research to action: enhancing crop yield through wild pollinators. Front Ecol Environ 12(8):439–447

Gibson RH, Nelson IL, Hopkins GW, Hamlett BJ, Memmott J (2006) Pollinator webs, plant communities and the conservation of rare plants: arable weeds as a case study. J Appl Ecol 43 (2):246–257

Godfray HCJ, Blacquiere T, Field LM, Hails RS, Potts SG, Raine NE, Vanbergen AJ, McLean AR (2015) A restatement of recent advances in the natural science evidence base concerning neonicotinoid insecticides and insect pollinators. Proc R Soc B Biol Sci 282(1818):6

Goulson D, Nicholls E, Botias C, Rotheray EL (2015) Bee declines driven by combined stress from parasites, pesticides, and lack of flowers. Science 347(6229):10

Hardman CJ, Norris K, Nevard TD, Hughes B, Potts SG (2016) Delivery of floral resources and pollination services on farmland under three different wildlife-friendly schemes. Agric Ecosyst Environ 220:142–151

Holland JM, Smith BM, Storkey J, Lutman PJW, Aebischer NJ (2015) Managing habitats on English farmland for insect pollinator conservation. Biol Conserv 182:215–222

Holzschuh A, Dainese M, Gonzalez-Varo JP, Mudri-Stojnic S, Riedinger V, Rundlof M, Scheper J, Wickens JB, Wickens VJ, Bommarco R, Kleijn D, Potts SG, Roberts SPM, Smith HG, Vila M, Vujic A, Steffan-Dewenter I (2016) Mass-flowering crops dilute pollinator abundance in agricultural landscapes across Europe. Ecol Lett 19(10):1228–1236

Iles DT, Williams NM, Crone EE (2018) Source-sink dynamics of bumblebees in rapidly changing landscapes. J Appl Ecol 55(6):2802–2811

Isaac NJB, van Strien AJ, August TA, de Zeeuw MP, Roy DB (2014) Statistics for citizen science: extracting signals of change from noisy ecological data. Methods Ecol Evol 5(10):1052–1060

Kessler SC, Tiedeken EJ, Simcock KL, Derveau S, Mitchell J, Softley S, Stout JC, Wright GA (2015) Bees prefer foods containing neonicotinoid pesticides. Nature 521(7550):74–U145

Kovacs-Hostyanszki A, Espindola A, Vanbergen AJ, Settele J, Kremen C, Dicks LV (2017) Ecological intensification to mitigate impacts of conventional intensive land use on pollinators and pollination. Ecol Lett 20(5):673–689

Kremen C, Williams NM, Thorp RW (2002) Crop pollination from native bees at risk from agricultural intensification. Proc Natl Acad Sci U S A 99(26):16812–16816

Le Feon V, Burel F, Chifflet R, Henry M, Ricroch A, Vaissiere BE, Baudry J (2013) Solitary bee abundance and species richness in dynamic agricultural landscapes. Agric Ecosyst Environ 166:94–101

Macgregor CJ, Williams JH, Bell JR, Thomas CD (2019) Moth biomass increases and decreases over 50 years in Britain. Nat Ecol Evol 3(12):1645–1649

Marini L, Ockinger E, Bergman KO, Jauker B, Krauss J, Kuussaari M, Poyry J, Smith HG, Steffan-Dewenter I, Bommarco R (2014) Contrasting effects of habitat area and connectivity on evenness of pollinator communities. Ecography 37(6):544–551

M'Gonigle LK, Ponisio LC, Cutler K, Kremen C (2015) Habitat restoration promotes pollinator persistence and colonization in intensively managed agriculture. Ecol Appl 25(6):1557–1565

Nieto A, Roberts SPM, Kemp J, Rasmont P, Kuhlman M, Garcia Criado M (2014) European red list of bees. Publication Office of the European Union, Luxembourg

Norton LR, Maskell LC, Smart SS, Dunbar MJ, Emmett BA, Carey PD, Williams P, Crowe A, Chandler K, Scott WA, Wood CM (2012) Measuring stock and change in the GB countryside for policy – Key findings and developments from the Countryside Survey 2007 field survey. J Environ Manag 113:117–127

Ollerton J, Erenler H, Edwards M, Crockett R (2014) Extinctions of aculeate pollinators in Britain and the role of large-scale agricultural changes. Science 346(6215):1360–1362

Osborne JL, Martin AP, Shortall CR, Todd AD, Goulson D, Knight ME, Hale RJ, Sanderson RA (2008) Quantifying and comparing bumblebee nest densities in gardens and countryside habitats. J Appl Ecol 45(3):784–792

Pe'er G, Zinngrebe Y, Hauck J, Schindler S, Dittrich A, Zingg S, Tscharntke T, Oppermann R, Sutcliffe LME, Sirami C, Schmidt J, Hoyer C, Schleyer C, Lakner S (2017) Adding some green to the greening: improving the EU's Ecological Focus Areas for biodiversity and farmers. Conserv Lett 10(5):517–530

Pocock MJO, Evans DM, Memmott J (2012) The robustness and restoration of a network of ecological networks. Science 335(6071):973–977

Potts SG, Biesmeijer JC, Kremen C, Neumann P, Schweiger O, Kunin WE (2010) Global pollinator declines: trends, impacts and drivers. Trends Ecol Evol 25(6):345–353

Potts SG, Imperatriz-Fonseca V, Ngo HT, Aizen MA, Biesmeijer JC, Breeze TD, Dicks LV, Garibaldi LA, Hill R, Settele J, Vanbergen AJ (2016) Safeguarding pollinators and their values to human well-being. Nature 540(7632):220–229

Powney GD, August TA, Harrower C, Outhwaite C, Isaac NJB (2017) UK Biodiversity Indicators 2017: D1c Status of pollinating insects. Technical background document. JNCC/Centre for Ecology and Hydrology, UK

Powney GD, Harrower C, Outhwaite C, Isaac NJB (2018) UK Biodiversity Indicators 2018: D1c Status of pollinating insects. Technical background document. JNCC/Centre for Ecology and Hydrology, UK

Powney GD, Carvell C, Edwards M, Morris RKA, Roy HE, Woodcock BA, Isaac NJB (2019) Widespread losses of pollinating insects in Britain. Nat Commun 10:6

Pywell RF, Heard MS, Woodcock BA, Hinsley S, Ridding L, Nowakowski M, Bullock JM (2015) Wildlife-friendly farming increases crop yield: evidence for ecological intensification. Proc R Soc B Biol Sci 282(1816):8

Redhead JW, Woodcock B, Pocock MJO, Pywell RF, Vanbergen AJ, Oliver TH (2018) Potential landscape-scale pollinator networks across Great Britain: structure, stability and influence of agricultural land cover. Ecol Lett 21(12):1821–1832

Richner N, Holderegger R, Linder HP, Walter T (2015) Reviewing change in the arable flora of Europe: a meta-analysis. Weed Res 55(1):1–13

Riedinger V, Mitesser O, Hovestadt T, Steffan-Dewenter I, Holzschuh A (2015) Annual dynamics of wild bee densities: attractiveness and productivity effects of oilseed rape. Ecology 96 (5):1351–1360

Robleno I, Storkey J, Sole-Senan XO, Recasens J (2018) Using the response-effect trait framework to quantify the value of fallow patches in agricultural landscapes to pollinators. Appl Veg Sci 21 (2):267–277

Rundlof M, Andersson GKS, Bommarco R, Fries I, Hederstrom V, Herbertsson L, Jonsson O, Klatt BK, Pedersen TR, Yourstone J, Smith HG (2015) Seed coating with a neonicotinoid insecticide negatively affects wild bees. Nature 521(7550):77–80

Sanchez-Bayo F, Wyckhuys KAG (2019) Worldwide decline of the entomofauna: a review of its drivers. Biol Conserv 232:8–27

Scheper J, Holzschuh A, Kuussaari M, Potts SG, Rundlof M, Smith HG, Kleijn D (2013) Environmental factors driving the effectiveness of European agri-environmental measures in mitigating pollinator loss – a meta-analysis. Ecol Lett 16(7):912–920

Scheper J, Bommarco R, Holzschuh A, Potts SG, Riedinger V, Roberts SPM, Rundlof M, Smith HG, Steffan-Dewenter I, Wickens JB, Wickens VJ, Kleijn D (2015) Local and landscape-level floral resources explain effects of wildflower strips on wild bees across four European countries. J Appl Ecol 52(5):1165–1175

Simon-Delso N, Amaral-Rogers V, Belzunces LP, Bonmatin JM, Chagnon M, Downs C, Furlan L, Gibbons DW, Giorio C, Girolami V, Goulson D, Kreutzweiser DP, Krupke CH, Liess M, Long E, McField M, Mineau P, Mitchell EAD, Morrissey CA, Noome DA, Pisa L, Settele J, Stark JD, Tapparo A, Van Dyck H, Van Praagh J, van der Sluijs JP, Whitehorn PR, Wiemers M (2015) Systemic insecticides (neonicotinoids and fipronil): trends, uses, mode of action and metabolites. Environ Sci Pollut Res 22(1):5–34

Siriwardena GM, Baillie SR, Buckland ST, Fewster RM, Marchant JH, Wilson JD (1998) Trends in the abundance of farmland birds: a quantitative comparison of smoothed Common Birds Census indices. J Appl Ecol 35(1):24–43

Siviter H, Koricheva J, Brown MJF, Leadbeater E (2018) Quantifying the impact of pesticides on learning and memory in bees. J Appl Ecol 55(6):2812–2821

Stanley DA, Raine NE (2017) Bumblebee colony development following chronic exposure to field-realistic levels of the neonicotinoid pesticide thiamethoxam under laboratory conditions. Sci Rep 7

Stanley DA, Garratt MPD, Wickens JB, Wickens VJ, Potts SG, Raine NE (2015) Neonicotinoid pesticide exposure impairs crop pollination services provided by bumblebees. Nature 528 (7583):548–550

Steckel J, Westphal C, Peters MK, Bellach M, Rothenwoehrer C, Erasmi S, Scherber C, Tscharntke T, Steffan-Dewenter I (2014) Landscape composition and configuration differently affect trap-nesting bees, wasps and their antagonists. Biol Conserv 172:56–64

Storkey J, Meyer S, Still KS, Leuschner C (2012) The impact of agricultural intensification and land-use change on the European arable flora. Proc R Soc B Biol Sci 279(1732):1421–1429

Storkey J, Macdonald AJ, Bell JR, Clark IM, Gregory AS, Hawkins NJ, Hirsch PR, Todman LC, Whitmore AP (2016) The unique contribution of Rothamsted to ecological research at large temporal scales In: Dumbrell AJ, Kordas RL, Woodward G (eds) Advances in ecological research, vol 55: large-scale ecology: model systems to global perspectives, vol 55, pp 3–42

van Geert A, van Rossum F, Triest L (2010) Do linear landscape elements in farmland act as biological corridors for pollen dispersal? J Ecol 98(1):178–187

van Strien AJ, van Swaay CAM, Termaat T (2013) Opportunistic citizen science data of animal species produce reliable estimates of distribution trends if analysed with occupancy models. J Appl Ecol 50(6):1450–1458

Vanbergen AJ, Baude M, Biesmeijer JC, Britton NF, Brown MJF, Brown M, Bryden J, Budge GE, Bull JC, Carvell C, Challinor AJ, Connolly CN, Evans DJ, Feil EJ, Garratt MP, Greco MK,

Heard MS, Jansen VAA, Keeling MJ, Kunis WE, Marris GC, Memmott J, Murray JT, Nicolson SW, Osborne JL, Paxton RJ, Pirk CWW, Polce C, Potts SG, Priest NK, Raine NE, Roberts S, Ryabov EV, Shafir S, Shirley MDF, Simpson SJ, Stevenson PC, Stone GN, Termansen M, Wright GA, Insect Pollinators Initiative (2013) Threats to an ecosystem service: pressures on pollinators. Front Ecol Environ 11(5):251–259

Vanbergen AJ, Heard MS, Breeze TD, Potts SG, Hanley N (2014) Status and value of pollinators and pollination services. Department for the Environment, Food and Rural Affairs, London

Wood TJ, Holland JM, Hughes WHO, Goulson D (2015) Targeted agri-environment schemes significantly improve the population size of common farmland bumblebee species. Mol Ecol 24 (8):1668–1680

Woodcock BA, Isaac NJB, Bullock JM, Roy DB, Garthwaite DG, Crowe A, Pywell RF (2016) Impacts of neonicotinoid use on long-term population changes in wild bees in England. Nat Commun 7:8

Woodcock BA, Bullock JM, Shore RF, Heard MS, Pereira MG, Redhead J, Ridding L, Dean H, Sleep D, Henrys P, Peyton J, Hulmes S, Hulmes L, Sarospataki M, Saure C, Edwards M, Genersch E, Knabe S, Pywell RF (2017) Country-specific effects of neonicotinoid pesticides on honey bees and wild bees. Science 356(6345):1393–1395

Woodcock BA, Garratt MPD, Powney GD, Shaw RF, Osborne JL, Soroka J, Lindstrom SAM, Stanley D, Ouvrard P, Edwards ME, Jauker F, McCracken ME, Zou Y, Potts SG, Rundlof M, Noriega JA, Greenop A, Smith HG, Bommarco R, van der Werf W, Stout JC, Steffan-Dewenter-I, Morandin L, Bullock JM, Pywell RF (2019) Meta-analysis reveals that pollinator functional diversity and abundance enhance crop pollination and yield. Nat Commun 10:10

Chapter 14
Designing Multifunctional and Resilient Agricultural Landscapes: Lessons from Long-Term Monitoring of Biodiversity and Land Use

Sabrina Gaba and Vincent Bretagnolle

Abstract In this chapter, we illustrate how the current challenges of agriculture - i.e. maintaining food security while preserving biodiversity, landscapes and ecosystem functions - have been addressed in a long-term social-ecological research site, the Zone Atelier Plaine & Val de Sèvre, operated since 1994. After a presentation of the study area and its research program, we present landscape changes over the past 25 years. Then, we review the most relevant studies conducted in this area that examined the effect of landscape on farmland biodiversity. We focus on biodiversity because it is a central tenet of sustainable agriculture, which delivers ecosystem services that benefit human well-being and ensure ecosystem stability and resilience. Finally, we show that approaches that sustain biodiversity can also support food production, the delivery of multiple ecosystem services and the economy of local stakeholders. One major result from studies in the LTSER ZA-PVS is that biodiversity decline is a severe threat to food security in itself: improving biodiversity can improve yields, gross margins, or both. Biodiversity in arable habitats is, therefore, required to maintain food security, especially through pollination, natural pest regulation and soil services. Long-term monitoring of biodiversity, ecosystem functioning and agricultural practices as well as land use revealed that the interplay between local and landscape factors can provide solutions to reach these goals of multifunctionality (biodiversity, food production and farmers' profitability) in

S. Gaba (✉)
Centre d'Etudes Biologiques de Chizé, UMR7372, CNRS & Université de La Rochelle, Villiers-en-Bois, France

USC 1339 Centre d'Etudes Biologiques de Chizé, INRAE, Villiers-en-Bois, France

LTSER Zone Atelier Plaine & Val de Sèvre, Villiers-en-Bois, France
e-mail: sabrina.gaba@inrae.fr

V. Bretagnolle
Centre d'Etudes Biologiques de Chizé, UMR7372, CNRS & Université de La Rochelle, Villiers-en-Bois, France

LTSER Zone Atelier Plaine & Val de Sèvre, Villiers-en-Bois, France

C. Hurford et al. (eds.), *The Changing Status of Arable Habitats in Europe*,
https://doi.org/10.1007/978-3-030-59875-4_14

agricultural landscapes. The next step is to engage stakeholders, in collaboration with scientists, and increase awareness of their mutual interdependencies with nature to foster the transition to more agroecological farming.

Keywords Zone Atelier Plaine & Val de Sèvre · Biodiversity · Organic farming · Ecosystem services · Multifunctional landscapes · Agri-environment schemes

1 Introduction

Over the past half century, agriculture has provided increasing food supply but at the same time, contributed massively to greenhouse gases emission, agrochemical pollution and soil degradation (Tilman et al. 2011; Del Grosso et al. 2012; Amundson et al. 2015). The expansion and intensification of agricultural activity is also threatening no less than 5407 animal species — 62% of those listed as threatened or near-threatened, placing agriculture second in the rank of the biggest threats to biodiversity (Maxwell et al. 2016). Apart from biodiversity loss in itself, this is particularly worrying for those crops for which yield depends on ecosystem functions and services such as pollination (Pywell et al. 2015; Tamburini et al. 2016). An increasing world population and concern about the sustainability of crop production has led to calls for producing food differently while creating agricultural systems that enhance biodiversity, promote closed-loop systems, increase soil health, and eliminate the dependence on external synthetic inputs (Altieri 1983; Tittonell 2014; Godfray 2015; Gliessman 2016). However, opinions are divided over the best strategic direction.

One option is to minimize farmland area and spare land for nature, hence reducing the negative environmental footprint of agriculture. In farmland areas, this is usually achieved by retaining on-farm semi-natural elements. Another option is to reduce external synthetic inputs (fertilisers and pesticides) by implementing alternative farming systems. This can be done at the field level, under the European agri-environment policy, MAEC in France, as part of the 2014–2020 reform of the second pillar of the Common Agricultural Policy (Pe'er et al. 2014). This can also be done at the farm level, i.e. by changing the agricultural system of the farm. Organic agriculture is the most popular alternative farming system (Willer and Lernoud 2019), with an increase of 23% of organic cropland between 2014 and 2015 in France (Source: Agence Bio/Organismes Certificateurs). Organic farming (OF) sustains agricultural productivity by avoiding or largely excluding synthetic fertilizers and pesticides (Lampkin 1990). Agroecology is another popular alternative to the massive use of synthetic chemicals (Altieri 1983; Gliessman 2016). Both agroecology and OF rely on the use of the natural functionality provided by the ecosystems to maintain crop production, protect crops against pests and pathogens, and improve the environment and farm profit (Bommarco et al. 2013; Gaba et al. 2014). Indeed, reducing applications of pesticides can maintain and promote natural enemies for pest control as well as pollinators - both major ecosystem services (ES) for global food production and agricultural businesses. On the one hand, invertebrate pests destroy 8–15% of global wheat, rice, maize, potato, soybean, and cotton production (Oerke 2006) and cause more than US$30billion in damage in the US each year (Pimentel et al. 1992). On the other hand, animal pollinators are essential for approximately 35% of global crop production, and 60–90% of all plant species are pollinator-dependent (Klein et al. 2007). The estimated worldwide value of insect-pollinated crops was €153 billion in 2009 (Gallai et al. 2009). However, the diversity and abundance of pests, their natural enemies and pollinators, all depend on both local management and landscape heterogeneity (here heterogeneity refers both to landscape composition and configuration, Fahrig et al. 2011). Therefore, strategies for enhancing pest control and pollination ESs to agriculture require a detailed understanding of the interactive effects of these two factors on predator and pollinator ecology, as well as between predator and pollinator biodiversity.

However, reducing or banning pesticides and fertilisers as well as making space for nature may result in lower yields per unit area (fuelling the long-standing land sharing/land sparing debate; Green et al. 2005, Phalan et al. 2011), and thus may require extending the area of land under production (Phalan et al. 2016). How, then, can we meet the current challenges that agriculture faces? In other words, which spatial arrangement of arable land and semi-natural habitats can make agricultural landscape multifunctional and resilient?

2 The Long-Term Social-Ecological Research Site, Zone Atelier Plaine & Val de Sèvre

2.1 Description of the ZA-PVS

The Long Term Socio-Ecological Research (LTSER) site, Zone Atelier Plaine & Val de Sèvre (ZA-PVS) is part of the French LTSER Research Infrastructure (Réseau des Zones Atelier, Lévêque et al. 2000). This network of 14 research platforms aims to analyse, understand and most importantly, implement management strategies that support socio-ecological system (SES) resilience and robustness. The LTSER ZA-PVS combines long-term monitoring of social-ecological components and their relations, supports interdisciplinary research, and engages stakeholders in transdisciplinary approaches (Bretagnolle et al. 2018b). The LTSER obtained its label in 2008, and has focused since 1994 on agroecology as an alternative agricultural model (*sensu* Altieri 1983), promoting nature-based solutions that integrate agricultural development and biodiversity conservation within resilient multifunctional landscapes. Its research program fits the unified operational framework of the Réseau des Zones Atelier (Bretagnolle et al. 2019) where the coupling between the social and ecological templates is represented by two loops: the "Ecosystem Services interface" (Haines-Young and Potschin 2010) and the

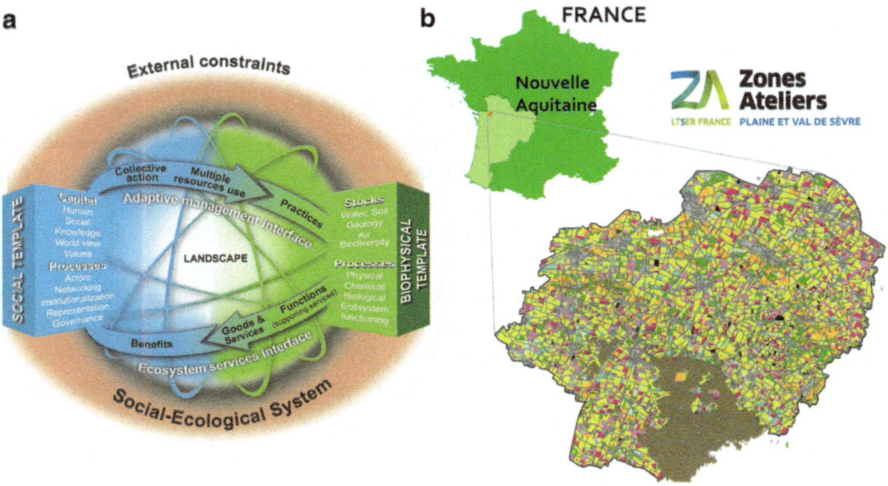

Fig. 14.1 (**a**) Conceptual framework of the French LTSER Zone Atelier Network (from Bretagnolle et al. 2019). The blue box shows the social template and the green one the biophysical template. The socio-ecosystem is represented with the two feedback loops acting in the landscape. (**b**) Location of the Zone Atelier Plaine & Val de Sèvre in West of France

"Adaptive Management interface" (Cumming et al. 2015) (Fig. 14.1). Using approaches such as monitoring and experimentation (and to a lesser extent, modelling), the LTSER ZA-PVS supports place-based knowledge production by both academic and non-academic participants, based on strong stakeholder collaboration (Angelstam et al. 2018; Bretagnolle et al. 2019). Its research program, that involves transdisciplinarity and interdisciplinarity, also aims to identify solutions to improve the capacity of the social-ecological system to reorganise itself in order to maintain its structure and functioning in response to local and global sustainability threats (e.g. food security, economic shocks, health) posed by climate change, agricultural pollution and biodiversity loss. This research program is organised around three main axes: agroecology (from food production to food consumption), ecohealth (joint analysis of human, animal and environment health in response to socio-ecosystem changes) and biodiversity conservation (participatory science and action research to maintain biodiversity, including endangered species).

The research strategy of the ZA-PVS combines three complementary approaches: (i) long-term monitoring of the social-ecological system, including its ecology (the whole trophic chain: plants, insects, spiders, small mammals and birds), landscape (configuration and composition, distribution of semi-natural habitats), and social aspects, such as land use, farmers' agricultural practices and public policies; (ii) experimenting with various stakeholders, management strategies at field, farm, landscape and village scales to foster food supply transition and global health within an agro-ecological and ecohealth frameworks; and (iii) engaging local stakeholders of different types in iterative and collaborative processes to produce place-based

Fig. 14.2 Various landscapes in LTSER Zone Atelier Plaine & Val de Sèvre. From left to right, on the top row a oilseed field and a relict vineyard in a cereal plain; on the bottom row a wheat field and a meadow

knowledge and pathways towards a resilient and multifunctional agricultural landscape.

The ZA-PVS is located south of the city of Niort, in the *Deux-Sèvres* department in the *Nouvelle-Aquitaine* Region, south-west France (Fig. 14.1b). This large study area (435 km^2) is typical of rural landscapes (Fig. 14.2) and is managed almost exclusively for arable and mixed farming. There are 24 administrative *communes* inside the ZA PVS and eight others (including the Niort urban district) partially within the ZA-PVS. The total human population of the ZA-PVS is around 29,000 inhabitants (excluding Niort city). More than half of the study site was designated as a NATURA 2000 site in 2003 under the Bird Directive (NATURA 2000 code FR5412007). Socioeconomic and ecological data have been collected continuously since 1994 (see Bretagnolle et al. 2018a, b for further details). Besides data collection, the ZA-PVS has been used to carry out experiments in real world conditions (i.e. socio-ecological experiments, Gaba and Bretagnolle 2020) at various spatial scales, from fine scale (~1 m^2), to habitats (crops) and landscapes (e.g. by manipulating the proportion of meadows or hedges using agri-environmental measures).

2.2 Temporal Variation of Landscape Composition

The ZA-PVS is an intensive farming area with mainly winter cereals (average 2009–2016: 41.5%). The most common crops are winter wheat (33.8%), maize (9.6%), sunflower (10.4%), oilseed rape (8.3%), pea (2%) and "grasslands" (13.5%), including permanent (meadows) and temporary grasslands (hay, clover, alfalfa). The land cover in the ZA-PVS has been monitored yearly at the field scale since 1994, twice a year (for winter and spring crops, respectively). The detailed knowledge of land cover (i.e. more than 30 categories of crops, plus build up areas, forest fragments, vineyards and hedges) has been used for setting up tightly targeted spatial designs for testing hypothesis on the effect of landscape on biodiversity as well as ecosystem functions.

From 1994 to 2020, field sizes increased and accordingly, the total number of fields fell from c. 19,000 to less than 12,000 today. Winter cereal cover has strongly increased from around 35% in 1995 to 45% in 2018 (Fig. 14.3), being present in almost 100% of the 1km^2 pixels in the ZA-PVS. At the same time, the meadow cover has strongly decreased, both in the long term (60% in 1970 down to 15% in 2016) and recently with a 30% decrease since 2007 with the abandonment of set-asides in the European Union common agricultural policy (CAP).

Organic farming increased in France from 1.7%, of farms in 2002 (1.87% of the surface of farmed area) to 9.46% in 2018 (7.55% of the farmed area; Agence BIO 2019). It also strongly increased in the ZA-PVS (Fig. 14.4a). Conversion to organic farming started in 1998 with only three farms. Currently the conversion rate is almost 15% of farmers with more than 12% of the ZA-PVS in organic farming. In some areas, organic farming represents almost 80% of cropland over 2 km^2 (e.g. Henckel et al. 2015). Other agri-environment schemes as well as organic farming have been implemented since 2004 in the entire study site by the CNRS research laboratory of Chizé (Bretagnolle et al. 2011a), covering up to one third of the study area in 2013. Six types of AESs have been implemented (Fig. 14.4b).

The high spatial heterogeneity in the LTSER ZA-PVS is an opportunity to analyse the effects of landscape gradients of interest (i.e. meadows, organic farming and woodland) on biodiversity and ecosystem functions.

3 Contrasting Effects of Landscape on Farmland Biodiversity and Ecosystem Functions in the LTSER ZA-PVS

3.1 Effects of Crop Type

Agricultural landscapes consist of a mosaic of habitat patches whose suitability may vary from beneficial to detrimental for any given species (Fahrig et al. 2011). Many organisms use multiple resources found in a variety of crops and non-crop habitats.

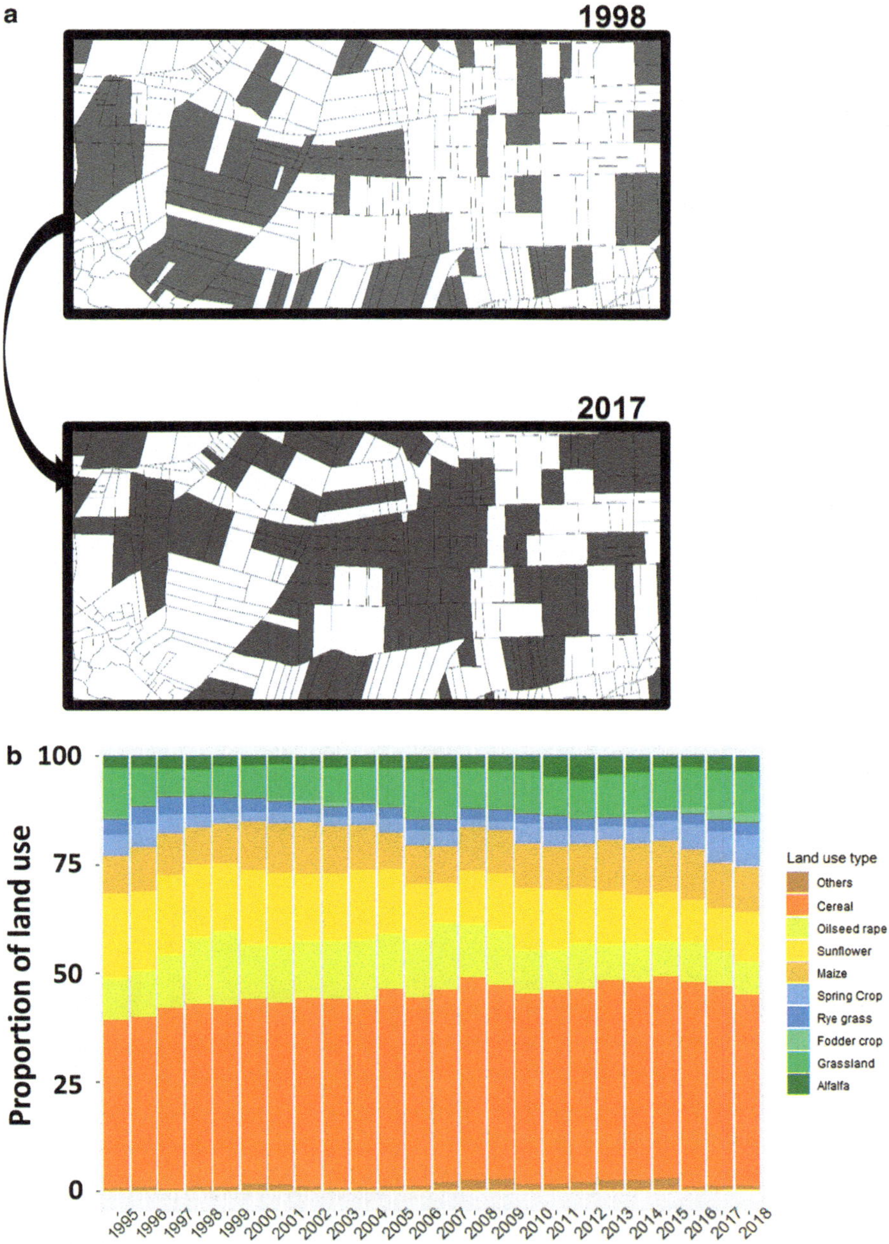

Fig. 14.3 (**a**) Fields in winter cereal (grey) in a part of the LTSER Zone Atelier in 1998 and 2017. (**b**) Temporal variation of land use from 1995 to 2018. Each colour shows a group of land use

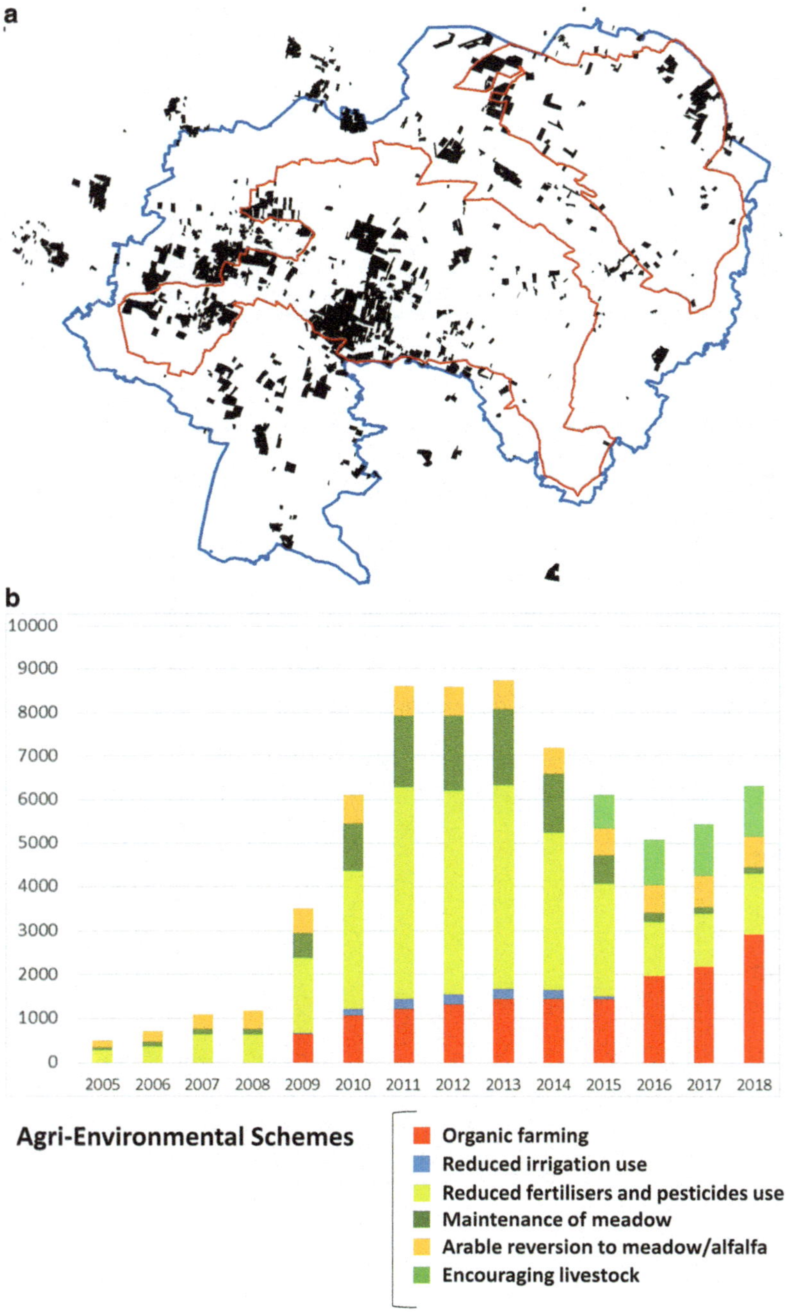

Fig. 14.4 (**a**) Location of the NATURA 2000 area (in red) and fields in organic farming (black polygons) on the LTSER Zone Atelier Plaine & Val de Sèvre. (**b**) Temporal variation of the amount of agri-environmental schemes in the LTSER ZA-PVS from 2005 and 2018. The implementation of AES started in 2004

The distribution in the landscape of these specific resources as well as the suitability of the habitat may be related to the diversity of crops (Vasseur et al. 2013; Schellhorn et al. 2015). Besides providing different environmental conditions and food availability, crop types differ in the intensity and timing of management practices, such as soil tillage, pesticides use and harvesting time. Considering the interplay between local and landscape effects, crop type can have a dual interactive effect on populations: an environmental filtering effect at local scale and a source-sink effect at the landscape scale.

In the ZA-PVS, weed species richness varies significantly between crop types, being highest in oilseed rape with on average, between 2014 and 2019, of 27.76 ($\pm$ 7.46sd) species per field; and lowest in maize (13.61 $\pm$ 6.99) (Bourgeois et al. 2020). Winter cereal and sunflower show intermediate numbers of weed species per field (19.00 $\pm$ 9.80 and 18.02 $\pm$ 6.89, respectively). Similar patterns are observed for carabid beetles (Marrec et al. 2015; Caro et al. 2016; Marrec et al. 2017): Wheat fields harboured on average 7.4 $\pm$ 3.5 species for an abundance activity of 64 $\pm$ 123, while alfalfa fields had 8.0 $\pm$ 4.0 (61.4 $\pm$ 78.8) and meadows 3.9 $\pm$ 2.7 (10.4 $\pm$ 13.2). *Poecilus cupreus* and *Brachinus sclopeta* are the two most abundant species overall in arable crops as well as meadows and grasslands. Oilseed rape supported higher abundances of *Poecilus cupreus* as well as *Anchomenus dorsalis* than winter wheat (0.46 and 1.09 times more, respectively). Despite the large quantity of insecticides used, the presence of oilseed rape all year round (from August the preceding year to July the following one) in the ZA-PVS, thus provides a suitable habitat over a long period, and may explain this diversity pattern for weeds and carabid beetles (Marrec et al. 2015). Effects of crop types are, however, less pronounced on birds and wild bees. In birds, crop type effects are far less significant than the presence of hedgerows and small woods in predicting species richness, whatever the spatial scale considered; at small spatial scales (200 m), only oilseed rape has a significant contribution, while at larger spatial scales grasslands contributes significantly to species richness (Henckel et al. 2019). However, when restricting analyses to open farmland habitats only (i.e., with no hedgerows), grasslands, alfalfa and to a lesser extent, rapeseed contribute strongly to predicting bird species communities (Henckel et al. 2019). Skylarks (*Alauda arvensis*) were also shown to depend on crop types, but even more on complementarity between crop types, typically cereals and grasslands (Marrec et al. 2017). Corn buntings (*Emberiza calandra*) are also strongly dependent on alfalfa, and to a lesser extent oilseed rape (Brodier et al. 2014). Bee-communities in mass-flowering crops (oilseed rape and sunflower) were three to four times less diversified than in semi-natural habitats (Rollin et al. 2015).

At the landscape scale, oilseed rape was shown to have a positive, negative or no effect on biodiversity depending on the species. The amount of oilseed rape in the landscape in a radius of 1000 m had a negative effect on the abundance of honeybee and *Lasioglossum sp.* supporting a dilution effect (Holzschuh et al. 2016). By contrast, and whatever the considered spatial scales (from 500 to 1000 m), the amount of oilseed rape in the landscape had a positive effect on *Poecilus cupreus* activity-density in oilseed rape. The effect was however negative in winter cereal

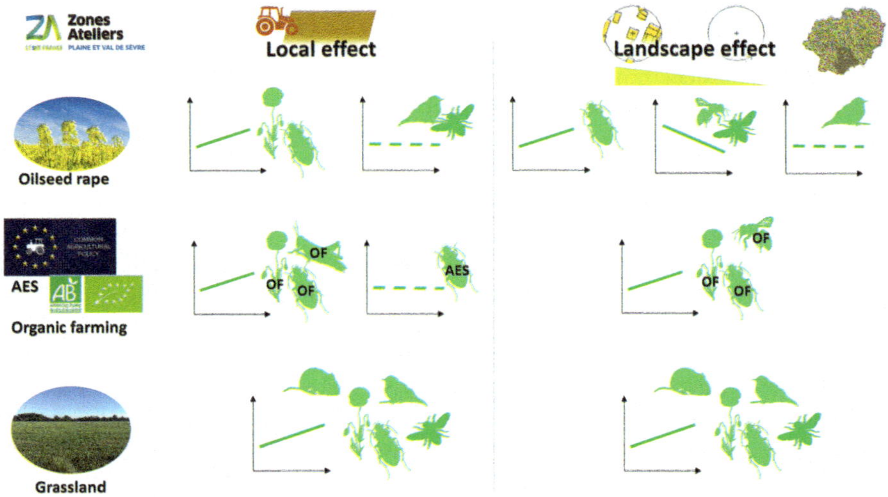

Fig. 14.5 Overview of the local and landscape effects of oilseed rape, grassland and organic farming on biodiversity in the LTSER Zone Atelier Plaine & Val de Sèvre

suggesting dilution-concentration effects (*sensu* Tscharntke et al. 2012) at play for this species when considering crops with different levels of attractiveness. For the carabid beetle *Anchomenus dorsalis*, the effect of oilseed rape in the surrounding landscape was only observed in winter cereals at the small scale (lower than 400 m buffer size) revealing no important long-distance movement for this species. In birds, community assembly is mainly predicted by hedgerows and woodlands, at any spatial scale from 200 to 1400 m, but the explained variation decreases with buffer size, especially in wooded landscapes; the contribution of crops also decreases with buffer size, whatever the species or the landscape types (Henckel et al. 2019). These results are synthesised in Fig. 14.5.

3.2 Landscape Effects of Grasslands on Biodiversity

In line with studies conducted in other study-sites (e.g. Chaplin-Kramer et al. 2011; Rusch et al. 2013; Haan et al. 2020), studies performed in the ZA-PVS revealed that the proportion of semi-natural habitat, such as temporary grasslands, meadows or hedgerows, can increase the abundance and the diversity of pest natural enemies, pollinators and to a lesser extent, birds in landscapes dominated by arable crops. Marrec et al. (2017) observed a positive, large-scale (from 500–700 to 1000 m) effect of grassland in the landscape on the activity-density of two main carabid species in oilseed rape and winter cereal; but the proportion of grassland within 250 m and 1000 m radius landscapes had no effect on spider activity-density in sunflower (Badenhausser et al. 2020). Grasslands are also important for common vole (*Microtus arvalis*) spatial and temporal dynamics (Bretagnolle et al. 2011a; Bonnet et al. 2013), and a similar effect was detected for weeds (Henckel et al. 2015). Birds in general are sensitive to the amount of grasslands at landscape scale (Bretagnolle et al. 2018c), and this is particularly the case for several flagship species such as little bustard (*Tetrax tetrax*) and Montagu's harrier (*Circus pygargus*) (Bretagnolle et al. 2011a, b). A strong effect of the proportion of semi-natural habitat in the landscape was also found on local wild bee richness (Rollin et al. 2019). The greatest landscape effect was predicted beyond a 20–25% threshold for herbaceous semi-natural habitats (within a 1000 m radius). However, the magnitude of the effect of semi-natural habitat proportion on bee richness varied with the local abundance of bees. The effect was higher at high compared to low bee abundance. This pattern can be explained by the dual function of semi-natural habitats, as a source of both feeding resources and nesting sites (Öckinger and Smith 2007; Goulson et al. 2010).

Increasing grasslands at the landscape scale further enhances several ecological functions. The proportion of meadows and temporary grasslands increased the predation rates of seeds and aphids in winter cereal (Perrot et al. in review) while this effect was only observed on aphid predation rate (not on seed predation) in sunflowers adjacent to plant-species rich meadows and temporary grasslands with few hedges in the neighbourhood (Badenhausser et al. 2020). These results support the need for using multiple prey sentinels when investigating the effect of landscape on pest control (McHugh et al. 2020). Higher abundance and diversity of pollinators in landscapes with large amounts of semi-natural habitat also improves pollination efficiency. The fruiting success was improved by increasing the contribution of large pollinators to pollination in landscapes rich in semi-natural habitats as shown by a bagged experiment. However, this positive effect was only observed during the flowering peak of oilseed rape suggesting that large pollinators were likely to be attracted by oilseed rape blooming.

3.3 Beneficial Effects of Reduced Agricultural Practices Intensity

Half of the ZA-PVS is a Natura 2000 site, where an extensive program of agri-environmental schemes (AES) has been set up since 2004. Many AES consist in reducing the intensity of agricultural practices, i.e. reducing pesticide inputs, mowing, frequency of cultivations (see for a list of AES, Brodier et al. 2014; Caro et al. 2016). There was little effect of AES on carabid beetles when excluding organic farming (see below), both locally or at landscape scale (Caro et al. 2016). Delayed cutting of alfalfa fields had a small positive effect on carabids, increasing their richness by 11.7%. This local effect was amplified in the landscape with large amount of organic farming, but at large spatial scales (larger than 750 m). The evaluation of the effect of a broad set of agri-environmental schemes (AESs) on carabid abundance-activity and species richness revealed that carabid species richness and abundance-activity only responded to the most restrictive AES. AES were also shown to affect positively grasshoppers, especially delayed cutting (Badenhausser et al. 2009; Bretagnolle et al. 2011a).

Among AES, organic farming has been shown to have the highest effect on biodiversity. Organic fields were found to act as key habitats for weed spatial dynamics (see also Gabriel et al. 2010). We found a strong landscape effect of organic farming in a 1 km buffer size on weed species richness in winter cereal fields (Henckel et al. 2015; Petit et al. 2016; Bourgeois et al. 2020). This effect was observed in both organic and conventional fields, although its magnitude was lower in conventional fields, especially in field cores highlighting the filtering effect of conventional management (especially chemical fertilisers and herbicides; Henckel et al. 2015). Differences in weed richness between organic and conventional fields were mostly explained by the higher diversity of less frequent species in organic fields. Using a structural equation model to quantify the effects of regional dispersal from source habitats in the landscape, local dispersal from field margin and their interplay with local conditions (crop type and field size), Bourgeois et al. (2020) showed that organic farming within the landscape predominantly acts indirectly via field margins. This highlights the key role of field margins in sustaining the local dynamics of weed communities through an improved connectivity between field core and the surrounding landscape; the magnitude of weed propagation from field margin to field core being however largely species-dependent as shown in Wilson and Aebischer (1995). Moreover, the strength of this two-step process (organic field cover plus field margin dispersal) was modulated by crop type, being especially pronounced in winter cereal compared to oilseed rape, sunflower and maize. This may be explained by the fact that most organic fields across the study area are sown with winter cereals. The strong positive effect of the amount of organic farming at the same scale was found to exceed the effect of local agricultural practices on weed richness (Petit et al. 2016). By contrast, the amount of organic farming in the landscape had no effect on weed abundance indicating that the nature and the scale at which the drivers act are not the same for weed abundance and weed richness. Similarly, we detected a strong positive effect of organic farming in winter cereal fields on carabid species richness and their abundance-activity (Caro et al. 2016). Carabid species richness increased by +20.2% in organic than in conventional winter cereal, in line with Tuck et al. (2014) who found an average increase of 25%. This local effect was further increased in landscapes with higher amount of organic farming. This effect was especially detected at small scales (i.e. at a radius lower than 500 m).

Honeybees were also investigated in the ZA-PVS, thanks to the presence of the ECOBEE platform (Odoux et al. 2014). Positive relationships at the local scale (300 m) between organic farming and worker brood area or number of honeybee adults were detected from a 6 year study with 60 apiaries (300 hives) located in landscapes varying in the proportion of organic farmland (Wintermantel et al. 2019). These authors found that organic fields impact colony size especially when they are nearby. There may be several pathways through which nearby organic farming may act on honeybee colonies, including reduce foraging efforts (i.e. lower energy consumption), insecticide reduction (i.e. lower impact on activity and survival), herbicide reduction (i.e. wider diversity of pollen sources) and provision of semi-natural elements. In the dearth period and at the beginning of the sunflower bloom, colonies exposed to organic farmland at the landscape scale had larger honey reserves (Requier et al. 2017; Wintermantel et al. 2019). Colonies in landscapes rich in organic farmland may benefit from increased availability of melliferous flowers after the oilseed rape bloom as well as a higher access to pollen demands more easily, which allowed them to forage more intensively on nectar sources.

3.4 Combined Effects of Past and Current Landscape Composition

Agricultural landscapes are highly dynamic in space and time because of crop rotation. Hence, they represent ephemeral habitats where environmental conditions only allow species to establish and successfully reproduce in a short window of time. Consequently, this temporal environmental heterogeneity directly affects the survival and the movement behaviour of species that use annual crops. In spite of its importance, the effect of temporal environmental heterogeneity on populations and species assemblies in agricultural landscape has been poorly investigated. Even fewer studies consider the combined effects of past and current landscape on population and community dynamics.

In the ZA-PVS, the activity-density of two common carabid beetles, *Poecilus cupreus* and *Anchomenus dorsalis*, in winter cereals and oilseed rape (Marrec et al. 2017) was significantly affected by both current and previous year landscape composition. These effects however varied with crop type and carabid species identity. The activity-densities of *P. cupreus* and *A. dorsalis* were both positively influenced by the amount of oilseed rape in the preceding year's landscape, but only *P. cupreus* activity-density responded positively to the amount of spring crops in the preceding year's landscape. *A. dorsalis* individuals overwinter in perennial habitats, particularly field margins (Marrec et al. 2015). Therefore, it is likely that *A. dorsalis* individuals move from previous oilseed rape fields to grasslands during the summer-fall period, rather than to an alternate annual crop such as spring crops. The colonisation of oilseed rape fields the next year may then occur from these perennial habitats. These results support the hypothesis that carabid beetles respond to the interannual redistribution of resources caused by crop rotation at the landscape level (Russon and Woltz 2015).

A similar response was also observed in pollinators. In agricultural landscapes, oilseed rape and sunflower provide resources during non-overlapping periods for pollinators. Using data collected in 172 fields over 5 years, we found that a high proportion of sunflower in the preceding landscape led to increases in *Lasioglossum* sp. and honeybee abundances in oilseed rape the following year. Sunflower is an important resource for wild bees (Todd et al. 2016) and honeybee (Requier et al.

2015). These results indicate that the management of landscape-scale patterns of mass-flowering crops accounting for delayed response of pollinators can ensure for wild pollinator persistence and crop pollination services.

4 Multifunctional Landscapes and Economic Performance

The benefits of grassland and AES, especially organic farming, for biodiversity on arable farmland are clear, with an average increase in species richness and abundance across taxa. However, the magnitude of these effects are strongly dependent on the guild or the taxon studied, as well as on landscape context. These conclusions are relevant to resolving the long-running argument between food security and the drive to ban chemical pesticides and fertilisers as exemplified by the organic/conventional farming and parallel land-sharing/land-sparing debate (e.g. Green et al. 2005; Phalan et al. 2011; Seufert et al. 2012; Ponisio et al. 2015; Seufert and Ramankutty 2017). Rather than simplifying the debate to "biodiversity vs. food production" (Bennett 2017), we sought in the ZA-PVS to consider agricultural landscapes as multifaceted and, importantly, multifunctional (Landis 2017; Turkelboom et al. 2018). We therefore explore the extent to which spatial arrangements of conventional farming, low-input farming (AES fields or farms) and organic farming, along with remnant semi-natural habitats, may best support multiple goals in multifunctional and resilient agricultural landscapes.

The important role that plant species richness (generally higher in semi-natural habitats) plays in maintaining ecosystem functions and, to some extent, multifunctionality has been highlighted previously (Allan et al. 2015; Soliveres et al. 2016; Wardle 2016; Pennekamp et al. 2018). In the ZA-PVS, we extend those results by revealing that weed diversity was a strong contributor to ecosystem multifunctionality especially when considering the simultaneous delivery of functions and when considering only aboveground ecological functions, i.e. pest control, pollination and bee richness as a proxy of biodiversity conservation (Gaba et al. 2020). This study also emphasizes the critical role of grasslands, including hay crops (particularly alfalfa) and meadows, for maintaining multifunctionality in agricultural landscape. Multifunctionality was greater in hay crops compared to winter cereal and oilseed rape fields. This pattern was mainly explained by less intensive management of hay crops as well as higher weed diversity. However, at the landscape scale, a modelling approach found that the loss of semi-natural habitat (such as grasslands) had contrasting effects on the delivery of three ESs: biodiversity decreases, pollination-independent crop production increases, while pollination-dependent crop production is maximized at an intermediate proportion of semi-natural habitat (Montoya et al. 2019). These results, in accordance with Pywell et al. (2015) and Tamburini et al. (2016), suggest that moderate increases in the amount of semi-natural habitat in simple agricultural landscapes (1–20% semi-natural habitat) allow ecosystem services essential for crop production to be maintained, which, in turn, increases the magnitude and stability of crop yield. This modelling approach also revealed that trade-offs involving crop pollination were strongly affected by the degree to which crops depend on pollination and by their relative requirement for pollinator densities. At the field scale, reducing herbicides and fertiliser inputs does not necessarily decrease yields in wheat (Gaba et al. 2016; Catarino et al. 2019b) nor in oilseed rape (Catarino et al. 2019a), increases average farmers' margins (Catarino et al. 2019b, a) and finally improves biodiversity (as shown by AES effects on biodiversity; see above) which increases yields, e.g. in oilseed rape and sunflower (Perrot et al. 2018, 2019).

4.1 Best Landscape Composition to Meet Various Stakeholders' Demands

The supply and demand components of ESs in agroecosystems are multifaceted because of the multiple ecological processes involved on the supply side and the multiple stakeholders that benefit from them on the demand side. Human demand is one side of the ecosystem-service equation and comprises different agents or stakeholders. These stakeholders vary in both their demand for and their valuation of different ecosystems services, thus have specific perspectives about how to manage efficiently agricultural landscapes (Fig. 14.6).

For example, farmers are mainly interested in maximizing crop yield per area of their cultivated land as this is mostly related to their profitability. Beekeepers, on the other hand, are looking for resource-rich landscapes to settle their beehives in order to maximise honey production and in turn their revenues. Agricultural unions or

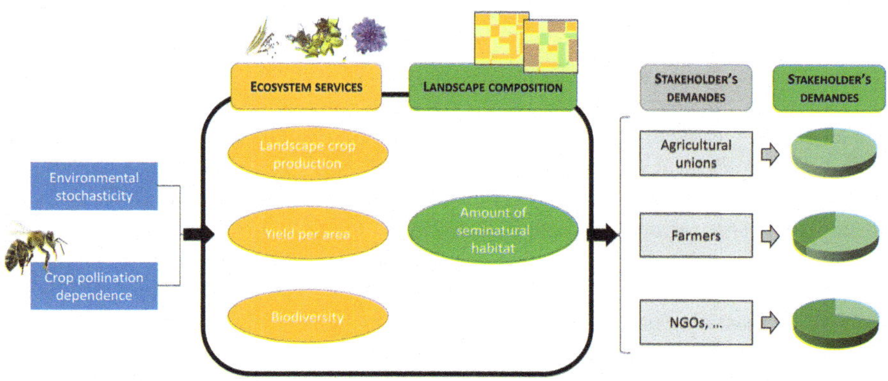

Fig. 14.6 Conceptual diagram of the modelling framework of Montoya et al. (2020) (modified from Montoya et al. 2020). Environmental stochasticity and crop pollination dependence effects the magnitude of three ecosystem services (landscape crop production, crop yield per area and biodiversity conservation) and the effects of landscape composition (amount of semi-natural habitat) associated with the delivery. The three groups of stakeholders are assumed to value each ecosystem service differently. The landscape compositions shows the best landscape compositions to fulfil their demands

cooperatives often aim at maximizing crop production at the regional level to deliver food security. NGO's, wildlife-friendly organizations and even residents expect agriculture not to jeopardise the future of farmland and nature in general. Considering a simple case-study model of three ESs and three groups of stakeholders (i.e. farmers, agricultural unions and NGO), Montoya et al. (2020) concluded that the ideal landscape composition, measured as the fraction of semi-natural habitat within the agricultural landscape, differs among stakeholders. In particular there is a trade-off between agricultural unions on one hand and farmers and NGOs on the other hand. The demands of farmers and NGOs were generally aligned at higher fractions of semi-natural habitat suggesting that farmers' best interest is that the landscape surrounding their land is not farmed. This supports empirical studies showing that higher levels of biodiversity are beneficial for farmers (Binder et al. 2018; Catarino et al. 2019a). However, the degree to which a crop depends on animal pollination and environmental stochasticity are two other main drivers of the best landscape composition. Higher pollination dependence of crops generally shifts the best landscape composition to larger fractions of semi-natural habitat, while high stochasticity, as expected under current global change, shifts maximum landscape production to lower fractions of semi-natural habitat and constrains the range of semi-natural habitat that maximizes farmers' profitability. The social average scenario, which considered multiple demands for ESs with a multifunctionality goal, was more robust to changes in crop type and stochasticity. Consequently, management for social average may be a better option for food security, livelihood opportunities, and biodiversity conservation, thus meeting various stakeholders' demands.

4.2 Policy Instruments for Multifunctional Landscape

In agricultural landscapes, stakeholders generally do not feel mutually dependent, i.e., they do not feel that they need one another to improve their situation. For instance, they typically ignore the cascade effects of their decision-making on the other stakeholders in the shared landscapes. Some ES can lead to mutual interest between ES beneficiaries. For example, pollination ES is of mutual benefit for beekeepers and farmers in agricultural landscapes. When visiting crop flowers, honeybees increase crop yields of mass-flowering crops (i.e. oilseed rape and sunflower; Perrot et al. 2018, 2019) and farmers' incomes (Catarino et al. 2019a). In return, mass-flowering crops provide abundant floral resources for honeybees, therefore ensuring honey production and overwinter survival success (Requier et al. 2017). Although these social interdependencies could contribute to collective action or elucidating potential divergences between related stakeholders' demands (Barnaud et al. 2018), it is not currently taken into account in any public policies. Using a bio-economic model (Fig. 14.7), Faure et al. (in review) explored how agricultural policies could act efficiently by playing on the mutual benefits for beekeepers and farmers of pollination ES to increase the delivery of multiple ESs (food and feed production, water quality and biodiversity conservation) and the

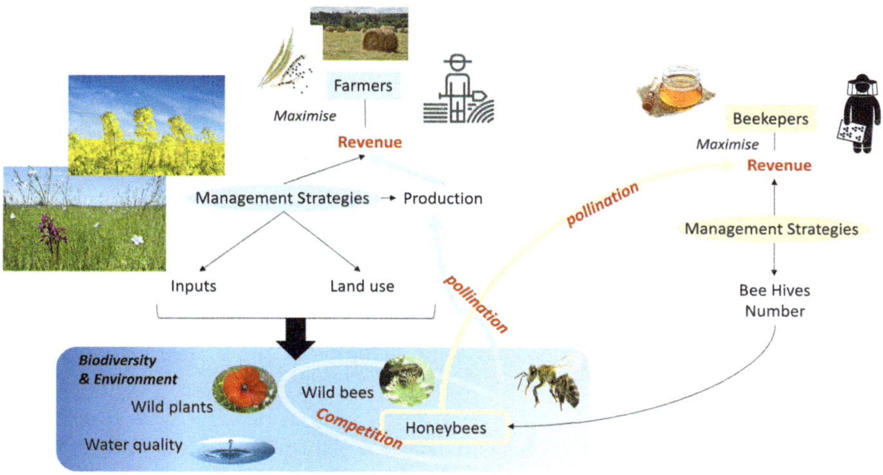

Fig. 14.7 Schematic representation of the bio-economic model of Faure et al. (in review)

economic performances at landscape scale, and both farmer and beekeeper profitability. Following the European beekeepers' recommendations (Breeze et al. 2019), the efficiency of two main incentive instruments were explored: (1) economic subsidies targeting beekeepers, and (2) taxes targeting farmers for reducing the use of agrochemicals. Supporting beekeeping as well as taxing pesticides both increased pollination. This generated a causal chain in the decision-making of the two stakeholders: implementing subsidies supporting beekeeping yielded an increase in the number of hives (change in the beekeeping strategy) resulting in a higher honeybee abundance, hence a higher OSR yield per ha. In turn, the farmers increased the OSR cultivated area and decreased their agrochemical use. Such positive feedback loops between the two stakeholder groups results in mutual benefit of farmers and beekeepers from the two policies. Furthermore, by linking the two stakeholders, the interdependency affected the delivery of ESs and the economy at the landscape scale. The subsidies benefited the economy (higher total wealth) but had low environmental performances (low wild bee abundance and increase of nutrient pollution in water). Tax, on the other hand benefited the environment but resulted in low economic benefits at landscape scale.

5 Conclusions

In conclusion, this chapter provides an overview of the research products from a Long Term Socio-Ecological research site, the Zone Atelier Plaine & Val de Sèvre, which aims to address the current challenges of agriculture, i.e. maintaining food security while preserving biodiversity, landscapes and ecosystems. One major result is that biodiversity decline is a severe threat to food security in itself; several results show that improving biodiversity improves yields, margins, or both; hence reducing biodiversity imperils crop production. Long-term monitoring of biodiversity, ecosystem functioning and agricultural practices, as well as land use, was revealed as a powerful tool for understanding the effect of human actions on biodiversity and the agricultural ecosystem. Furthermore, it paves the way for producing knowledge, shared by various stakeholders, which may contribute to tackling difficult problems such as food security or environmental sustainability in a more efficient way. Here, we show that the interplay between local and landscape factors can provide solutions to maintain biodiversity, crop production and farmers' profitability. Engaging stakeholders in collaboration with scientists, as well as increasing the awareness of stakeholders' mutual interdependencies with nature, is now needed to foster agroecological transition.

Acknowledgments The LTSER Zone Atelier "Plaine & Val de Sèvre" is supported by CNRS/INEE, ALLENVI and INRAE. The research in this chapter has been supported by the French Ministry of Ecology project (2017-2020 "Pollinisateurs"), the ANR AGROBIOSE (ANR-13-AGRO-286 0001), the ANR IMAGHO (ANR-18-CE32-0002), the ERANET ECODEAL (FP7 ERA-NET BiodivERsA/FACCE-JPI), ERA-Net BiodivERsA FARMLAND (ANR-11-EBID-0004), the Région Nouvelle Aquitaine 'Birdland' and 'Harmony' project and the INRAe project "ESPACE" (MP SMACH). We would like to express our thanks to Jean-Luc Gautier, Marilyn Roncoroni, Alexis Saintilan and hundreds of field workers and students over the 27 years who contributed to the monitoring programs. We sincerely thank the local stakeholders, especially farmers, for their collaboration and involvement on the research programs.

References

Agence BIO (2019) Agence bio. Available at: https://www.agencebio.org/wp-content/uploads/2019/06/DP-AGENCE_BIO-4JUIN2019.pdf

Allan E, Manning P, Alt F, Binkenstein J, Blaser S, Blüthgen N et al (2015) Land use intensification alters ecosystem multifunctionality via loss of biodiversity and changes to functional composition. Ecol Lett 18:834–843

Altieri MA (1983) Agroecology: the scientific basis of alternative agriculture. Division of Biological Control, University of California, Berkeley

Amundson R, Berhe AA, Hopmans JW, Olson C, Sztein AE, Sparks DL (2015) Soil and human security in the 21st century. Science 348:1261071

Angelstam P, Manton M, Elbakidze M, Sijtsma F, Adamescu MC, Avni N et al (2018) LTSER platforms as a place-based transdisciplinary research infrastructure: learning landscape approach through evaluation. Landsc Ecol 34:1–24

Badenhausser I, Amouroux P, Lerin J, Bretagnolle V (2009) Acridid (Orthoptera: Acrididae) abundance in Western European grasslands: sampling methodology and temporal fluctuations. J Appl Entomol 133:720–732

Badenhausser I, Gross N, Mornet V, Roncoroni M, Saintilan A, Rusch A (2020) Increasing amount and quality of green infrastructures at different scales promotes biological control in agricultural landscapes. Agric Ecosyst Environ 290:106735

Barnaud C, Corbera E, Muradian R, Salliou N, Sirami C, Vialatte A et al (2018) Ecosystem services, social interdependencies, and collective action: a conceptual framework. Ecol Soc 23:15

Bennett EM (2017) Changing the agriculture and environment conversation. Nat Ecol Evol 1:1–2
Binder S, Isbell F, Polasky S, Catford JA, Tilman D (2018) Grassland biodiversity can pay. Proc Natl Acad Sci U S A 115:3876–3881
Bommarco R, Kleijn D, Potts SG (2013) Ecological intensification: harnessing ecosystem services for food security. Trends Ecol Evol 28:230–238
Bonnet T, Crespin L, Pinot A, Bruneteau L, Bretagnolle V, Gauffre B (2013) How the common vole copes with modern farming: insights from a capture-mark-recapture experiment. Agric Ecosyst Environ 177:21–27
Bourgeois B, Gaba S, Plumejeaud C, Bretagnolle V (2020) Weed diversity is driven by complex interplay between multi-scale dispersal and local filtering. Proc. R. Soc. London. Ser. B, Biol. Sci. 287:20201118
Breeze TD, Boreux V, Cole L, Dicks L, Klein A, Pufal G et al (2019) Linking farmer and beekeeper preferences with ecological knowledge to improve crop pollination. People Nat 1:562–572
Bretagnolle V, Gauffre B, Meiss H, Badenhausser I (2011a) The role of grassland areas within arable cropping systems for the conservation of biodiversity at the regional level. In: Lemaire G, Hodgson J, Chabbi A (eds) Grassland productivity and ecosystem services. CAB International, Cambridge, MA, pp 251–260
Bretagnolle V, Villers A, Denonfoux L, Cornulier T, Inchausti P, Badenhausser I (2011b) Rapid recovery of a depleted population of Little Bustards Tetrax tetrax following provision of alfalfa through an agri-environment scheme. Ibis (Lond 1859) 153:4–13
Bretagnolle V, Berthet E, Gross N, Gauffre B, Plumejeaud C, Houte S et al (2018a) Description of long-term monitoring of farmland biodiversity in a LTSER. Data Br 19:1310–1313
Bretagnolle V, Berthet E, Gross N, Gauffre B, Plumejeaud C, Houte S et al (2018b) Towards sustainable and multifunctional agriculture in farmland landscapes: lessons from the integrative approach of a French LTSER platform. Sci Total Environ 627:822–834
Bretagnolle V, Siriwardena G, Miguet P, Henckel L, Kleijn D (2018c) Local and landscape scale effects of heterogeneity in shaping bird communities and population dynamics: crop-grassland interactions. In: Agroecosystem diversity: reconciling contemporary agriculture and environmental quality. Elsevier Science, San Diego, pp 241–243
Bretagnolle V, Benoit M, Bonnefond M, Breton V, Church JM, Gaba S et al (2019) Action-orientated research and framework: insights from the French long term social-ecological research network. Ecol Soc 24:10
Brodier S, Augiron S, Cornulier T, Bretagnolle V (2014) Local improvement of skylark and corn bunting population trends on intensive arable landscape: a case study of the conservation tool Natura 2000. Anim Conserv 17:204–216
Caro G, Marrec R, Gauffre B, Roncoroni M, Augiron S, Bretagnolle V (2016) Multi-scale effects of agri-environment schemes on carabid beetles in intensive farmland. Agric Ecosyst Environ 229:48–56
Catarino R, Bretagnolle V, Perrot T, Vialloux F, Gaba S (2019a) Bee pollination outperforms pesticides for oilseed crop production and profitability. Proc R Soc B Biol Sci 286:20191550
Catarino R, Gaba S, Bretagnolle V (2019b) Experimental and empirical evidence shows that reducing weed control in winter cereal fields is a viable strategy for farmers. Sci Rep 9004
Chaplin-Kramer R, O'Rourke ME, Blitzer EJ, Kremen C (2011) A meta-analysis of crop pest and natural enemy response to landscape complexity. Ecol Lett 14:922–932
Cumming GS, Allen CR, Ban NC, Biggs D, Biggs HC, Cumming DHM et al (2015) Understanding protected area resilience: a multi-scale, social-ecological approach. Ecol Appl 25:299–319
Del Grosso SJ, Parton WJ, Adler PR, Davis SC, Keough C, Marx E (2012) Daycent model simulations for estimating soil carbon dynamics and greenhouse gas fluxes from agricultural production systems. In: Managing agricultural greenhouse gases. Elsevier, San Diego, pp 241–250
Fahrig L, Baudry J, Brotons L, Burel FG, Crist TO, Fuller RJ et al (2011) Functional landscape heterogeneity and animal biodiversity in agricultural landscapes. Ecol Lett 14:101–112

Faure J, Gaba S, Mouysset L (in review) Designing multifunctional landscapes: the hidden consequences of the interdependency between the stakeholders through pollination. *In revision*
Gaba S, Bretagnolle V (2020) Social-ecological experiments to foster agroecological transition. People Nat. 2(2):317-327
Gaba S, Bretagnolle F, Rigaud T, Philippot L (2014) Managing biotic interactions for ecological intensification of agroecosystems. Front Ecol Evol 2:29
Gaba S, Gabriel E, Chadœuf J, Bonneu F, Bretagnolle V (2016) Herbicides do not ensure for higher wheat yield, but eliminate rare plant species. Sci Rep 6:30112
Gaba S, Cheviron N, Perrot T, Piutti S, Gautier JL, Bretagnolle V (2020) Weeds enhance multifunctionality in arable lands in South-West of France. Front Sustain Food Syst 4:71
Gabriel D, Sait SM, Hodgson JA, Schmutz U, Kunin WE, Benton TG (2010) Scale matters: the impact of organic farming on biodiversity at different spatial scales. Ecol Lett 13:858–869
Gallai N, Salles JM, Settele J, Vaissière BE (2009) Economic valuation of the vulnerability of world agriculture confronted with pollinator decline. Ecol Econ 68:810–821
Gliessman S (2016) Transforming food systems with agroecology. Agroecol Sustain Food Syst 40:187–189
Godfray HCJ (2015) The debate over sustainable intensification. Food Secur 7:199–208
Goulson D, Lepais O, O'Connor S, Osborne JL, Sanderson RA, Cussans J et al (2010) Effects of land use at a landscape scale on bumblebee nest density and survival. J Appl Ecol 47:1207–1215
Green RE, Cornell SJ, Scharlemann JPW, Balmford A (2005) Farming and the fate of wild nature. Science 307:550–555
Haan NL, Zhang Y, Landis DA (2020) Predicting landscape configuration effects on agricultural Pest suppression. Trends Ecol Evol 35:359–367
Haines-Young R, Potschin M (2010) The links between biodiversity, ecosystem services and human well-being. In: Raffaelli D, Frid C (eds) Ecosystem ecology: a new synthesis. Cambridge University Press, Cambridge, pp 110–139
Henckel L, Börger L, Meiss H, Gaba S, Bretagnolle V (2015) Organic fields sustain weed metacommunity dynamics in farmland landscapes. Proc R Soc B Biol Sci 282:1–9
Henckel L, Meynard CN, Devictor V, Mouquet N, Bretagnolle V (2019) On the relative importance of space and environment in farmland bird community assembly. PLoS One 14:e0213360
Holzschuh A, Dainese M, González-Varo JP, Mudri-Stojnić S, Riedinger V, Rundlöf M et al (2016) Mass-flowering crops dilute pollinator abundance in agricultural landscapes across Europe. Ecol Lett 19:1228–1236
Klein A-M, Vaissière BE, Cane JH, Steffan-Dewenter I, Cunningham SA, Kremen C et al (2007) Importance of pollinators in changing landscapes for world crops. Proc Biol Sci 274. 66, 95–96, 191
Lampkin N (1990) Farming systems and ecological diversity. In: Organic Farming. Farming Press, Ipswich, pp 574–579
Landis DA (2017) Designing agricultural landscapes for biodiversity-based ecosystem services. Basic Appl Ecol 18:1–12
Lévêque C, Pavé A, Abbadie L, Weill A, Vivien F-D (2000) Les zones ateliers, des dispositifs pour la recherche sur l'environnement et les anthroposystèmes. Natures, Sci. Sociétés 8:44–52
Marrec R, Badenhausser I, Bretagnolle V, Börger L, Roncoroni M, Guillon N et al (2015) Crop succession and habitat preferences drive the distribution and abundance of carabid beetles in an agricultural landscape. Agric Ecosyst Environ 199:282–289
Marrec R, Caro G, Miguet P, Badenhausser I, Plantegenest M, Vialatte A et al (2017) Spatiotemporal dynamics of the agricultural landscape mosaic drives distribution and abundance of dominant carabid beetles. Landsc Ecol 32:2383–2398
Maxwell SL, Fuller RA, Brooks TM, Watson JEM (2016) Biodiversity: the ravages of guns, nets and bulldozers. Nature 536:143
McHugh NM, Moreby S, Lof ME, Van der Werf W, Holland JM (2020) The contribution of semi-natural habitats to biological control is dependent on sentinel prey type. J Appl Ecol 57:914–925

Montoya D, Haegeman B, Gaba S, de Mazancourt C, Bretagnolle V, Loreau M (2019) Trade-offs in the provisioning and stability of ecosystem services in agroecosystems. Ecol Appl 29:e01853

Montoya D, Gaba S, de Mazancourt C, Bretagnolle V, Loreau M (2020) Reconciling biodiversity conservation, food production and farmers' demand in agricultural landscapes. Ecol Model 416:108889

Öckinger E, Smith HG (2007) Semi-natural grasslands as population sources for pollinating insects in agricultural landscapes. J Appl Ecol 44:50–59

Odoux JF, Aupinel P, Gateff S, Requier F, Henry M, Bretagnolle V (2014) ECOBEE: a tool for long-term honey bee colony monitoring at the landscape scale in West European intensive agroecosystems. J Apic Res 53:57–66

Oerke EC (2006) Crop losses to pests. J Agric Sci 144:31–43

Pe'er G, Dicks LV, Visconti P, Arlettaz R, Báldi A, Benton TG et al (2014) EU agricultural reform fails on biodiversity. Science 344:1090–1092

Pennekamp F, Pontarp M, Tabi A, Altermatt F, Alther R, Choffat Y et al (2018) Biodiversity increases and decreases ecosystem stability. Nature 563:109–112

Perrot T, Gaba S, Roncoroni M, Gautier JL, Bretagnolle V (2018) Bees increase oilseed rape yield under real field conditions. Agric Ecosyst Environ 266:39–48

Perrot T, Gaba S, Roncoroni M, Gautier JL, Saintilan A, Bretagnolle V (2019) Experimental quantification of insect pollination on sunflower yield, reconciling plant and field scale estimates. Basic Appl Ecol 34:75–84

Perrot T, Rusch A, Coux C, Gaba S, Bretagnolle V (2020) Proportion of grassland drives biological pest control services in agricultural landscapes. Front Ecol Evol. *in review*

Petit S, Gaba S, Grison A-L, Meiss H, Simmoneau B, Munier-Jolain N et al (2016) Landscape scale management affects weed richness but not weed abundance in winter wheat fields. Agric Ecosyst Environ 223:41–47

Phalan B, Balmford A, Green RE, Scharlemann JPW (2011) Minimising the harm to biodiversity of producing more food globally. Food Policy 36:S62–S71

Phalan B, Green RE, Dicks LV, Dotta G, Feniuk C, Lamb A et al (2016) How can higher-yield farming help to spare nature? Science 351:450–451

Pimentel D, Acquay H, Biltonen M, Rice P, Silva M, Nelson J et al (1992) Environmental and economic costs of pesticide use. Bioscience 42:750–760

Ponisio LC, M'gonigle LK, Mace KC, Palomino J, De Valpine P, Kremen C (2015) Diversification practices reduce organic to conventional yield gap. Proc R Soc B Biol Sci 282:20141396

Pywell RF, Heard MS, Woodcock BA, Hinsley S, Ridding L, Nowakowski M et al (2015) Wildlife-friendly farming increases crop yield: evidence for ecological intensification. Proc R Soc B Biol Sci 282:20151740

Requier F, Odoux JF, Tamic T, Moreau N, Henry M, Decourtye A et al (2015) Honey bee diet in intensive farmland habitats reveals an unexpectedly high flower richness and a major role of weeds. Ecol Appl 25:881–890

Requier F, Odoux JF, Henry M, Bretagnolle V (2017) The carry-over effects of pollen shortage decrease the survival of honeybee colonies in farmlands. J Appl Ecol 54:1161–1170

Rollin O, Bretagnolle V, Fortel L, Guilbaud L, Henry M (2015) Habitat, spatial and temporal drivers of diversity patterns in a wild bee assemblage. Biodivers Conserv 24:1195–1214

Rollin O, Pérez-Méndez N, Bretagnolle V, Henry M (2019) Preserving habitat quality at local and landscape scales increases wild bee diversity in intensive farming systems. Agric Ecosyst Environ 275:73–80

Rusch A, Bommarco R, Jonsson M, Smith HG, Ekbom B (2013) Flow and stability of natural pest control services depend on complexity and crop rotation at the landscape scale. J Appl Ecol 50:345–354

Russon H, Woltz J (2015) Movement patterns of carabid beetles between heterogenous crop and noncrop habitats. Mich Entomol Soc 47:186

Schellhorn NA, Gagic V, Bommarco R (2015) Time will tell: resource continuity bolsters ecosystem services. Trends Ecol Evol 30:524–530

Seufert V, Ramankutty N (2017) Many shades of gray—the context-dependent performance of organic agriculture. Sci Adv 3:e1602638

Seufert V, Ramankutty N, Foley JA (2012) Comparing the yields of organic and conventional agriculture. Nature 485:229–232

Soliveres S, Manning P, Prati D, Gossner MM, Alt F, Arndt H et al (2016) Locally rare species influence grassland ecosystem multifunctionality. Philos Trans R Soc B Biol Sci 371:20150269

Tamburini G, De Simone S, Sigura M, Boscutti F, Marini L (2016) Soil management shapes ecosystem service provision and trade-offs in agricultural landscapes. Proc R Soc B Biol Sci 283:20161369

Tilman D, Balzer C, Hill J, Befort BL (2011) Global food demand and the sustainable intensification of agriculture. Proc Natl Acad Sci U S A 108:20260–20264

Tittonell P (2014) Ecological intensification of agriculture-sustainable by nature. Curr Opin Environ Sustain 8:53–61

Todd KJ, Gardiner MM, Lindquist ED (2016) Mass flowering crops as a conservation resource for wild pollinators (Hymenoptera: Apoidea). J Kansas Entomol Soc 89:158–167

Tscharntke T, Tylianakis JM, Rand TA, Didham RK, Fahrig L, Batáry P et al (2012) Landscape moderation of biodiversity patterns and processes – eight hypotheses. Biol Rev 87:661–685

Tuck SL, Winqvist C, Mota F, Ahnström J, Turnbull LA, Bengtsson J (2014) Land-use intensity and the effects of organic farming on biodiversity: a hierarchical meta-analysis. J Appl Ecol 51:746–755

Turkelboom F, Leone M, Jacobs S, Kelemen E, García-Llorente M, Baró F et al (2018) When we cannot have it all: ecosystem services trade-offs in the context of spatial planning. Ecosyst Serv 29C:566–578

Vasseur C, Joannon A, Aviron S, Burel F, Meynard JM, Baudry J (2013) The cropping systems mosaic: how does the hidden heterogeneity of agricultural landscapes drive arthropod populations? Agric Ecosyst Environ 166:3–14

Wardle DA (2016) Do experiments exploring plant diversity-ecosystem functioning relationships inform how biodiversity loss impacts natural ecosystems? J Veg Sci 27:646–653

Willer H, Lernoud J (2019) The world of organic agriculture. Statistics and emerging trends 2019. FiBL, Frick, and, IFOAM – Organics International, Bonn

Wilson PJ, Aebischer NJ (1995) The distribution of dicotyledonous arable weeds in relation to distance from the field edge. J Appl Ecol 32:295–310

Wintermantel D, Odoux JF, Chadœuf J, Bretagnolle V (2019) Organic farming positively affects honeybee colonies in a flower-poor period in agricultural landscapes. J Appl Ecol 56:1960–1969

Part IV
Farmland Bird Case Studies

Chapter 15
Historic Overview and Conservation Perspectives of the Czech Grey Partridge (*Perdix perdix*) Population

Miroslav Šálek and Václav Zámečník

Abstract In the Czech Republic, the grey partridge is a champion of diverse farmland. Data from the mid seventeenth century shows a close connection between its abundance and agricultural practices. The number of partridges peaked at more than 2 million birds between 1933 and 1937. Thereafter, from the mid 1950s, the consolidation of the land into large monocultures and the onset of chemicals in agriculture led to a population decline. These factors, alongside the high predation of nests and females, caused the Czech population to decline to only 8–16,000 pairs in 2014–2017. The partridge is highly sedentary and movements of >10 km are uncommon. Home range size is related to habitat quality and varies from 0.5 ha to 12 ha. At present, the species prefers unmanaged vegetated habitats that are sparsely distributed through the landscape. Local hunting associations tend to support the partridges by releasing hand-reared birds, but without any evidence of a long-term positive effect. Since 2004, there has been a special agri-environmental measure available to farmers called 'bio-belts' to increase food supply for farmland species from late summer until spring. We consider that the large-scale creation of suitable habitats through a subsidy system is the best and most promising means of providing long-term support for the grey partridge populations in the present Czech agricultural landscape.

Keywords Agricultural landscape structure · Agricultural intensification · Agri-environmental measures · Bio-belts · Czech Republic · Common Agricultural Policy (CAP) · Grey partridge Perdix perdix · Habitat requirements · Pesticide use

M. Šálek (✉)
Department of Ecology, Faculty of Environmental Sciences, Czech University of Life Sciences Prague, Praha Suchdol, Czechia
e-mail: salek@fzp.czu.cz

V. Zámečník
Czech Society for Ornithology, Prague, Czechia
e-mail: zamecnik@birdlife.cz

C. Hurford et al. (eds.), *The Changing Status of Arable Habitats in Europe*,
https://doi.org/10.1007/978-3-030-59875-4_15

1 History and Current Status

The grey partridge ('partridge') is naturally widespread in the Czech Republic (present over an area 80,000 km^2), preferring agricultural areas with lower altitudes up to 500 m (Sekera 1966). There was a documented increase in numbers in the country from the mid seventeenth century, in keeping with increases in deforestation and expanding agriculture. The numbers peaked during the 1930s when, in 1935, the total Czech population was estimated at 2.4 million pairs (Hudec and Šťastný 2005). Average annual hunt totals achieved an estimated 674,000 birds between 1924 and 1929, rising further to 2.2 million between 1933 and 1937 (Komárek 1945). After 1948, the annual hunt totals dropped significantly, e.g., to 586,000 between 1948 and 1953, as did the population density, with average spring numbers falling to ca.758,000 individuals between 1966 and 1968 (unpublished hunting statistics, Forest Management Institute, Brandýs nad Labem, Fig. 15.1). A clear decline continued during the next period in the Czech Republic: rough estimates refer to 9–18,000 pairs in 1985–1989, 11–22,000 pairs in 2001–2003 (Šťastný et al. 2006) and 8–16,000 pairs in 2014–2017 (K. Šťastný, unpubl. data). The species remained a

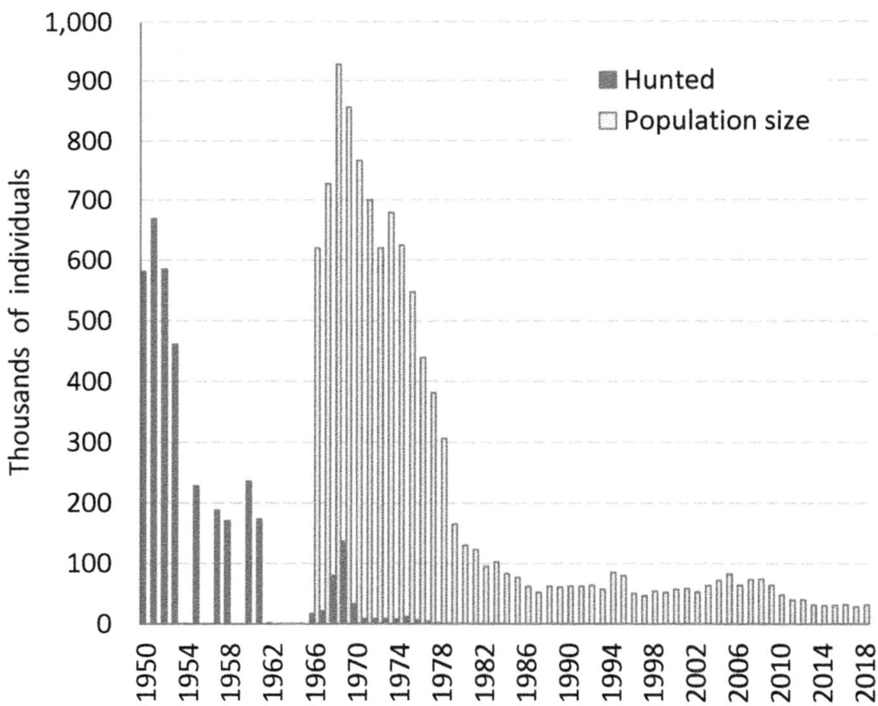

Fig. 15.1 Hunting statistics from the 1950s and population numbers of grey partridges from 1966 in the Czech Republic according to unpublished data by Hunting Evidence, The Forest Management Institute (ÚHÚL), Brandýs nad Labem

popular game bird until 1971, at which point hunting declined dramatically. In fact, it is believed that the Czech partridge population has declined by as much as 99% over the past 80 years, despite having been awarded the status of a specially protected species in 1992.

Long-term data on hunting and spring abundance in Třeboň Basin, South Bohemia (82 km^2 of agricultural land), refer to the influence of human activities on partridge numbers between 1666 and 1920 (Nováková and Hanzl 1966). At the beginning of this period, breeding density achieved no more than 2 pairs/km^2. After the first large-scale drainage and introduction of potato-growing around 1775, the density increased to ca.6 pairs/km^2. Subsequent meadow cultivation, the extension of fodder cropping, crop rotation with regular fallowing and the introduction of more intense natural fertilizing significantly contributed to further population growth, which increased to ca.9 pairs/km^2 after 1830.

A further wave of massive drainage in favour of improving field cultivation, together with the introduction of deeper ploughing and more widespread beet production, led to a further increase of up to ca.13 pairs/km^2 after 1850. Although harsh winters and several wars negatively influenced the dynamics of partridge populations in this period, the numbers always regenerated within a few years. Note that species management did not change during the three centuries; it included occasional release of reared birds (documented since 1668), winter capturing and chambering (recorded from 1684) as well as predator control (mandatory hunting of foxes, mustelids and large raptors since 1703). However, the authors emphasize that these measures did not exert a long-term positive effect on partridge numbers, being unable to support the populations without the provision of suitable habitat.

Spring densities of 13 or more pairs/km^2 were common in the period from 1850 until the 1960s. For example, the mean partridge density in an area of 382 km^2 at Beroun, central Bohemia, between 1958 and 1962 reached 11–20 pairs/km^2, and locally up to 40 pairs/km^2 (Nováková and Hanzl 1966). Such densities can still be found, albeit rarely (see below), indicating the potential for the population to grow if local habitat conditions are favourable.

2 Habitat

The partridge requires sufficient vegetation to provide shelter for adults, chicks and nests, as well as a good view of the surroundings to provide an early escape from predators. The seeds of arable weeds are a key food component for adult partridges, while chicks require an abundance and variety of insects (Janda 1966). During winter green leaves are important, mainly of winter cereals or other crops. Stubble fields are attractive as they supply both food and shelter (Buner et al. 2005). In the early days of agriculture, most open landscapes met these requirements, including cultivated fields, but modern farmland has mostly lost these attributes. At present, the conditions can be partially met in wastelands, such as unmanaged weed places on

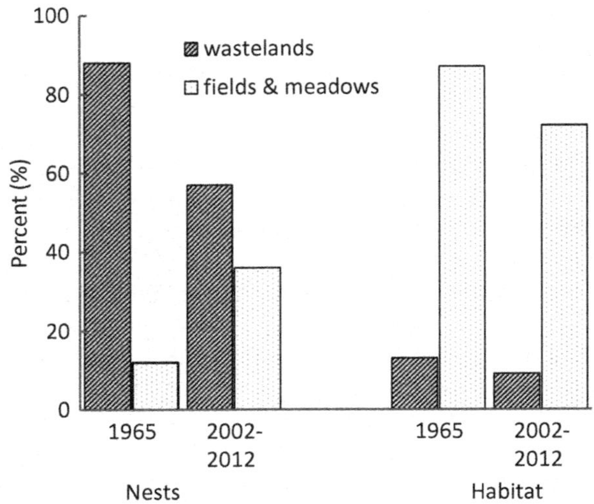

Fig. 15.2 Grey partridges have a strong preference for nesting in unmanaged habitats, e.g., wastelands, including grassy strips. There is a limited area of these habitats in the current agricultural landscape (1965: east of Prague, total area 1.6 km^2, 76 nests; Bouchner and Fišer 1966; 2002–2012: Písek, suburban Prague and Milešín, total area 31 km^2, 94 nests; own unpublished data)

urban fringes, landfill sites, grassy strips along fields and dirt roads, fallow fields, grassy hillsides and groves. (Figs. 15.2 and 15.3)

3 Causes of the Recent Decline

Evidence suggests that the two main causes of decline in the Czech grey partridge population are agricultural changes and predation. We discuss both of these in the sections below.

3.1 Agriculture

While the early days of small-scale farming triggered the golden era of the Czech partridge population, modern agriculture is responsible for its subsequent dramatic decline. First, the change of the political system, post 1948, resulted in extensive socialist planning that supported large-scale state farming, leading to the consolidation of most small and, until then, privately owned lands.

Consequently, small fields abundantly lined with grassy strips were transformed into extremely large and uniform field units. This led to a massive loss of two essential habitat features for grey partridges – food and shelter (Figs. 15.4 and 15.5). The country lost an estimated 120,000 km of dirt roads, 800,000 km of grassy boundaries, 30,000 km of linear mosaics with diverse plant, shrub and tree cover and 35,000 ha of groves (Klápště and Franková 2015). Písek, in South Bohemia, provides a good example of the dramatic changes that were so detrimental to the grey

Fig.15.3 (**a**) and (**b**) The modern Czech landscape offers few opportunities for the safe placement of nests during the main egg laying period in May. Most of the available locations are in linear landscape components, i.e., field hedges and narrow herbaceous or grassy strips with sparse shrubs along watercourses, paths, and between fields. Partridges prefer to nest here but they are easy prey for predators in these linear structures (**a**). Less productive habitat patches are essential for partridges (**b**). In suburban Prague, these areas are crucial for supporting a secure partridge population into the future. (Photos: Miroslav Šálek (**a**) and Václav Zámečník (**b**))

Fig. 15.4 (**a**) and (**b**) Nesting partridges are much safer and more difficult to find in wider field boundaries and edges with a high and dense cover of herbs than in narrow strips of sparse grass. (Photos: Miroslav Šálek)

partridge. Here, an area of 26 km^2 at an altitude of 300–400 m comprised 84% and 77% of arable land (including meadows and pastures) in 1952 and 1995, respectively. The length of edge lines in this area (except along forests) declined from a

b

Fig. 15.5 (**a**) and (**b**) Even in low vegetation it is easy to overlook an incomplete clutch as the female carefully covers the eggs with dry stalks (a). There might be up to 22 eggs laid at daily intervals in the clutch, followed by another 25 days of incubation, making the total exposure time of the nest and the female to the risk of predation almost 50 days. (Photos: Miroslav Šálek)

320 m to 277 m per hectare (i.e., a loss of about 13%) and the area of arable units increased 6.6 times from 0.8 to 5.3 hectare. Moreover, in 1952 almost 21% of arable land was fallow, in 1995 no fallow land was present (Horajsová 2001). A second factor in the decline was the introduction of toxic pesticides and other chemical products in agriculture, which is responsible for increased chick and adult mortality (Potts 1986), either through the accumulation of toxins in the tissues or through reductions in the availability of food (Kuijper et al. 2009). However, no relevant studies are available in the country.

3.2 *Predation*

Partridges are very susceptible to predation at all stages of their life cycle. Ground nests are exposed to many predator species over extremely long egg-laying periods (sometimes >20 days) as well as during incubation periods of 25 days or more (Jenkins 1961, own observations). Nest failures were always high despite systematic predator control in the past (Slaný 1947). Bouchner and Fišer (1966) who monitored 76 nests in lowland farmland east of Prague in 1965 found that 46% of nests were predated (25% by corvids, 21% by mammals) even though most nests were situated in preferred wasteland habitats. Nest predation is also responsible for the vast majority of nest losses in the modern agricultural landscape, where the nests are situated in various habitats (including wastelands as previously). Between 2002 and 2012, in three different areas (agricultural lowland at Písek, South Bohemia, suburban Prague and Milešín, Vysočina Hills), of 98 nests found, the majority were predated, with only 26% successfully hatching (own unpublished data).

Partridge adults and chicks rely on camouflage in vegetation, freezing on the ground or, in an emergency, on rapid escape flights. Although very inconspicuous, they became frequent prey for various mammalian and avian predators. We monitored 168 radio-tagged adult partridges in three areas (Písek, Prague, Milešín) between 2002 and 2010 and found that their survival was not generally longer than 3–4 months (median = 87 days, mean = 126 ± 8 se days; Rymešová et al. 2012). The results prove that predation is the major threat for partridges. Of 93 known causes of death, 83% were due to avian or mammalian predators in similar proportions. In 12% of cases, the death was due to other reasons, with no indications of predator attack, i.e. probably due to illness, parasites or poisoning. Two birds (2%) died after collisions with cars or power lines and two incubating females (2%) were killed by machinery while the meadow was being harvested. We found a significantly lower adult mortality in suburban Prague (42%), in an area with a higher proportion of wastelands compared with either Písek (80%) or Milešín (83%). Surprisingly, the partridges survived better during the covey period (autumn and winter) than during breeding time. Significantly more females than males died throughout the year (27% of 74 females vs 14% of 94 males), and particularly during the breeding season (26% of females while 8% of males), as the females are more vulnerable to predators during laying and incubation.

4 Population Movements and Dynamics

Despite considerable population dynamics, the partridge is highly sedentary and movements of >10 km are scarce (6% of birds; Cepák et al. 2008). Between 1960 and 1962, data on the movement of 191 from 473 ringed partridges documented that up to 70% of birds remained within 1 km of the release point, with a mean movement of ca.850 m (Fišer and Bouchner 1966). Our telemetry results from 2001 to 2004 agreed with these findings. Most of the pre-breeding movements executed by the pairs were associated with the occupation of breeding territories. In general, in Písek, where there is a lower availability of wastelands, the pairs dispersed further (on average >500 m) than their Prague counterparts (<500 m). We showed that the drop from winter to pre-breeding densities is mostly not due to predation but due, in part at least, to pre-breeding dispersal. In addition, we found that unpaired males (representing 8–14% of the February population; Šálek et al. 2002) are nomadic and travel the greatest distances (Šálek and Marhoul 2008). Specifically, four males that remained unpaired after the regular pairing-off period in February moved on average 1828 m (±sd 636 m). Except for one male that was predated by an eagle owl (*Bubo bubo*) during late April, the other males paired off and entered into nesting, indicating their role in successful gene flow between sub-populations. On the other hand, paired males in Prague were the most sedentary cohort, suggesting that most males prefer to stay near to their wintering sites, not abandoning them unless resources become limited.

Habitat supply and predation rates are essentially intertwined drivers of the partridge population dynamics, spatial activity and spring density. Specifically in today's landscape, limited habitat supply enhances spatial activity and predation risk, which reduces population density. We compared the seasonal dynamics of two different populations between 1997 and 1999 in a typical agricultural landscape (Písek, South Bohemia) and suburban rich-wasteland Prague (Šálek et al. 2004). The pre-breeding (April) densities strongly differed, with 2–5 pairs per km^2 in Písek and 24–33 pairs in Prague (Šálek and Marhoul 1999) but the pre-breeding densities in both areas were very similar to the respective 1965 levels (Hudec and Černý 1977). Seasonal fluctuations were strong in both areas, with two obvious phenomena: increasing covey size from summer to winter and a decreasing density of pairs from early spring to the beginning of breeding period at the end of April. The growth of covey size from summer to winter was due to the clustering of surviving members from different families, which arose from 5 to 10 individuals per covey in Písek and from 7–8 to 14 in Prague. As documented on radio-tagged birds, this increase is partly due to post-breeding movements, achieving on average 905 m (± sd 537 m) in Písek and 388 m (± sd 256 m) in Prague (Šálek and Marhoul 2008). The most likely cause of the birds actively searching for alternative resources over the landscape and for the concentration of the coveys at the most favourable sites is the dramatic reduction in food and shelter after crop harvesting in late summer and autumn (Putaala and Hissa 1998). The birds also look for suitable unrelated breeding partners in this period (Cramp and Simmons 1982). The winter (December, January) densities were 15–30 and 95–145 individuals per km^2 in Písek and Prague, respectively, i.e. the highest throughout the year. However, a sharp pre-spring drop followed afterwards, returning to the previous pre-breeding densities.

4.1 Home Range Size

We revealed a high variation in the size of breeding home ranges (primarily May to June) both between and within the localities (own unpublished data). The home ranges in Prague were smaller (min 1.1 ha, average ± sd 3.7 ± 3.4 ha, n = 27) than in Milešín (min 0.5 ha, mean 6.0 ± 5.2 ha, n = 63) and Písek (min. 3.3 ha, mean 7.6 ± 4.5 ha, n = 20). Most of the home ranges comprised winter wheat and oilseed, two dominant crops in the current Czech field landscape. Wastelands were present on average in 14% of home ranges, and other suitable habitats, e.g., temporarily fallow fields, artificial bio-strips, sparse tree patches and gardens, represented in 6%. Although we have not shown a direct positive relationship between home range size and amount of wasteland therein at the level of individual pairs, in terms of their availability within the study areas (Prague 28%, Milešín 8%, Písek 4%), there appears to be an obvious positive effect.

5 Conservation Considerations and Advice

In the Czech Republic, the partridge is a former champion of diverse farmland. It always has a high profile and still maintains a high attraction for hunters as potential game, for farmers for its economic value in pest control and for scientists as a model species for research and popularisation of farmland biodiversity. Before 1949, there were more than 500 registered articles on this species in the former Czechoslovakia, mostly in hunter's popular periodicals (Hudec 1966). The species was listed as specially protected and vulnerable in the Czech Republic in 1992 according to Decree 395/1992 Coll. and classified as near threatened in the Czech Red list (Šťastný et al. 2017).

Local hunting associations tend to support the partridge populations by releasing hand-reared birds. It is a relatively complicated process, as it requires the approval of the responsible nature protection body. More than 57,000 birds were released in the period from 2000 until 2018, at an average of 3030 birds a year (T. Kunca, Ministry of Agriculture of the Czech Republic, pers. comm.). In some cases, these releases are not allowed due to the inappropriate origin of birds or because no habitat improvements are made but, as no general guidance exists, each nature protection body makes its own criteria. The efforts of hunters in predator control vary widely among hunting associations and hunted species. For example, the red fox (*Vulpes vulpes*) with sustained growth in the 1990s still maintains a high population density despite e.g., 80,000 foxes being killed in the country in 1998 (Řehák 2000). Similarly, the stone marten (*Martes foina*) has an increasing population trend despite 1500 bags in 1998 in the country (Koubek et al. 2000). Illegal hunting takes place in avian predators (corvids, raptors, owls), but no accurate data exist.

Current, non-systematic, natural history observations of partridges are stored in national databases (Avif, NDOP), providing the potential for future analyses and conservation. However, there are no systematic long-term research projects on this species. Some local protection actions happen on the ground, but without precise description or feedback. Most conservationists and ornithologists consider that the best way to protect partridges is to raise the carrying capacity of the environment. They also believe that, as the partridge is an r-strategist, there is no point in preserving special areas for the long-term maintenance of the population. On the contrary, it is advisable to provide temporary early succession sites connected by stepping stones to enable pairs to disperse through the landscape. This can be ensured by providing e.g., areas sown with special seed mixtures without using chemical products, that offer food and shelter in addition to the wastelands set in and alongside the fields. Computer modelling indicates that optimal habitat requirements are a combination of 5% of arable area as insect-rich brood-rearing habitat and over 6 km per km^2 of linear nesting cover (Winspear and Davies 2005) in the form of wild-flower and hedge strips optimally at least 6 m wide (Buner et al. 2005). Alternatively, in some situations, instead of using linear elements, it might be better to create blocks of suitable habitat of a minimal size of 0.3 ha and minimal width of 20 m (Sotherton 1998).

Unsprayed strips along field edges may increase both food availability and chick survival (Rands 1985) and serve as stepping stones. In principle, increasing the carrying capacity of the environment would reduce dispersal, predation and home range size while enhancing survival and population density. At the same time, the species can efficiently serve as an umbrella for other farmland species with similar habitat requirements (Potts 2010).

6 Conservation Perspectives

So far, the most promising opportunity for partridge protection arose from the Common Agriculture Policy (CAP) after joining EU, especially the second CAP pillar called Rural Development Programme (RDP). The agri-environmental (AE) measures are an important part of the RDP, as they promote the ways of using agricultural land that is consistent with the protection and improvement of the environment. The partridge has been considered as a target species already during the preparation of new AE measures in Czechia. Specifically, the experiences of game hunters from Milešín, Vysočina Hills (Mühlhansl et al. 2002, Fig. 15.6) were interpreted to inform a specific measure called 'bio-belts', which was officially adopted in 2004. The objective of this measure is to increase the food supply for partridges and other species associated with field habitats. Conditions of the measure are to:

1. Establish 6–12 m (since 2015 up to 24 m) wide strips that are located at the edge of arable land or within fields (inside fields in ploughing direction);
2. Sow the belts from 1 April to 15 June with seed mixtures of proven quality, consisting of dominant spring cereals, and other compulsory crops as buckwheat (*Fagopyrum esculentum*), millet (*Panicum* sp.) and cabbage (*Brassica oleracea*). In addition, at least two voluntary crops must be sown, selected from sunflower (*Helianthus annuus*), canary grass (*Phalaris canariensis*), bunch or bundle (*Phacelia tanacetifolia*), flax (*Linum usitatissimum*), white lupine (*Lupinus albus*) or legumes (*Pisum sativum* and/or *Vicia faba*);

Fig. 15.6 (**a**) and (**b**) 'Bio-belts' with a mixture of herbs provide food throughout the summer to early spring, not only for partridges but also for other game species and pollinators (**a**). Since 2015, there has been an extension of bio-belts to aid pollinator populations, these have also improved conditions for other animals during spring and summer (**b**). (Photos: Miroslav Šálek (**a**) and Anna Šrámková (**b**))

3. Leave the belts without any management until the end of March of the following year;
4. Prohibit the application of pesticides, entry for agricultural machinery and do not use as headlands; and
5. Include the option to change the position of the bio-belts during the five-year period depending on the crop rotation.

The experience of local hunters confirms that those bio-belts placed in suitable habitat (e.g. within large field units and further from the forest) are favoured by the partridges and can support their survival through late summer until winter (M. Bažant, pers. comm.). However, the total area of bio-belts in the country was only 2000 ha in 2016 (Ministry of Agriculture, pers. comm. 2017). Considering that a bio-belt has a width of 12 m and that the total length of bio-belts was 1600 km, compared to the 800,000 km of grassy boundaries that we have lost (Klápště and Franková 2015), the area of bio-belts now represents less than 0.1% of the total arable land area. Unfortunately, isolated habitat patches embedded in a poor surrounding landscape can attract the attention of many animals together with their predators, so that these sites can paradoxically act as ecological traps for partridges due to excessive predation pressure (Bro et al. 2004; Rantanen et al. 2010). Within the programming period of CAP after 2020, a new measure is being prepared aimed specifically at partridges. It suggests combining two parallel strips at least 6 m wide, the clover-grass and the bio-strip, supplemented with sorghum to improve the potential for hiding and feeding.

After the collapse of the communist system in 1989, land privatisation and restitution enabled landowners to access their property and re-establish the landscape structure including the construction of new dirt roads, wetlands, erosion-controlling grassy strips, windbreaks and infiltration polders. In late 2019, this land consolidation had been achieved on about one third of agriculture land, with the establishment of 3400 km of dirt (in part paved) roads and 790 ha of erosion control measures, such as grassy strips and new tree plantations (L. Růžková, State Land Office of the Czech Republic, pers. comm.). However, there has been no evaluation of the possible benefits of this habitat diversification for the partridges.

Organic farming is known to provide great potential for birds, including partridges (Štefanová and Šálek 2014). There is a fundamental difference in the approaches to farming applied by the family-owned farms and organic farms on one side and by the cooperatives and large enterprises farming on rented land on the other (Sklenicka et al. 2015). The owners of smaller farms and organic farms tend to re-establish a functional landscape structure. Several of them use agricultural subsidies and other funds to recreate landscape features that keep the ecosystems more stable. An example of good practice is the organic farm managed by Petr Marada in Šardice, South Moravia (48°57′52.1″N 17°01′37.8″E). Within his 73 ha farm, he established 29 ha of orchards, 6.5 ha of a grassy valley with bushes and trees, 3.5 ha of permanent grasslands and one small wetland. Other grassland he cultivates on 13 ha out of 34 ha of arable land and 3.2 ha of arable land he left as bio-belts. The partridge numbers within the Šardice cadastral territory increased from

100 individuals (counted by hunters before spring dispersal) in 2007 to 150 - individuals in 2018 (P. Marada, pers. comm.).

Higher densities of partridges have also been recorded locally in specific open areas such as airports, industrial zones or photovoltaic power plants maintained to sustain dominant herbal habitats without regular agricultural practices or the use of chemicals. For example, breeding densities inside the industrial zone of the car producer TPCA Ovčáry near to Kolín (central Bohemia 50°04′11.6″N 15°13′23.4″E) achieved 8 pairs/10 ha in 2017 (own unpublished data). These areas may have great potential as local refuges, so their importance needs to be popularised through wide-ranging conservation organizations such as the Czech Society for Ornithology or the Czech Union for Nature Conservation.

Other activities such as releasing of reared partridges and predator control need very careful consideration. Several methods have been applied in these projects, such as:

- Large groups of adults being released immediately into the field;
- Small groups being released from stationary aviaries after several days of stay; or by
- Families, raised in aviaries, being installed in the field and released after chick fledging.

However, we have no feedback on the efficiency of these approaches in the Czech partridge populations. There is still no clear evidence to date that releasing birds can establish vital populations in the country. According to our findings carried out in Vysočina Highlands, hand-reared birds have smaller home ranges than the wild ones, wild females showed no interest in pairing with hand-reared males and wild males preferred wild females instead of hand-reared females (Rymešová et al. 2013). Releasing the more conspicuous hand-reared birds, which are easier to catch, can even have a negative impact on wild populations by attracting the attention of predators. Potts (2010) documented that the partridge densities on his Sussex farm increased after changing the management approach, i.e. from the releasing of farmed birds to making habitat improvements. In any case, habitat evaluation and, if necessary, habitat improvement, is required before sanctioning release projects. Although the populations of some predator species, such as foxes or stone martens, have demonstrably increased in recent decades, poorly managed predator control is controversial. At a local level, hunters often see it as the only way to protect partridge populations, but it is an ethically controversial tool with uncertain effects and its application should be considered carefully in the context of other conservation measures.

7 Summary

The grey partridge is a naturally widespread and historically attractive game species in the Czech landscape, whose rapid decline during the last half century is due to the increased intensity of agriculture. Its demands for vegetation shelter, a good view of surroundings, and the availability of weed seeds and insects make this species a suitable indicator of diverse agricultural habitats and an umbrella species for other animals with similar habitat requirements.

The grey partridge is sedentary and relies strictly on unstable early succession sites dominated by mosaics of wild herbs, an extremely scarce commodity in the present landscape. The lack of these habitats results in longer dispersal, greater spatial activity within larger home ranges, higher predation risk on adults, nests and chicks, which all reduces breeding density and total population numbers.

Local Czech partridge populations show strong seasonal dynamics, which indicates possible growth in abundance under favourable conditions. Therefore, we need a national strategy for systemic species support to create a widespread supply of suitable habitat and increase the carrying capacity of the environment, which can then facilitate safer dispersion, higher survival rates and the successful fledging of chicks.

After joining EU, the carrying capacity of the environment with suitable habitat in the country could be increased through the Common Agriculture Policy (CAP), which opened up opportunities for supporting the appropriate agri-environmental measures. To be effective, any measure needs to achieve a compromising balance between the species requirements, the farmers' willingness/resistance to sustaining unwanted herbs (weeds) and a sufficiently motivating amount of financial subsidy. A popular measure with relatively favourable results currently applied by the Czech Ministry of Agriculture is the establishment and maintenance of bio-belts with seed mixtures of proven quality.

Other grey partridge populations are supported through organic farming and in specific areas (such as airports, industrial zones, photovoltaic power plants and other non-agricultural land maintained to sustain dominant vegetation communities) without regular agricultural practice or the use of chemicals. On the other hand, locally applied non-systemic releasing of partridges from farms accompanied with predator control has yielded inconsistent results, making the recommendation of these measures questionable.

In summary, we consider the large-scale creation of suitable habitats through a subsidy system the best and most promising form of long-term support for grey partridge populations in the present Czech agricultural landscape. However, we have not resolved issues related to monitoring the operative optimisation and effectiveness of these measures yet and, to date, farmers have shown little interest in the recommended and available measures. So now the state institutions and non-governmental organizations need a high profile publicity campaign to increase awareness of and promote the appropriate agri-environmental measures.

References

Bouchner M, Fišer Z (1966) Hnízdění koroptví a ztráty na snůškách [Nesting of partridges and clutch losses]. Symposium o koroptvi. Proceedings of the symposium on partridge, Prague, October 1965. VÚLHM & ČMS, pp. 72–81. [in Czech with summary in English]

Bro E, Mayot P, Corda E, Reitz F (2004) Impact of habitat management on grey partridge populations: assessing wildlife cover using a multisite BACI experiment. J Appl Ecol 41:846–857

Buner F, Jenny M, Zbinden N, Naef-Daenzer B (2005) Ecologically enhanced areas – a key habitat structure for re-introduced grey partridges *Perdix perdix*. Biol Conserv 124:373–381

Cepák J, Formánek J, Hořák D, Jelínek M, Klvaňa P, Schröpfer L, Škopek J, Zárybnický J (2008) Atlas migrace ptáků České a Slovenské republiky [Czech and Slovak bird migration atlas]. Aventinum, Praha. [in Czech with summary in English]

Cramp S, Simmons KEL (eds) (1982) The birds of the Western Palearctic, vol 2. Oxford University Press, Oxford

Fišer Z, Bouchner M (1966) Rozptyl koroptví po vypuštění do honiteb [Dispersion of partridges after release into hunting grounds]. Symposium o koroptvi. Proceedings of the symposium on partridge, Prague, October 1965. VÚLHM & ČMS, pp. 82–87. [in Czech with summary in English]

Horajsová D (2001) Analýza dlouhodobých změn ve struktuře zemědělské krajiny s využitím GIS [Analysis of long-term changes in the structure of agricultural landscape using GIS]. Diploma thesis, Faculty of Forestry, Czech University of Life Sciences Prague, 52 pp.

Hudec K (1966) Několik poznámek k výzkumu koroptve v ČSSR [Some notes on the research of partridge in Czechoslovakia]. Symposium o koroptvi. Proceedings of the symposium on partridge, Prague, October 1965. VÚLHM & ČMS, pp. 40–44. [in Czech with summary in English]

Hudec K, Černý W (eds) (1977) Fauna ČSSR, Ptáci 2 [Fauna of Czechoslovakia, Birds 2]. Academia, Praha

Hudec K, Šťastný K (eds) (2005) Fauna ČR, Ptáci – Aves II/1. [Fauna of the Czech Republic, Birds II/1]. Academia, Praha

Janda J (1966) Přirozená potrava koroptve polní *Perdix perdix* (Linné) v přírodě [Natural foods of Grey partridge Perdix perdix Linné in nature]. Symposium o koroptvi. Proceedings of the symposium on partridge, Prague, October 1965. VÚLHM & ČMS, pp. 93–99. [in Czech with summary in English]

Jenkins D (1961) Population control in protected partridges (*Perdix perdix*). J Anim Ecol 30:235–258

Klápště J, Franková L (2015) Sucho [Drought]. Ochrana přírody 6:23–27. [in Czech]

Komárek J (1945) Myslivost v českých zemích [Hunting in the Czech lands]. Čin, Praha. [in Czech]

Koubek P, Červený J, Bufka L (2000) Současné poznatky o rozšíření, populační dynamice a potravní ekologii některých lasicovitých v České republice [Current knowledge on distribution, population dynamics and feeding ecology of some mustelids in the Czech Republic]. Proceedings of the Conference 'Predátoři v myslivosti 2000', Hranice 1-2 September 2000, pp. 40–51. [in Czech with abstract in English]

Kuijper DPJ, Oosterveld E, Wymenga E (2009) Decline and potential recovery of the European Grey Partridge (*Perdix perdix*) population – a review. Eur J Wildl Res 55:455–463

Mühlhansl I, Havlát F, Holý J, Jobánek P, Nagy O, Odehnal J, Poledna M (2002) Úpravy biotopu pro koroptev polní (*Perdix perdix* L.) v rámci podpory zachování genofondu vzácného a ohroženého druhu zvěře v k. ú. Milešín, Heřmanov, Nová Ves a Vlkov [Habitat modifications for partridge (*Perdix perdix* L.) in support of conservation of gene pool of rare and endangered game species in cadastral areas Milešín, Heřmanov, Nová Ves and Vlkov]. Unpublished report, Hunting Club Borovina-Milešín

Nováková E, Hanzl R (1966) Vliv anthropogenních změn prostředí na početní stavy koroptví v třeboňské pánvi a na Berounsku [Influence of anthropogenic changes in the environment on the numbers of partridges in the Třeboň basin and in the Beroun region]. Symposium o koroptvi. Proceedings of the symposium on partridge, Prague, October 1965. VÚLHM & ČMS, pp. 45–71. [in Czech with summary in French]

Potts GR (1986) The partridge: pesticides, predation and conservation. Collins, London

Potts GR (2010) Long term partridge studies in Sussex. BOU proceedings: lowland farmland birds 3: delivering solutions in an uncertain world

Putaala A, Hissa R (1998) Breeding dispersal and demography of wild and hand-reared grey partridges *Perdix perdix* in Finland. Wildl Biol 4:137–145

Rands MRW (1985) Pesticide use on cereals and the survival of grey partridge chicks: a field experiment. J Appl Ecol 22:49–54

Rantanen E, Buner F, Riordan P, Sotherton N, Macdonald D (2010) Habitat preferences and survival in wildlife reintroductions: an ecological trap in reintroduced grey partridges. J Appl Ecol 47:1357–1364

Řehák L (2000) Analýza vývoje početních stavů lišky obecné v ČR a několik slov k jejímu současnému postavení v myslivosti [Analysis of the development in population numbers of the red fox in the Czech Republic and a few words on its current position in hunting]. Proceedings of the Conference 'Predátoři v myslivosti 2000', Hranice 1-2 September 2000, pp. 69–87. [in Czech]

Rymešová D, Šmilauer P, Šálek M (2012) Sex- and age-biased mortality in wild Grey Partridge (*Perdix perdix*) populations. Ibis 154:815–824

Rymešová D, Tomášek O, Šálek M (2013) Differences in mortality rates, dispersal distances and breeding success of commercially reared and wild grey partridges in the Czech agricultural landscape. Eur J Wildl Res 59:147–158

Šálek M & Marhoul P (1999) Sezónní dynamika a příčiny ztrát koroptve polní (*Perdix perdix*): výsledky sčítání a telemetrického sledování v letech 1997–1999 [Seasonal dynamics and causes of loss in the Grey Partridge (*Perdix perdix*): results of counts and telemetry observations in 1997–1999]. Sylvia 35: 55–67. [in Czech with summary in English]

Šálek M, Marhoul P (2008) Spatial movements of grey partridge *Perdix perdix*: male-biased spring dispersal and effect of habitat quality. J Ornithol 149:329–335

Šálek M, Marhoul P, Pintíř J (2002) Rychlost párování, poměr pohlaví a věková struktura pražské populace koroptve polní (*Perdix perdix*) během února 2000–2002 [Pairing rate, sex ratio and age structure of a Grey Partridge (*Perdix perdix*) population in Prague, February 2000–2002]. Sylvia 38: 29–40. [in Czech with summary in English]

Šálek M, Marhoul P, Pintíř J, Kopecký T, Slabý L (2004) Importance of unmanaged wasteland patches for the grey partridge *Perdix perdix* in suburban habitats. Acta Oecol 25:23–33

Sekera J (1966) Problém koroptví v Českolsovensku [Problems of partridges in Czechoslovakia]. Symposium o koroptvi. Proceedings of the symposium on partridge, Prague, October 1965. VÚLHM & ČMS, pp. 5–17. [in Czech with summary in English]

Sklenicka P, Janeckova Molnarova K, Salek M, Simova P, Vlasak J, Sekac P, Janovska V (2015) Owner or tenant: who adopts better soil conservation practices? Land Use Policy 47:253–261

Slaný J (1947) Koroptev a orebice jako lovní ptáci [Partridges like game birds]. Zář, Brno

Sotherton NW (1998) Land use changes and the decline of farmland wildlife: an appraisal of the set-aside approach. Biol Conserv 83:259–268

Šťastný K, Bejček V, Hudec K (2006) Atlas hnízdního rozšíření ptáků v České republice [Atlas of breeding birds distribution in the Czech Republic 2001–2003]. Aventinum, Praha. [in Czech with summary in English]

Šťastný K, Bejček V, Němec M (2017) Červený seznam ptáků České republiky [The red list of birds of the Czech Republic]. Příroda, Praha 34: 107–154

Štefanová M, Šálek M (2014) Effects of integrated farming on herbal and bird species diversity in Czech agricultural landscapes. Pol J Ecol 62:161–176

Winspear R, Davies G (2005) A management guide to birds of lowland farmland. The RSPB, Sandy

Chapter 16
Perspectives on the Declining Ortolan Bunting (*Emberiza hortulana*) Population in Northern Sweden

Christer Olsson

Abstract The ortolan bunting (*Emberiza hortulana*) population in northern Sweden, which represents about 50% of the Swedish population, historically bred in cultivated farmland habitats. However, in the 1970s, it became evident that the ortolan bunting population had begun to move away from the farming landscape into forest clear-cut areas, which currently host the strongholds of the species in Sweden. The key factor that these two habitats share is the availability of exposed soil for feeding. Recent studies on the declining ortolan bunting population in the mid-Sweden farming landscape indicate that the six most important feeding microhabitats are unsown rows in the tracks of tractor wheels; sparsely sown or small bare patches in the crops; supplemental oat feeders; the canopies of trees and bushes; uncultivated field borders; and ditches containing eroded soil. In the forest clear-cuts, exposed soil occurs where vehicles have disturbed the surface layer during the logging process. These areas of bare soil persist for about 5–10 years before being fully vegetated by trees and bushes. The logging industry is now cooperating in a project with bird ecologists to create suitable habitats for longer periods and over larger areas. Evidence from their West African wintering area suggests that the greatest losses in the population are due to habitat changes in Sweden and not due to factors on the wintering grounds.

Keywords Changing agricultural practices · Cultivated habitat degradation · Methyl mercury · Ortolan bunting *Emberiza hotulana* · Forest clear-cuts · Pesticides · Population decline · Yellowhammer *Emberiza citrinella*

C. Olsson (✉)
Fågelperspektiv, Umeå, Sweden

C. Hurford et al. (eds.), *The Changing Status of Arable Habitats in Europe*,
https://doi.org/10.1007/978-3-030-59875-4_16

1 Historical Background

The ortolan bunting (*Emberiza hortulana*) has been documented as a regular breeding bird in the province of Västerbotten in northern Sweden since at least the eighteenth century. There are no sources available for finding well documented records of this species before then. This implies that, even at that time, the ortolan bunting was present in the community of Umeå, where I have studied this species over the last 50 years.

In the late 1940s, there was a major survey of the ortolan bunting distribution in Sweden, when it was said to be quite common throughout the region of Västerbotten. Thereafter, during the 1950s and early 1960s, there was a steep decline in the ortolan bunting population across the whole of Sweden, mainly due to the widespread use of methyl mercury, which was traditionally used in agricultural chemicals as a fungicide or pesticide. This metal was a lethal physical threat to ortolan buntings (Svensson et al. 1999).

In the early 1960s many studies focused on the brutal effects of methyl mercury on the farmland avifauna. As a result of this research, the substance was prohibited from use in the agricultural landscape of the country in 1966.

The ortolan bunting populations responded quickly to this, especially in northern Sweden (Fig. 16.1), where they became increasingly abundant until the mid 1970s. At this point, there was a sudden change in the structure of the farming landscape which led to increasing areas of monocultures, where the same types of crop were intensively cultivated over huge areas, with few feeding areas available for the ortolans, which were accustomed feeding on open patches of soil along the edges of these fields.

Fig. 16.1 A male from one of the last seven pairs of ortolan bunting in its former agricultural stronghold in northern Sweden: Röbäcksslätten, in the vicinity of Umeå in June 2019. (Photo by Christer Olsson)

However, despite the changing agricultural practices, the ortolan population remained in a relatively stable state for about a decade in the rural areas of northern Sweden. This was primarily because the species underwent a dramatic change in habitat preferences during this period, moving from their former crop areas to clear-cut areas made by the forest industry; a habitat that was increasing rapidly at that time (Olsson and Wiklund 1999).

There were no further reports of changes in the ortolan bunting population until the end of the 1980s when we heard the first warnings of a new decline. Initially, a few monospecific studies were carried out on discrete local populations. Among the first to notice the decrease were those counting the numbers of passerines on autumn migration. They detected an obvious decrease of this species at migration studying stations throughout the country (Kjellén 2002). Subsequent investigations during the breeding season revealed a dramatic decline in the Swedish ortolan bunting population; a decline that is ongoing (Sondell 2013).

By the turn of the millennium the ortolan bunting was a scarce breeding species in Sweden. For example, in the entire farming landscape area of Röbäcksslätten (in the vicinity of Umeå), a survey in June 1972 recorded 125 singing males. In this same area in June 2019 only seven singing males remained (https://www.artportalen.se/ViewSighting/ViewSightingList). The last reliable estimation of the entire Swedish population is from 2012, when a survey estimated the total Swedish population to comprise about 3700 pairs, of which c.1800 were found in the county of Västerbotten.

2 Ecology

The ortolan bunting has a long history as a typical inhabitant of a farming landscape that nowadays has almost disappeared. The birds cannot cope in today's intensively farmed areas, without any ditches, rows of trees for the males to sing from and without possibly the most important feature, i.e. open soils without crops or vegetation. Neither can the ortolan bunting find suitable habitats in areas abandoned from agricultural use for more than ten years. In the region of Västerbotten, most former arable sites are now either highly industrialised or have been out of active cultivation for decades. As such, it is reasonable to say that the species has now become obsolete in its original and favoured habitat.

Ortolan buntings continue to decline rapidly in Sweden, mainly due to changes in agricultural practices. Landscape simplification, due to the removal of semi-natural elements, leads to a lack of breeding habitats. Furthermore, due to intensive artificial fertiliser applications and advanced agricultural equipment and technology, crops grow taller and denser than 50 years ago, so this species has faced many challenges and threats (Sondell et al. 2018).

3 Changing Habitat Preferences

The ortolan bunting changed its habitat preferences, from arable regions to forest clear-cuts, almost completely within a timeframe of 20 years from the mid 1970s. This was a response to the changes in cultivation practices and large scale turmoil in the forest industry, leading to the creation of huge areas of virgin open landscape.

In the region of Västerbotten, and elsewhere in Sweden, the 1970s was a period when the old form of commonly appearing open ditches were transformed to underground culverts, leading the water away from the fields. With the open ditches gone, the rows of bushes or trees for singing males to sit in disappeared. The era of huge areas of monocultures in the farming landscape was the basic model for the decade and for those to come.

In the beginning, there were no alarming signals to suggest that the population of the ortolan bunting as a whole was in a phase of decline in Västerbotten. We estimated that the number of buntings disappearing from the crop areas were equal to those settling in the forest clear-cut areas. This kind of habitat is definitely of interest to ortolan buntings during the primary years, when the vegetation is low and there are many open areas with lots of exposed soil and sand created through disturbance by tractors and other huge machines driving over sensitive ground.

Typically, clear-cut areas are of interest to ortolan buntings for a period of five to ten years (Fig. 16.2), before they are overgrown by dense vegetation (Fig. 16.3) of young birches (*Betula* sp.), sallow (*Salix* sp.), aspen (*Populus tremula*), rowan trees (*Sorbus aucuparia*), planted pine (*Pinus* sp.) or spruce (*Picea* sp.). As this happens, the buntings disappear and move to adjacent areas with their preferred stage of habitat development. In the absence of these early successional areas, the species will undergo further declines.

3.1 Current Agricultural Habitats

Habitat changes, particularly in the agricultural areas (Wretenberg et al. 2007), environmental toxins and a generally high mortality during migration and in wintering areas are possible explanations for the decline of the Scandinavian ortolan bunting populations (Menz and Arlettaz 2012; Naturvårdsverket 2017). In Finland, a population crash in the early 1990s was specifically “associated with the amount of small-scale environmental heterogeneity and with agricultural practices” (Vepsäläinen et al. 2005). Ortolans are today present in the Nordic countries in a range of different breeding environments: farmland, forest clear-cuts in northern areas, wild-fire zones, peat bogs, bioenergy forests, and gravel extraction quarries, but also in industrial areas including runways and railroads (Ottvall et al. 2008). This diverse habitat use has confounded attempts to characterise the current preferred/optimal habitats for the species.

Fig. 16.2 A forest clear-cut area with lots of exposed soil and sand. Since the decline in the cultivated habitats, this is typical breeding habitat for ortolan buntings in northern Sweden. (Photo by Christer Olsson)

Breeding ortolans forage mainly on the ground and require bare soil for searching for prey (Menz et al. 2009). Ortolans also eat green larvae found in tree crowns (Stolt 1974), as well as seeds. Menz et al. (2009) determined that without small patches of open soil, ortolans find food (usually dark invertebrates) difficult to locate and extract. This suggests that a key factor in the decline of this species may be that the foraging prerequisites are not good enough to support a reproductive output capable of maintaining the population size (Sondell et al. 2018).

In a survey carried out in the central Swedish region of Närke (Sondell et al. 2018), only 18% of all observed foraging was outside the crop fields, and then mainly in natural grass, at oat feeders, and in the tree canopies. There was only one observed instance of successful hunting of insects in flight. The majority of foraging happened in crop fields; in fact a single field with two unsown rows was the favoured location for 41% of the foraging trips.

The study published by Sondell, J. et al. (2018) also found that the foraging habitats or specific food sources used by ortolans in mid-Sweden fall into six main categories:

1. Unsown rows in tractor wheel tracks;
2. Sparsely sown or small bare patches in the crops;
3. Supplemental oat feeders;
4. Tree or bush canopies;

Fig. 16.3 One pair of ortolan buntings still bred in this forest clear-cut in June 2019, but the future is bleak. Birch (*Betula*) and spruce trees (*Picea abies*) are recolonising strongly and the amount of exposed soil is at a very low level. (Photo by Christer Olsson)

5. Uncultivated field borders; and
6. Other local sources, such as ditches containing eroded soil (blown there from the field surface by strong winds in dry weather), natural grass close to the nests and flying insects.

Few, if any, of these are available in highly cultivated areas nowadays.

3.2 The Forest Clear-Cut Areas

In northern Sweden, the ortolan bunting now breeds mainly in forest clear-cut areas. In the county of Västerbotten, the birds spend most of their time in this habitat. In a study by Sondell (2013), ortolans were present in 48 of 123 forest clear-cut study areas. Investigations of these forest clear-cuts found that the birds prefer very open areas with exposed soil (Fig. 16.2) and that they avoid areas with lots of moist ground. Another important requirement of these clear-cuts is that they provide scattered, uncut trees and that exposed soil exceeds 10% of the total area: the area of open soil is typically created during forestry operations. Furthermore, almost all of the forest clear-cuts holding nesting pairs exceeded 30 acres in area (Naturvårdsverket 2017).

4 The Migration and the Wintering Grounds

An important requirement of the research into the long-term decline of the Swedish ortolan bunting population is to detect the migration routes, stopover points and wintering grounds of the birds. In recent years, there has been a study aiming to shed light on these issues; this involved equipping ortolans with light geolocators (Selstam et al. 2015). In all, 45 ortolan buntings were fitted with geolocators, 39 of them from northern Sweden, mainly from the Umeå area. This study found that the birds started their autumn migration during August (median 24 August) and that most birds had stopovers for about a week north of the 50th parallel. All of the birds made stopovers on the Iberian Peninsula or in Morocco, lasting from 6 to 32 days. The analysis also indicated that several of the birds passed through the French county of Les Landes, south of Bordeaux. On arriving to the south of the Sahara, all of the birds stayed at one or two locations for up to 35 days before arriving at the main wintering grounds, where all six birds with complete data (five of them from the Umeå region) stayed for approximately 5½ months.

The autumn migration lasted on average for 56 days from the beginning of the migration in Sweden until arriving at the first wintering/stopover location south of Sahara. After arriving in west Africa, one bird wintered at the same location until spring migration. The other birds visited one or two locations for up to 61 days before arriving at their main wintering locations. The birds stayed at the main wintering locations in West Africa (south-western Mali and northern Guinea) for 150–182 days. The return migration in spring started in late March (22nd) or the first half of April (median 7 April). All birds arrived at stopover points in Morocco or Spain a few days later, and remained for 5–18 days. It is uncertain whether one of the birds passed Sahara before its first stopover or not. Several other birds stayed at two locations in Morocco or Spain for up to 25 days. Two birds also stopped in central Europe (6 days) and Finland (14 days). On average, the spring migration took 40 days. The seven males all wintered in West Africa within a limited area of some 40,000–50,000 km^2.

The wintering area is located at the northern slopes of the Fouta Djallon mountain range. According to aerial photos, the ground is generally sparse savannah forest interrupted by brook ravines and cultivated areas, with part of the ground burnt periodically. The mountains are also the source of three large rivers: Senegal, Gambia and Niger and precipitation is relatively high. The two largest flocks of ortolan buntings reported from West Africa were at the airport of Labé, central Guinea, where the birds were on partly burnt grassland close to the air strip. All of the tagged ortolan males winter in areas situated between 50 and 350 km from Labé.

The large number of ortolan buntings wintering in West Africa suggests that the declines in the Swedish population are mostly due to habitat changes in Sweden and not on factors occurring during migration or on the wintering grounds (Selstam et al. 2015).

4.1 The Influence of Trapping on Ortolan Bunting Populations

Sondell et al. (2011) identified the three most likely causes of the decline in the Swedish ortolan bunting populations as: (1) changes in the agricultural breeding habitat influencing the production of offspring; (2) hunting or trapping along the migration route, especially in France; and (3) factors affecting the survival in the wintering area and migration routes for ortolan buntings. In a small region in SW France between the Bay of Biscay, the Pyrenees and the pine forests south of Bordeaux situated on an ortolan bunting migration route, there is a restricted agricultural area that still traps the species. The aim is to trap them and then fatten them for human consumption. Before the EU banned legal hunting in 1999 some 50,000 birds were caught annually (Selstam et al. 2015). Today, no catching figures are available, but it is probable that the catch is still substantial (bulletins from Ligue pour la protection des oiseaux, LPO).

5 Current Conservation Measures for Ortolan Buntings in Sweden

There are two conservation programmes designed specifically to preserve the declining ortolan bunting breeding population Sweden, these are described briefly below.

5.1 Measures to Improve Agricultural Habitats for Ortolan Buntings

The classic environment for the ortolan bunting in Sweden, fields cultivated with barley and oat, needs small areas of exposed soil to provide optimum habitat. Initially, to achieve this, it is essential to inform the farmers in the breeding areas that they host a threatened species on their land. The best chances of positive results are where the farmers are willing to cooperate with ecologists, because leaving parts of a field proposed for growing crops with patches of exposed soil will incur financial losses for the farmer.

The regional government in the county of Örebro in Central Sweden has contracted landowners of cultivated fields hosting ortolan buntings and provided financial aid to sow only the half of the normal quantity of crops. Thus the farmers are leaving strips (approx. 50 cm in width) of exposed soil both in the middle of the fields and at their borders. Here the birds can find optimal feeding habitat and protection. These strips are also commonly used by skylarks (*Alauda arvensis*) and yellowhammers (*Emberiza citrinella*), other species declining as a consequence of changing agricultural practices. It is critical that these strips of exposed soil are not too narrow, as the birds need to land and take off without any obstacles from the surrounding vegetation (the wing-length of the ortolan bunting is approx. 26 cm). It is also important that the strips are not treated with pesticides.

5.2 Measures to Improve Forest Clear-Cut Habitats for Ortolan Buntings

There is also a conservation programme developed specifically to preserve the ortolan bunting population in the forest clear-cut areas in northern Sweden; this was designed by the responsible authority in Sweden (Naturvårdsverket 2017). Their focus is to create optimal conditions in as many forest clear-cuts as possible where ortolan buntings still breed. This work has concentrated on creating larger areas of disturbed ground with exposed soil and on removing tree saplings to maintain an open landscape. The forest industry has at least partially recognised that there is opportunity to generate some goodwill from this work and is cooperating with the organisers of the national ortolan bunting project, which was founded in 2017 and is ongoing until at least 2021.

The hope is that the work of these programmes will preserve the remnants of the ortolan bunting population in Sweden.

References

Artportalen (2019) (https://www.artportalen.se/ViewSighting/ViewSightingList) (In Swedish)

Kjellén N (2002) Sträckräkningar i Falsterbo från förr och nu. Meddelande nr 212 från Falsterbo fågelstation. Anser 114–123. (In Swedish)

Menz MHM, Arlettaz R (2012) The precipitous decline of the ortolan bunting: time to build on scientific evidence to inform conservation management. Oryx 46:122–129. https://doi.org/10.1017/s0030605311000032

Menz MHM, Mosimann-Kampe P, Arlettaz R (2009) Foraging habitat selection in the last Ortolan Bunting *Emberiza hortulana* population in Switzerland: final lessons before extinction. Ardea 97(3):323–333. https://doi.org/10.5253/078.097.0308

Naturvårdsverket (2017) Åtgärdsprogram för ortolansparv *Emberiza hortulana* 2017–2021. Rapport 6781. (In Swedish with English summary)

Olsson C, Wiklund, J (1999) Västerbottens fåglar. 665 pages. Umeå. ISBN 91-973576-0-X (In Swedish)

Ottvall R, Green M, Lindström Å, Svensson S, Esseen P-A, Marklund L (2008) Ortolansparvens Emberiza hortulana förekomst och habitatval i Sverige. Ornis Svecica 18:3–16. (In Swedish)

Selstam G, Sondell J, Olsson P (2015) Wintering area and migration routes for Ortolan Buntings *Emberiza hortulana* from Sweden determined with light-geologgers. Ornis Svecica 25:3–14

Sondell J (2013) Aktuellt om ortolansparven 2012–2013. Fåglar i Kvismaren 28:8–13. (In Swedish)

Sondell J, Brookes C, Persson M (2011) Ortolan Bunting *Emberiza hortulana* at Kvismaren, Central Sweden – breeding studies and suggested management. Ornis Svecica 21:167–178

Sondell J, Durà C, Persson M (2018) Breeding prerequisites for Ortolan Bunting *Emberiza hortulana* in Swedish farmland, with special focus on foraging. Ornis Svecica 29:5–25

Stolt B-O (1974) Gulsparvens *Emberiza citrinella* och ortolansparvens *Emberiza hortulana* förekomst vid Uppsala under 1960-talet. Vår Fågelvärld 33:210–217. (In Swedish)

Svensson S, Svensson M, Tjernberg M (1999) Svensk fågelatlas. Vår fågelvärld, supplement 31. Stockholm. (In Swedish)

Vepsäläinen V, Pakkala T, Phia M, Tiainen J (2005) Population crash of the Ortolan Bunting *Emberiza hortulana* in agricultural landscapes in southern Finland. Ann Zool Fennici 42:91–107

Wretenberg J, Lindström Å, Svensson S, Pärt Å (2007) Linking agricultural policies to population trends of Swedish farmland birds in different agricultural regions. J Appl Ecol 44:933–941

Chapter 17
The Impact of Land-Use Change on Arable Plant Habitats and Wintering Farmland Birds on a Farm in South-West Wales, UK

Clive Hurford

Abstract Gupton Farm supported a nationally important arable flora and fauna until the summer of 2013, when the landowners ended the tenancy agreement and took over the management of the site. Thereafter the landowners abandoned cultivation across large areas of the farm, including those areas hosting a diverse and abundant arable flora and a wintering population of c.12,300 farmland birds. Although similar land-use changes have been applied widely across Europe in the name of agricultural sustainability, sites of known national or international conservation importance have rarely been deliberately degraded or destroyed in the name of nature conservation.

The landowner, one of the four largest landowners in the UK, made the land-use changes at Gupton Farm to enable 'natural processes' and 'sustainable agriculture' at the expense of the biodiversity interest. The relevant national conservation agency and leading nature conservation NGOs in the UK supported this decision. This case study describes the impact of the land-use changes made in April 2013 on the flora and fauna of Gupton Farm; it also discusses issues relating to good practice in nature conservation management.

Keywords Arable habitat loss · Arable flora · Biodiversity losses · *Glebion segetum* · Land-use change impacts · Important Arable Plant Area (IAPA) · Wintering farmland birds

1 Introduction

Gupton Farm is located on the Castlemartin Peninsula in south-west Wales and covers an area of approximately 127 ha. The farm has medieval origins (Owen 1918) and, until recently, supported a nationally/internationally important arable plant flora and one of the largest wintering farmland bird populations in Wales. For example,

C. Hurford (✉)
Serapias (Ecological Monitoring Consultancy), Ty Llaeth, Y Fraich, Fishguard, UK
e-mail: clive.hurford@serapias.net

C. Hurford et al. (eds.), *The Changing Status of Arable Habitats in Europe*,
https://doi.org/10.1007/978-3-030-59875-4_17

there were extensive populations of corn marigold (*Glebion segetum*), field woundwort (*Stachys arvensis*) and corn spurrey (*Spergula arvensis*), which are all in rapid decline in Wales. The farm also supported large numbers of wintering birds including c.3700 lapwings (*Vanellus vanellus*), over 3400 golden plover (*Pluvialis apricaria*) and more than 1600 linnets (*Carduelis cannabina*). Sections 3 and 4 provide a detailed account of the flora and fauna.

1.1 Recent Historical Background

From 1974 to April 2013 the tenant farmers managed the land relatively consistently. In contrast to many other farms in the UK, the tenants applied a low intensity approach to farming, which allowed the flora and fauna associated with the site in the 1960s to persist and develop until the tenancy ended. The decision to end the tenancy agreement in April 2013 prompted a series of baseline surveys to define the biodiversity interest of the site and inform management decisions ahead of the proposed land-use changes.

These baseline surveys took place in the period from July 2012 to March 2013 and focussed on describing:

- The land-use that was responsible for creating such an exceptional flora and fauna;
- The arable plant flora, which qualified as nationally/internationally important; and
- The wintering farmland bird population, which was exceptional for modern-day farmland in Wales.

This chapter describes the changes in the status of the flora and avifauna resulting from the land-use changes implemented in the summer of 2013.

2 Land-Use Changes at Gupton Farm

In the period from 1974 to 2012, the combination of cereal cultivation in summer and cattle grazing in winter created the conditions to support a rich arable flora and fauna. The sections below describe this land-use regime, and the subsequent changes to it.

2.1 Land-Use by the Tenant Farmers in the Period from 1975 to 2012

From 1975 to 2006, the tenant farmers applied a relatively consistent summer land-use regime, comprising the cultivation of a mix of fodder crops, i.e., maize *Zea mays* and cultivars of spring barley *Hordeum vulgare* undersown with grass, and permanent grass. In the early days the main crop was spring barley and grass, with maize restricted to the lower, sandier fields in the west of the site. The fodder mix of maize, spring barley and grass continued in the period from 2006 to 2012, but with an increase in the extent of maize fodder (ca.52 ha) at the expense of spring barley (ca.11 ha): this equated to 4000 tons of maize and grass and 25 tons of spring barley. Herbicide treatments, e.g. of Atrazine, were applied annually until 2007, typically at the time of planting (in mid to late May), but no insecticides were used. There were no subsequent treatment applications after 2007.

The fields along the spine of the farm were permanent grass and, until 2006, were grazed by a resident herd of up to 400 cattle. Thereafter, the only summer grazing involved a small number of cattle in the fen meadows and one or two ponies in a field near the farmhouse (Fig. 17.1). In this period, there were three silage cuts from the permanent grass fields each year.

The winter land-use during this period focused on maintaining the resident dairy herd, which increased from c.200 animals in the mid-1970s to c.400 animals in the period from the late 1970s to 2006. These cattle spent the winter months in relatively small groups in the fields on sandier soils in the west of the site (Fig. 17.4). The herd was fed on local maize, barley and grass fodder. After the tenants stopped dairy

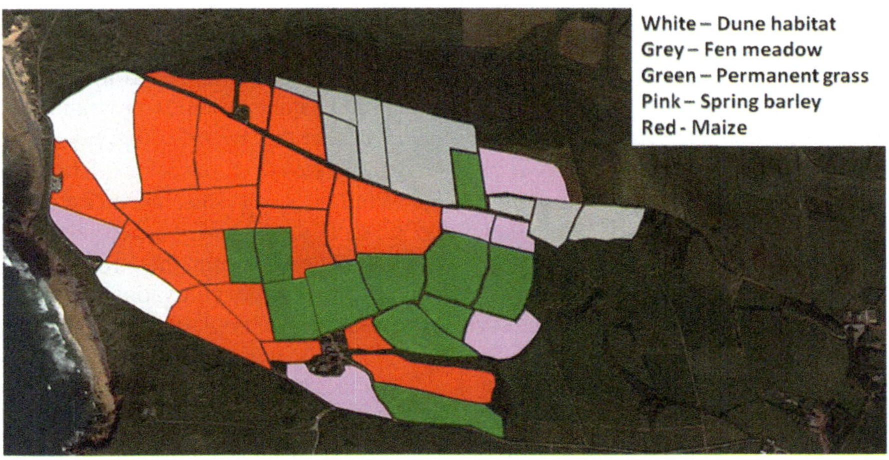

Fig. 17.1 This shows the summer land-use at Gupton Farm in August 2012. The few cattle on the farm spent the summer grazing the areas of fen meadow in the north of the site

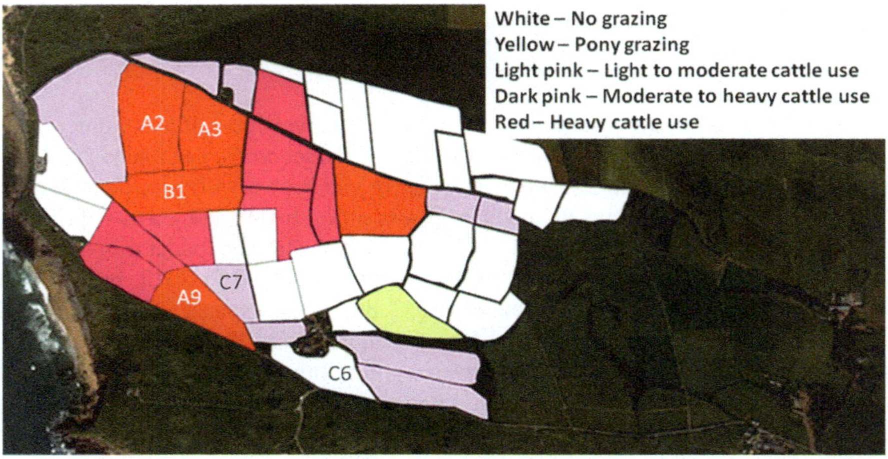

Fig. 17.2 This map shows the land-use at Gupton Farm in the winter of 2012/13. For the purposes of the wintering farmland bird survey, we gave each field a number code; the numbered fields supported the largest numbers of farmland birds

farming in 2006, cattle from other local farms were wintered on the site in increasing numbers until 2012/13, peaking at c.1000 animals in the 2012/13 winter. Again, these animals wintered in the sandy-soiled fields in the west of the site (Fig. 17.2) and fed mostly on fodder grown at Gupton.

2.2 Land-Use Changes Introduced in 2013

In April 2013, the landowners introduced land-use changes that aimed to reinstate natural processes on the farm, primarily by initiating a dune reversion programme in the west of the site. The underlying rationale was that this would offset any future risks to the farmland arising from climate change-related sea-level rise. The only cultivated areas on the farm in summer 2013 were four small fields in the east of the site comprising c.1.6 ha of spring barley and 9.3 ha of spring barley undersown with grass (Fig. 17.3). Agricultural activity in the west of the site during the 2013 summer was restricted to light cattle-grazing in fields previously under cultivation. Section 4 describes the impact that these land-use changes had on the wintering farmland bird population.

By the summer of 2018, the area of land under arable cultivation had stabilised at around c.28 ha of spring barley, of which only c.15 ha was in fields in the west of the site (Table 17.1). While this was an increase on the area under cultivation compared to the summer of 2013, it was still c.37 ha (40%) less than in the 40 years up to and including 2012 (Fig. 17.4). Less than 2 ha (6%) was cultivated in the area that had

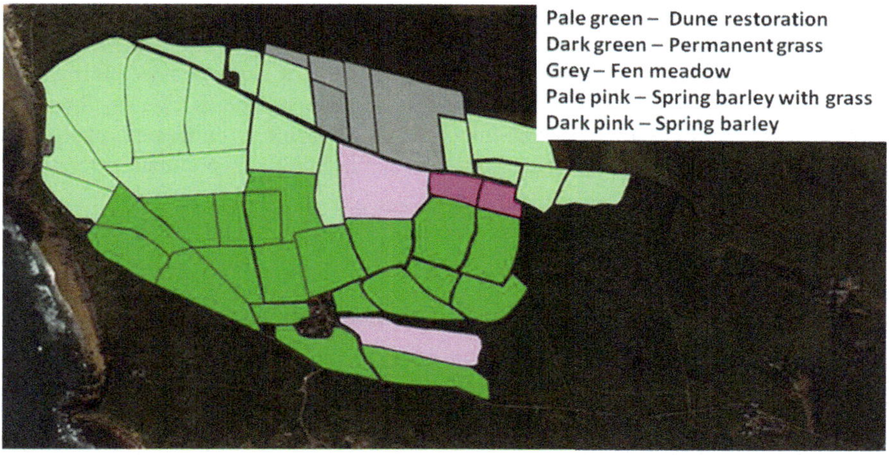

Fig. 17.3 A map of Gupton Farm showing the land-use in the summer of 2013, when cultivation in the west of the site was abandoned in favour of a dune reversion policy. The area under cultivation declined from c.63 ha in 2012 to 11 ha in 2013. Some of the dune restoration and permanent grass enclosures were lightly grazed by a herd of c.20 cattle

Table 17.1 The area (ha) of maize and spring barley cultivated in the years before and after the land-use changes at Gupton Farm (the figures in bold are post land-use change)

	West Gupton			East Gupton		
Crop type	2012	**2013**	**2018**	2012	**2013**	**2018**
Maize (ha)	52	**0**	**0**	0	**0**	**0**
Spring barley (ha)	0	**0**	**15**	11	**11**	**13**
Total	52	**0**	**15**	11	**11**	**13**

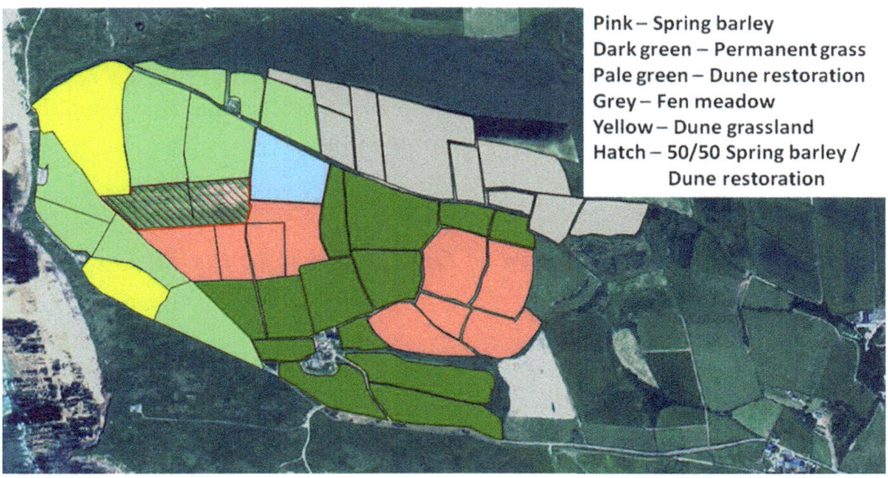

Fig. 17.4 This map illustrates the land-use at Gupton Farm in the summer of 2018. It shows an increase in the area of land under cultivation of c.19 ha compared to 2013, but a loss of 38 ha on the area cultivated in 2012. The 'blue' field was a sacrificial bird crop comprising a mix of mostly non-native plant species

supported the greatest diversity and abundance of arable plants and the vast majority of the wintering farmland birds.

Following the end of the tenancy in April 2013, winter land-use was minimal and comprised light grazing of the grassland and dune reversion areas by a small herd of 20–30 cattle. In the dune reversion areas that were formerly cultivated, one of the stated aims was to graze out the weeds, i.e. the arable plant population (National Trust 2014).

3 The Arable Plant Flora at Gupton Farm

The arable flora at Gupton Farm included many archaeophyte species, i.e. species introduced to Britain before 1500. Many of these probably arrived as contaminants of cereal seed introduced by the Romans (Wilson and King 2003) and can still be found growing inside the Colosseum and Roman Forum (Deakin 1855). However, modern farming methods have had a dramatic impact on the UK archaeophyte flora, and species associated with cultivated land now form c.60% of the most rapidly declining species in the UK (Preston et al. 2002).

In response to the threat to the UK arable flora, Plantlife developed a scoring system to identify areas of local, national and international importance for arable plants (Byfield and Wilson 2005). Using the 'Important Arable Plant Area' (IAPA) scoring system for free-draining sandy soils, Gupton Farm scored 63 points and qualified as a nationally important area (Table 17.2): it failed to meet the criteria for international importance by only seven points, i.e. by a single species.

If refound, the small-flowered catchfly *Silene gallica* (previously recorded on the sandy soils in the west of the site) would have added another seven points to the total and placed the farm in the 'Internationally important' category. Other notable IAPA scoring species recorded at Gupton Farm include henbane *Hyoscyamus niger*, bur chervil *Anthriscus caucalis* and weasel's snout *Misopates orontium*. Ideally, however, any IAPA/SSSI would have combined Gupton Farm with the adjoining Newton and Neath Farms to form a more robust protected area that would have comfortably qualified as of international importance.

3.1 The Arable Plant Flora of Gupton Farm in the Period to April 2013

In the summer of 2012, the arable plant flora at Gupton Farm was of considerable conservation importance. At this time, some 63 ha (c.50%) of the farm was under cultivation, comprising 52 ha of maize and 11 ha of spring barley (Fig.17.1). The remainder of the farm was a mix of permanent grass (used for silage or summer cattle

Table 17.2 The IAPA scoring species of arable plant recorded at Gupton Farm, including the species classed as 'vulnerable' and 'of local concern'

IAPA scoring plant species recorded at Gupton Farm			
Vulnerable			
Species	Score	Species	Score
Silene gallica	7	*Spergula arvensis*	6
Sinapis arvensis	7	*Stachys arvensis*	7
Species of local concern			
Anchusa arvensis	2	*Lamium hybridum*	2
Anthriscus caucalis	4	*Melilotus altissimus*	2
Brassica nigra	1	*Misopates orontium*	3
Chenopodium rubrum	2	*Papaver dubium*	1
Glebion segetum	2	*Papaver rhoeas*	1
Coronopus squamatus	1	*Raphanus raphanistrum* ssp. *marit.*	2
Erodium moschatum	4	*Raphanus raphanistrum* ssp. *raph.*	2
Fumaria bastardii	2	*Sherardia arvensis*	1
Fumaria capreolata	2	*Solanum nigrum*	1
Hyoscyamus niger	4	*Thlaspi arvense*	2
Lamium amplexicaule	2		
		Total	70
Species at Gupton Farm in the list of "100 species showing the greatest decline between 1930–69 and 1987–99"			
Species	Gupton status	Species	Gupton status
Silene gallica	Rare	*Hyoscyamus niger*	Rare
Spergula arvensis	Locally abundant	*Fallopia convolvulus*	Locally common
Glebion segetum	Abundant	*Stachys arvensis*	Locally common
Sinapis arvensis	Locally common	*Lamium purpureum*	Locally common

grazing), fen meadow and small areas of dune grassland. During the subsequent winter months (2012/13), the farm held up to 1000+ cattle, fed mostly on fodder crops grown at Gupton.

The area of Gupton Farm that qualified as being of national, and very probably international, importance was the cultivated land on sandy soils in the west of the site. At least 22 IAPA scoring species occurred on these soils, compared with only 13 on the deeper soils in the east of the site. The respective IAPA totals further underline the relative conservation value of these two areas: the fields in the west of the farm had an IAPA score of 63, compared to a score of only 36 for the fields in the east (Fig.17.5).

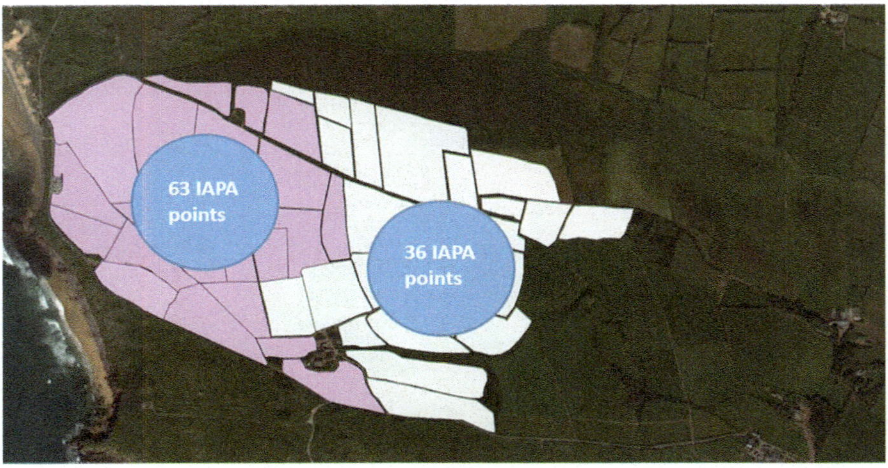

Fig. 17.5 The pink area highlights the key fields for arable plants at Gupton Farm pre-2013. This coincides with the area of cultivated sandy soils

However, the IAPA scoring system takes no account of species abundance, so a field with a single plant of a rare species will score the same as a field with a thousand plants. At Gupton Farm, it was both the diversity and abundance of the arable plant species classed as 'vulnerable' or 'of local concern' that defined the conservation value of its flora. Species such as corn marigold *Glebion segetum*, corn spurrey *Spergula arvensis* and common stork's-bill *Erodium cicutarium* dominated large areas of land, with purple swathes of henbit deadnettle *Lamium amplexicaule* also adding colour to the landscape.

3.2 The Impact of the 2013 Land-use Changes on the Arable Plant Flora at Gupton Farm

The land-use changes implemented in April 2013 resulted in a decline of c.52 ha (83%) on previous summers. In the west of the site, which had formerly supported the highest diversity and abundance of arable plants, no land was cultivated.

By 2018, the most visible impact of the land-use changes on the arable flora was the restricted distribution and abundance of the commoner arable plants such as corn marigold *Glebion segetum*, scentless mayweed *Tripleurospermum inodorum* and corn spurrey *Spergula arvensis*. The array of colour formerly associated with the landscape (Fig.17.6) was no longer evident; a direct consequence of the reduced area of land under cultivation at Gupton in 2018, some 38 ha less than in 2012.

Fig. 17.6 A field margin at Gupton Farm during harvesting in August 2012. Often, in the fields in the west of the farm, the abundance of corn marigold (*Glebion segetum*) and other arable plants extended well into the crop. Photo: Clive Hurford

Furthermore, by this time, many of the scarcer arable plant species that were previously locally abundant in the west of the site, e.g. *Lamium amplexicaule* and musk stork's-bill *Erodium moschatum*, were difficult to find.

4 Wintering Farmland Birds at Gupton Farm

After the landowners announced the land-use changes proposed for April 2013, a team of local ornithologists organised monthly bird counts at Gupton Farm in the 2012/13 winter to establish the size of the wintering farmland bird population. The counts took place on 26 Oct, 30 Nov, 04 Jan and 30 Jan (adverse weather deferred the December count to early January). For analysis purposes, I have focussed on the mid-winter counts, i.e. those carried out in November, December and January. The wintering population is more stable in these months, whereas the October and February counts are more prone to distortion by influxes of migrating birds and early/late arrival and departure dates.

In the 2012/13 winter, the number of birds using Gupton Farm peaked at 12,300 on 04 January. The main species contributing to this total were strongly associated with farmland, namely lapwing (*Vanellus vanellus*), golden plover (*Pluvialis*

Fig. 17.7 Thousands of wintering lapwings and golden plovers associated with the cattle in the sandy-soiled areas of Gupton Farm in the decades leading up to the 2012/13 winter: this was taken in November 2012. Photo: Clive Hurford

apricaria), skylark (*Alauda arvensis*), chaffinch (*Fringilla coelebs*) and linnet (*Carduelis cannabina*). Other farmland species present included chough (*Pyrrhocorax pyrrhocorax*), curlew (*Numenius arquata*), greenfinch (*Chloris chloris*), goldfinch (*Carduelis carduelis*), house sparrow (*Passer domesticus*), meadow pipit (*Anthus pratensis*), pied wagtail (*Motacilla alba*), reed bunting (*Emberiza schoeniclus*), rook (*Corvus frugilegus*) and starling (*Sturnus vulgaris*). A wintering population of 12,300 birds is exceptionally high for a farm in Wales. The vast majority of these birds wintered on the sandy, free-draining soils that were cultivated in the summer and cattle-grazed in winter (Fig.17.7).

However, the subsequent loss of cultivation from the areas that had supported large numbers of wintering farmland birds had a dramatic impact on the wintering bird population. The key species affected by the changes were lapwing, golden plover, skylark, chaffinch and linnet. The following sections describe the impact of the land-use changes on each of these species.

4.1 Lapwing

The lapwing is on the Birds of Conservation Concern 4 Red List (Eaton et al. 2015) and is a 'Species of Principal Importance' in Section 42 of the NERC Biodiversity

Duty (NERC 2006). The Lapwing diet mainly comprises soil invertebrates, supplemented by perhaps 10% grass and arable seeds (Cramp et al. 1977–1994).

The wintering population at Gupton Farm prior to, and including, the 2012/13 winter showed a strong preference for the sandy-soiled fields under cattle use in the west of the site, especially Fields A2, A3 and B1, which together could hold up to 2000 birds (Fig. 17.2). Elsewhere on the farm, flocks of up to 350 birds used the sandy-soiled fields in the south of the site, particularly Fields A9 and C7. Invariably, the lapwings were closely associated with the cattle grazing in the sandy-soiled fields (Fig. 17.6); their preference for these fields is most likely linked to:

- The historic lack of insecticide treatments in these fields;
- The high volume of seeds in the soil from the abundant arable plant flora; and
- Constant soil disturbance by cattle providing a readily available source of both soil invertebrates and arable plant seeds.

4.1.1 Winter 2012/13 Survey Results

During the 2012/13 winter, the lapwing population at Gupton Farm peaked at 3773 birds on 4 January (Fig. 17.8), indicating that Gupton was an important farmland wintering site for the species in Wales. The mean number of lapwings recorded over the three mid-winter counts in Nov, Dec and Jan was 2592, which is a better indicator of the numbers wintering on the farm.

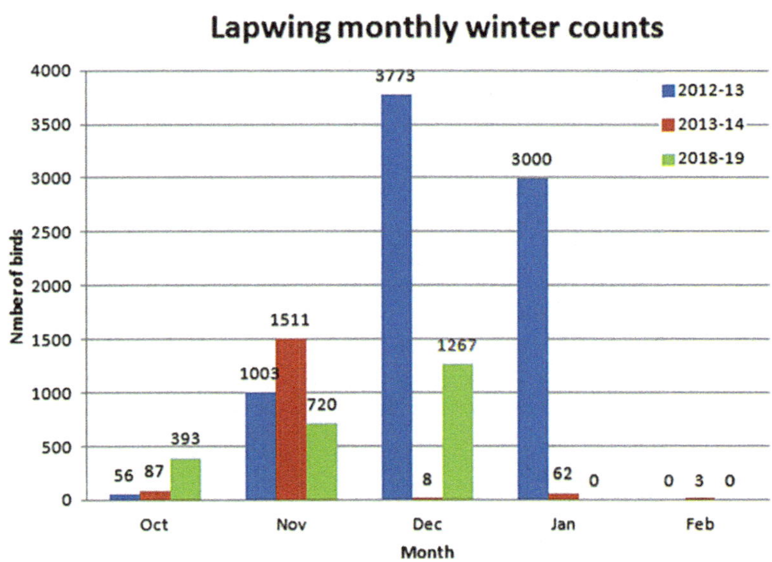

Fig. 17.8 The count data for lapwings in the 2012/13, 2013/14 and 2018/19 winters. Note that, in 2013/14, the lapwings arrived in good numbers in November before abandoning the site

4.1.2 Winter 2013/14 Survey Results

In the first winter after the landowner introduced the land-use changes, the lapwings started to arrive, as usual, in November, when 1511 were present. However, this was the peak count of the winter, with the farm, to all intents and purposes, abandoned as a feeding area before the December count (Fig. 17.8). The only lapwings seen subsequently during the counts that winter were eight birds in December 2013 and 62 in January 2014. The mean of the mid-winter counts fell from 2592 in 2012/13 to 527 in 2013/14, a decline of 2065 (c.–80%).

4.1.3 Winter 2018/19 Survey Results

The 2018/19 winter was the fifth after the land-use changes at Gupton Farm. By this time, the land management at Gupton Farm was relatively stable with no major modifications planned. During this winter period, the peak mid-winter lapwing count was 1267 in December 2018 and the mean of the three mid-winter counts was 662: a 69% decline on the mean recorded in the 2012/13 winter. However, this is misleading because it suggests that lapwings continued to be a regular wintering species at Gupton Farm, which is not the case. Flocks of up to c.1300 birds would shelter in the fields at Gupton occasionally, but only when disturbed from feeding grounds elsewhere. No lapwings were present during the counts in January and February 2019 at Gupton.

4.2 Golden Plover

The golden plover is on the Birds of Conservation Concern Wales 3 Red List (Johnstone and Bladwell 2016) and a 'Species of Principal Importance' in Section 42 of the NERC Biodiversity Duty (NERC 2006).

The diet of the golden plover consists of a wide variety of invertebrates, mostly beetles and worms (Newton 2017). Most of the golden plovers at Gupton Farm wintered in the fields in the west and south-west of the site, particularly Fields A2, A9, B1 and C7: these areas hosted large numbers of out-wintering cattle. Their preference for these fields was probably due to the historic lack of insecticide treatments and the ready availability of soil invertebrates in the sandy-soils disturbed by cattle.

4.2.1 Winter 2012/13 Survey Results

During the 2012/13 winter, the golden plover flock started to arrive in November, and peaked at c. 3450 birds on the December count (Fig. 17.9). This mid-winter

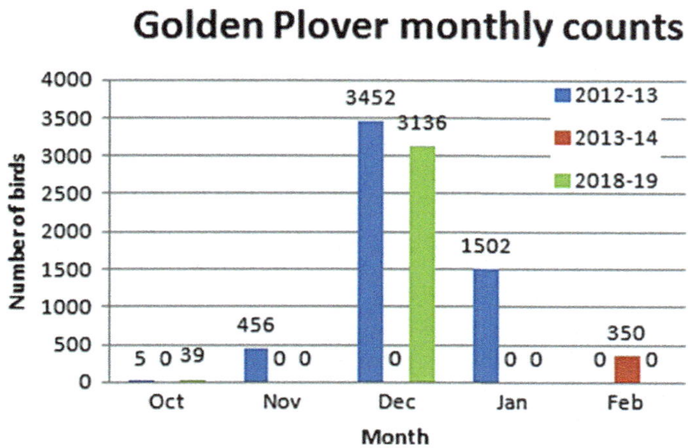

Fig. 17.9 This shows the monthly count data for golden plovers over the 2012/13, 2013/14 and 2018/19 winters. Note that, after the 2012/13 winter, the species no longer wintered at Gupton Farm and was present on only three of the ten subsequent count dates

count indicated that Gupton was an important wintering site for golden plover in Wales. The mean number of birds over the three core mid-winter counts was c.1800.

4.2.2 Winter 2013/14 Survey Results

In the 2013/14 winter, following the introduction of the land-use changes, there were nil counts for golden plovers in October, November, December and January. The species was present at Gupton only on the February count date, when c.350 birds were there. The mean core mid-winter counts fell from 1800 in 2012/13 to zero in 2013/14, a 100% decline in the core wintering population.

4.2.3 Winter 2018/19 Survey Results

In the 2018/19 winter, five years after the introduction of the land-use changes, golden plovers no longer wintered at Gupton Farm, though the species was present on two of the five count dates. A small flock of 39 passage birds was there in October 2018 and a large flock of 3136 birds was present on the day of the December count, but there were no other reports of this flock. This suggests that these birds were part of winter movement, perhaps on their way to Ireland. The mean of the November, December and January counts in 2018/19 was 1045, which is a 41% decline on the mean mid-winter count of 1800 in 2012/13.

At this time, a much-reduced and mobile flock of c.1000 golden plovers was wintering on the Castlemartin peninsula, spending most of its time feeding on land adjacent to Gupton Farm. On occasions, these birds would be disturbed from their feeding grounds, e.g. by a raptor, and disperse to Gupton, where they would shelter until the danger had passed. Golden plovers abandoned Gupton Farm as a feeding area after the land-use changes in 2013.

4.3 Skylark

The skylark appears on the Birds of Conservation Concern 4 Red List (Eaton et al. 2015) and is a 'Species of Principal Importance' in Section 42 of the NERC Biodiversity Duty (NERC 2006).

Although the diet of the skylark varies and includes insects, leaf material and seeds, in the winter months it has a strong preference for arable plant seeds (Cramp et al. 1977–1994). The wintering distribution of skylarks at Gupton Farm reflected this, with most of the birds favouring the sandy-soiled fields in the west of the site, where cattle trampling helped to replenish seed availability, i.e. in fields A3 and A9 (Fig. 17.2).

In summer, these fields supported an abundant arable plant flora with a high vegetation cover of common stork's-bill *Erodium cicutarium* and corn marigold *Glebion segetum*.

4.3.1 Winter 2012/13 Survey Results

During the 2012/13 winter, skylarks began to arrive at Gupton Farm in October, with numbers building throughout November and peaking at c.750 birds in early January (Fig. 17.10). This was a large skylark wintering population for a farm in Wales: elsewhere, numbers seldom exceeded 150 birds. The mean count during the core mid-winter months of November, December and January was 518.

4.3.2 Winter 2013/14 Survey Results

During the 2013/14 winter, the first after the land-use changes, skylarks began to arrive at Gupton as usual in October, with numbers peaking at 276 in November. Thereafter, the population declined month on month as the winter progressed. The peak of 276 birds represented a decline of c.63% on the previous winter. The mean count during the core mid-winter months of November, December and January was 173, a decline of 67% on the 2012/13 midwinter mean.

4.3.3 Winter 2018/19 Survey Results

During the 2018/19 winter, five years after the introduction of the land-use changes, skylark numbers peaked at 409 birds in the January count, a decline of 45% on the

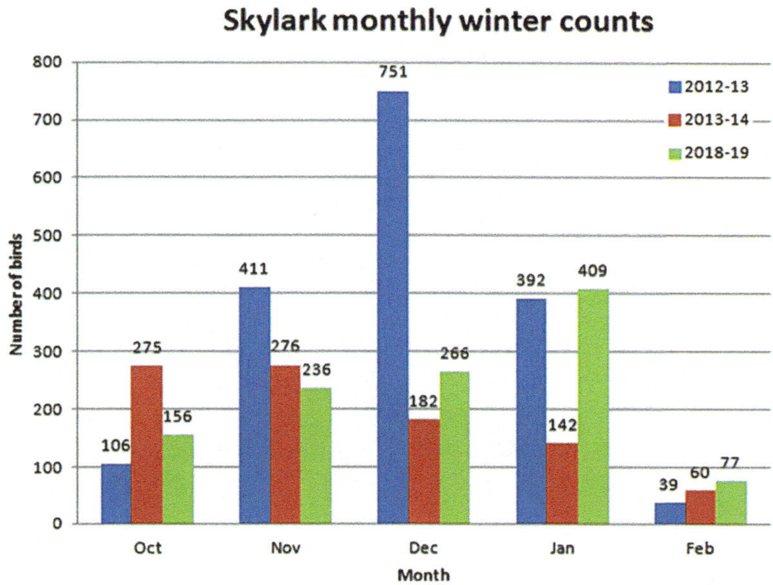

Fig. 17.10 The monthly count data for skylarks at Gupton Farm over the 2012/13, 2013/14 and 2018–19 winters

2012/13 peak. The mean count during the core mid-winter months of November, December and January was 304 which, although an increase on the mid-winter mean of 171 in 2013/14, was a decline of 41% on the 2012/13 mid-winter mean.

4.4 *Chaffinch*

The chaffinch is a common and widespread breeding species in the UK and is categorised as of 'Least Concern' by the International Union for Conservation of Nature. However, Gupton Farm consistently supported large flocks of chaffinches in the years leading up to the land-use changes in April 2013 and I have included it here to underline the impact that land-use change can have on even the commoner seed-eating species.

4.4.1 Winter 2012/13 Survey Results

During the 2012/13 winter, chaffinches began to arrive at Gupton Farm in October, with numbers building throughout November and peaking at c.1841 birds in the December count (Fig. 17.11). This was one of the largest wintering

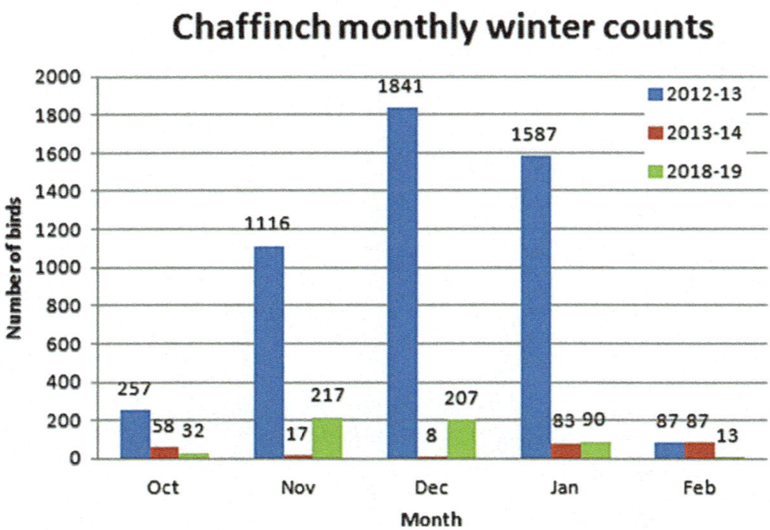

Fig. 17.11 The monthly count data for chaffinches at Gupton Farm over the 2012/13, 2013/14 and 2018/19 winters

populations on farmland in Wales. The mean count during the mid-winter months of November, December and January was 1514.

4.4.2 Winter 2013/14 Survey Results

During the 2013/14 winter, the first after the introduction of the land-use changes, chaffinch numbers peaked at 87 in February, a decline of 1754 (–95%) on the peak count in the 2012/13 winter. The mean count during the core mid-winter months of November, December and January was 36, a decline of 98% on the 2012/13 mean.

4.4.3 Winter 2018/19 survey results

During the 2018/19 winter, five years after the introduction of the land-use changes at Gupton Farm, the chaffinch wintering population peaked at 217 birds in the November count, a decline of 1624 (–88%) on the 2012/13 peak. The mean of the mid-winter counts was 171, a decline of 1343 birds (–89%) on the mean mid-winter count in 2012/13.

4.5 Linnet

The linnet is on the Birds of Conservation Concern 4 Red List (Eaton et al. 2015) and is a 'Species of Principal Importance' in Section 42 of the NERC Biodiversity Duty (NERC 2006).

The linnet is primarily a seed-eating species that relies less on invertebrate food than any other Western Palaearctic finch (Cramp et al. 1977–1994). The distribution of the wintering linnet population at Gupton Farm reflected this and was restricted to fields that had held maize or spring barley during the summer months. As with the other specialist farmland birds at Gupton, the highest concentrations of feeding birds were found on the free-draining sandy soils disturbed by cattle grazing, e.g. Fields B1, A2 and A3 (Fig. 17.2). These fields had an abundant arable flora during the summer, typified by a high vegetation cover of common stork's-bill *Erodium cicutarium*, corn marigold *Glebion segetum* and corn spurrey *Spergula arvensis*.

4.5.1 Winter 2012/13 Survey Results

Linnets started to arrive at Gupton Farm in late October 2012, with numbers building throughout December and peaking at c.1670 in January (Fig. 17.12): an exceptionally high winter count for farmland in Wales. The mean of the mid-winter counts taken in the months of November, December and January was 1195.

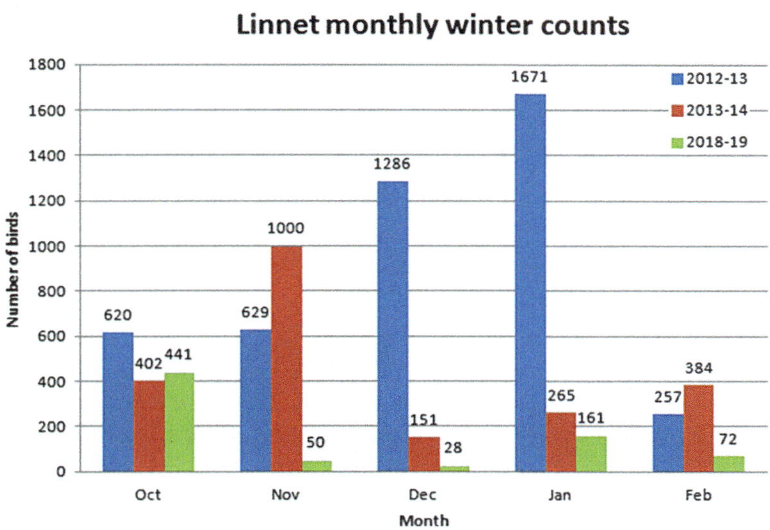

Fig. 17.12 The monthly count data for linnets at Gupton Farm over the 2012/13, 2013/14 and 2018/19 winters

4.5.2 Winter 2013/14 Survey Results

In 2013/14, the first winter after the land-use changes were introduced, c.400 linnets were present in October, which increased to 1000 birds in November. This was the peak count of the winter and represented a decline of 671 (c.–40%) on the 2012/13 winter peak. The mean of the mid-winter counts fell from 1195 in 2012/13 to 472 in 2013/14, a decline of 723 (c.–61%).

4.5.3 Winter 2018/19 Survey Results

In the 2018/19 winter, five years after the land-use changes were introduced, c.440 linnets were present in October, which declined to only 28 birds on the mid-winter count in December: a decline of 1643 birds (–98%) on the number of birds present in December 2012. The mean of the mid-winter counts fell from 1195 in 2012/13 to 239 in 2018/19, a decline of 956 (c.–80%).

5 Discussion

This case study has identified several key actions that could help to avoid a repeat of the biodiversity losses seen at Gupton Farm. These include:

- Early stakeholder engagement;
- Collating existing data and identifying knowledge gaps;
- Gathering relevant survey data;
- Prioritisation; and
- Mitigation.

These issues are discussed under the appropriate headings below.

5.1 Early Stakeholder Engagement

Early engagement with stakeholders is a key phase of the management planning cycle for nature protection areas (Jordan 2015; Westberg and Waldenström 2016). In the case of Gupton Farm, the landowners engaged with the stakeholders very late in proceedings, i.e. only 10 months before they implemented the land-use changes.

Despite the late stakeholder engagement, ecologist stakeholders generated two reports drawing attention to the likely consequences of the proposed land-use changes at Gupton Farm (Wilson 2013; Hurford 2013): the landowners had access to these reports before the tenancy agreement ended in April 2013. The first report (Wilson 2013) emphasised that the exceptional arable plant flora and wintering bird

Fig. 17.13 A typical mix of spring barley and arable plants at Gupton Farm before the land-use changes were made in April 2013, with abundant corn marigolds (*Glebion segetum*) and a wide range of associated arable plant species including borage (*Borago officinalis*) and wild radish (*Raphanus raphanistrum*) as seen here. This field was taken out of cultivation, sown with a *Lolium perenne* and *Trifolium repens* seed mix and sheep-grazed in an attempt to eradicate the 'weeds'

populations at Gupton Farm were the result of (a) "growing spring maize and barley for silage without agrochemical inputs" and (b) "the overwintering of large numbers of young cattle on the fields". The report also explained that this management is likely to have resulted in the accumulation of a very large arable plant seedbank and a large population of soil organisms on which the birds feed. The report went on to state that the "continuation of the current farming system is likely to be crucial to the maintenance of the biodiversity of the Castlemartin Valley."

The second report (Hurford 2013) summarised the results of the 2012/13 wintering farmland bird survey, collated the historical records of the arable flora, and described the land-use that was responsible for producing the exceptionally diverse arable flora and fauna at Gupton Farm. The report also analysed the distribution of the flora and fauna to inform decisions associated with mitigation.

Earlier stakeholder engagement would have allowed more time to collate existing data and identify knowledge gaps (see 5.2), and made it possible to gather additional data on key elements of the flora and fauna (Fig. 17.13).

5.2 Collating Existing Data and Identifying Knowledge Gaps

In summer 2012, when the landowners presented their plans for the management changes at Gupton Farm, the only survey data on the flora and fauna of the farm were *ad hoc* natural history records collected by, or submitted to, the respective county recorders. The landowners had commissioned hydrological and geomorphological surveys, but the latter failed to take into account that every dune system in Wales had been in a phase of stabilisation for a century or more. This means that the hind dunes have been stabilising faster than the fore-dunes have been accreting sand, leading to a net loss of bare sand and a diminishing area of species-rich, successionally-young dune habitats. At Kenfig, for example, open sand habitats diminished from c.45% of the site in the 1940s to less than 2% in 2010 (Carrington et al. 2010). Google Earth images show that Broomhill Burrows (the dune system adjacent to Gupton) was in a similarly stable state, particularly on the landward side of the road where the farm is situated. As such, the most likely consequence of the proposed land-use changes would be a loss of nationally important arable habitats in favour of closed, species-poor grassland.

Had a data collation exercise taken place, it would have revealed a lack of structured biodiversity data for the land.

5.3 Data Gathering to Fill Knowledge Gaps

At the time of announcing the planned land-use changes in summer 2012 the landowners had not surveyed any components of the flora and fauna. This was despite being aware of the important arable flora; the shrill carder bee (*Bombus sylvarum*) population (and abundant pollinator fauna); and the large wintering farmland bird population. By then, however, it was already too late to carry out a breeding bird survey or undertake comprehensive surveys of the flora and invertebrate fauna. There was only time to carry out a presence/absence survey of some of the more interesting arable plants on the farm (Shellswell 2013) and to organise a survey of the wintering farmland bird population (Hurford 2013). The results from these surveys informed Sections 3 and 4 above.

A report by Wilson (2013) recommended that further work should include a more comprehensive arable plant survey in 2013–2015, including an estimate of the availability of arable plant seeds to wintering birds. This report also recommended that there should be a full survey of soil organisms and other invertebrates, both to determine the conservation importance of the invertebrate fauna and to ascertain its importance to the wintering birds. The landowner ignored these recommendations, and the data gathered subsequent to stakeholder engagement, and implemented the land-use changes, without any modifications in April 2013.

5.4 *Prioritisation*

Identifying the conservation priorities for a site should be a key component of any designation process or management planning exercise on any site of high nature conservation value (Hurford 2006). From a biodiversity perspective, the conservation priority at Gupton was the nationally/internationally important arable flora and fauna. Had the landowners engaged with the stakeholders earlier in the planning process, it would have been clear from the outset that there was a conflict between the proposals to reinstate 'natural processes' and promote 'sustainable agriculture' and the need to support the high biodiversity associated with the arable habitats on the site.

5.5 *Mitigation*

If developers are planning a project that will damage or destroy a site of national or international nature conservation importance in the UK, the likelihood is that either permission will be refused or that mitigation will be needed to offset the expected conservation losses. During the discussions with the landowners ahead of the land-use changes, the stakeholders suggested a mitigation proposal that would have limited the damage to the farmland flora and fauna; this proposal involved two phases:

(1) To continue cultivating and cattle-grazing the six fields that supported the highest density of arable plants and more than 90% of the wintering farmland bird population, at least until there was evidence that the plans for dune restoration were being successful; and
(2) To use this time to ensure that there was a sufficiently large area of arable habitat (with a suitably rich and abundant arable flora) elsewhere on the Castlemartin peninsula capable of supporting the large wintering farmland bird population being displaced from Gupton Farm.

The landowners received a report providing details of the measures in the Glastir agri-environment scheme that could contribute to funding the cultivation of the key fields at Gupton while the mitigation options were explored (Wilson 2013). The stakeholders also gathered both the pre- and post-2013 land-use data included in this report, initially in the hope of informing a mitigation package that never materialised and to allow assessment of the scale of future biodiversity losses in the context of the land-use changes.

In the event, the landowner implemented the land-use changes at Gupton Farm in April 2013, as originally planned, without any mitigation to offset the biodiversity losses.

6 Conclusions

The land-use changes at Gupton Farm had a predictably devastating impact on the flora and fauna of the arable habitats at Gupton Farm. This is hardly a unique occurrence, however, as landowners have made similar land-use changes with similar consequences widely across Europe and beyond. What is unique is that these land-use changes were made a) in the name of nature conservation and b) with the backing of the national conservation agency and leading UK conservation NGOs.

In the months leading up to the end of the tenancy agreement, two reports were prepared for the landowners (Hurford 2013; Wilson 2013) warning of the biodiversity losses likely to ensue if the land-use changes went ahead as planned. These reports also suggested a number of reasonable measures that could have mitigated the losses caused by the land-use changes, but the landowner rejected these proposals. There is no doubt that Gupton Farm qualified for notification as a Site of Special Scientific Interest and that the relevant authority, the national conservation agency, should have designated it as a protected area. However, the agency chose not to do this and supported the landowner's planned land-use changes, as did the other large UK conservation NGOs . Throughout the stakeholder meetings there was a strong sense of indifference shown towards the biodiversity of arable habitats, both by the landowners and by the individuals representing the conservation bodies. This indifference was expressed through a number of statements in support of the land-use changes, e.g. that Gupton Farm was 'just a big bird table', implying that after the restoration of the dune habitats the birds would disperse more evenly across the other farms in the region. However, a survey of the arable fields in the region would have revealed that there were no other farms with an extensive and abundant arable flora, nor indeed an arable field not treated with pesticides. Away from Gupton Farm, arable plants occurred only in small patches in field gateways - so there were no alternative feeding areas for wintering farmland birds on the Castlemartin peninsula.

Equally concerning was that the conservation NGOs seemed incapable of seeing the bigger picture, i.e. the plant conservation NGO only considered the rare plant interest on the site and the bird conservation NGO only considered the breeding bird interest. For example, one of the leading plant conservation NGOs was happy to support the land-use changes on the basis that there were more important sites for arable plants in Wales. This may be true if we consider a field with a population of c.30 cornflowers (*Centaurea cyanus*) to be of more ecological importance than an abundant and diverse arable plant flora and fauna covering more than 41 ha of land. In reality, however, if you factor in:

(a) The abundance of the arable flora;
(b) The abundance of the wintering farmland bird population;
(c) The abundant pollinator population;
(d) The presence of the shrill carder bee (*Bombus sylvarum*); and
(e) The extensive area of arable habitat present on the farm,

there is a strong likelihood that Gupton Farm hosted one of the two most important areas of arable habitat in Wales.

Unfortunately, the lack of concern shown for the arable flora and fauna at Gupton Farm by the representatives of the nature conservation bodies reflects a general attitude among conservationists towards arable habitats. At the time of writing, of more than 5000 SSSI in England and Wales, only four are designated to protect arable habitats: three in England and one in Wales. It is a similar story at the European level, with the EC Habitats Directive also failing to provide any protection for arable habitats or species. Consequently, the rich biodiversity historically associated with arable habitats, which has been part of the European cultural landscape for thousands of years, is virtually unprotected.

This lack of protection stems from the fact that no country considers arable habitats to be 'natural' nor the flora associated with them to be 'native'. Farmers often shoulder the blame for the disappearance of these exceptionally rich and diverse habitats, but the indifference of conservationists towards arable habitats is equally responsible. The 'native' and 'natural' concepts present a very real barrier to the effective conservation of biodiversity in the UK.

That the scale of biodiversity losses at Gupton Farm took place over such a short space of time provides an insight to the scale of biodiversity losses that must have been associated with the historic eradication of the arable flora and fauna across the wider countryside. The solutions to this problem must centre on increased awareness and increased levels of protection. Unfortunately, we could not prevent the losses at Gupton Farm but I would like to think that we can learn from them.

Acknowledgements Thanks to Terry and Michael Watkins, not only for their stewardship of such exceptional farmland habitats over the term of their tenancy, but also for their invaluable help in providing the land-use information included in this report. Thanks also to those who helped with collecting and providing data, notably: to Bob Haycock (the BTO representative for Pembrokeshire) for his support with organising and carrying out the wintering bird surveys; to Stephen Evans (BSBI Recorder for Pembrokeshire) for providing the historic plant records; to Cath Shellswell (Plantlife) for producing the arable plant report; and to Phil Wilson for his report on the most cost-effective and efficient options for managing the key arable areas on the site. Finally, thanks to Phil Wilson and to Chris Millican for their comments and recommendations on the drafts of this chapter.

References

Byfield AJ, Wilson PJ (2005) Important Arable Plant Areas: identifying priority sites for arable plant conservation in the United Kingdom. Plantlife International, Salisbury

Carrington D, Hurford C, Jones A, Pankhurst T (2010) The Fen Orchid – a species on the brink. Br Wildl 22(1):1–8

Cramp S et al (1977–1994) Handbook of the birds of Europe the Middle East and North Africa – the birds of the Western Palearctic. Oxford University Press, Oxford (A 9 Vol multi-author work edited by Cramp and Cramp and Perrins)

Deakin R (1855) Flora of the Colosseum of Rome. Oxford University, Oxford

Eaton M, Aebischer N, Brown AF, Hearn R, Lock L, Musgrove AJ, Noble D, Stroud D, Gregory R (2015) Birds of conservation concern 4: the population status of birds in the UK, Channel Islands and Isle of Man. Br Birds 108:708–746

Hurford C (2006) Identifying the conservation priority: using limited resources to best effect. In: Hurford C, Schneider M (eds) Monitoring nature conservation in cultural habitats. Springer, Dordrecht, pp 61–64

Hurford C (2013) Gupton farm: a report on the land-use, arable plant flora and wintering farmland bird population, July 2012 – March 2013. Unpublished report

Johnstone I, Bladwell S (2016) Birds of conservation concern in Wales 3: the population status of birds in Wales. Birds Wales 13(1):3–32

Jordan T (2015) Konfliktkunskap för rovdjursförvaltningen. Rapport 6674, Naturvårdsverke. ISBN 978-91-620-6674-1. ISSN 0282-7298

National Trust (2014). https://www.youtube.com/watch?v=iZNZtBIzy2U

NERC (2006) NERC biodiversity duty, 814 Section 40 of the Natural Environment and Rural Communities Act 2006

Newton I (2017) Farming and birds. William Collins, London

Owen H (ed) (1918) Calendar of Pembroke shire records. London

Preston CD, Pearman DA, Dines TD (2002) New atlas of the British and Irish flora. Oxford University Press, Oxford

Shellswell C (2013) Gupton farm – arable plant survey. Plantlife, Salisbury

Westberg L, Waldenström C (2016) How can we ever create participation when we are the ones who decide? On natural resource management practice and its readiness for change. J Environ Policy Plann

Wilson PJ (2013) Options for the management of Gupton farm, Castlemartin. Unpublished report

Wilson PJ, King M (2003) Arable plants – a field guide. WILDguides, Hampshire

Chapter 18
The Conservation of the Grey Partridge (*Perdix perdix*) on Farmland in Southern England

N. W. Sotherton, N. J. Aebischer, and J. A. Ewald

Abstract The grey partridge (*Perdix perdix*) was once a common breeding bird in Britain but has declined by over 90% over the last 50 years. Beginning in 1968, the Game & Wildlife Conservation Trust (GWCT) has investigated the causes of this decline in Britain and identified three main factors involved in it. These are the loss of nesting habitat, loss of insect-rich brood-rearing habitat and increased predation during the breeding season. The conservation of a once widespread species of farmed land such as the grey partridge requires bottom-up intervention, so remedial measures need to be appropriate for farmers and their advisors. Therefore, such solutions needed to be costed, ideally supported by subsidy (agri-environmental schemes), pragmatic and simple. This has resulted in a GWCT grey partridge management package based on scientifically proven recommendations. We now need to persuade farmers and land managers to adopt the management package nationally and reverse the decline of this species. An important factor in achieving adoption is attracting subsidy support for the management options and providing demonstration sites where visiting farmers can see how appropriate and practical management leads to successful recovery of grey partridges. For example, at a landowner-led demonstration site in West Sussex, southern England, partridges went from near extinction to densities providing enough birds in autumn to reinstate a wild grey partridge shoot. The detailed results are reported here, and offer hope for more widespread recovery of the species across the UK.

Keywords Grey partridge *Perdix perdix* · Farmland birds · Biodiversity · Sustainable use · Sussex study · Agri-environment schemes

N. W. Sotherton (✉) · N. J. Aebischer · J. A. Ewald
Game & Wildlife Conservation Trust, Hampshire, UK
e-mail: nsotherton@gwct.org.uk; naebischer@gwct.org.uk; jewald@gwct.org.uk

C. Hurford et al. (eds.), *The Changing Status of Arable Habitats in Europe*,
https://doi.org/10.1007/978-3-030-59875-4_18

1 Introduction

Numbers of the grey partridge (*Perdix perdix*) (Fig. 18.1) a breeding farmland bird found across Europe and a traditional quarry species, have declined by 92% over the last 50 years (Massimino et al. 2019) (Fig. 18.2). Across Europe, grey partridges have shown a long–term decline amounting to 94% over the last 35 years (PECBMS 2017). In the UK the species appears on the Red List of Birds of Conservation Concern (Eaton et al. 2015). It was also a priority species identified by the UK Biodiversity Action Plan process (Anon 1995) and is one of 19 species included in the Farmland Bird Indicator, which is a UK government biodiversity indicator for the natural environment (DEFRA 2011).

Concerns generated by the grey partridge decline have prompted considerable research in the UK and elsewhere (Potts 1986, 2012; Aebischer et al. 2000; Aebischer and Ewald 2012; Bro et al. 2005). Beginning in 1968, the Game & Wildlife Conservation Trust (GWCT) has investigated the causes of the UK decline and identified three main factors involved in it, linked to agricultural intensification: the destruction of nesting habitat, resulting in poor holding capacity; a pesticide-induced reduction in chick-food insects in crops, leading to poor chick survival; increased predation pressure on the remaining habitat, leading to adult and nest losses (Potts 1986, 2012; Aebischer et al. 2000; Aebischer and Ewald 2012). Similar factors have been identified elsewhere in Europe. Nevertheless, the declines have continued.

Fig. 18.1 Covey of grey partridges in a field margin in the Sussex Study Area. (Copyright: Peter Thompson)

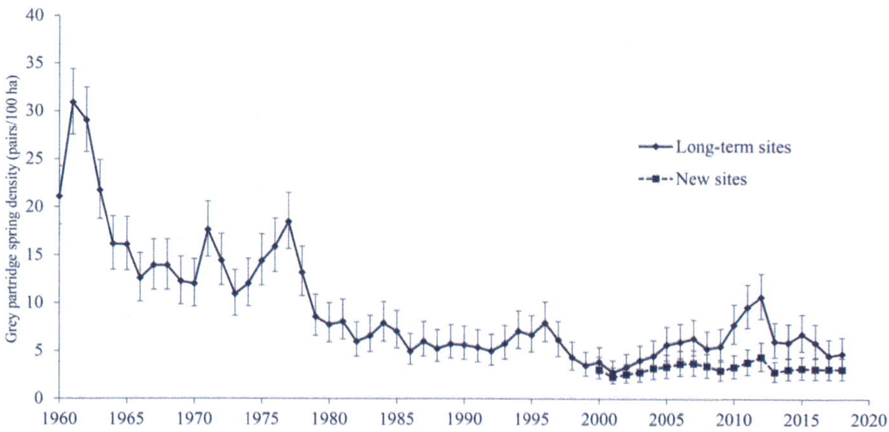

Fig. 18.2 Trends in grey partridge spring pair density counted through the GWCT's Partridge Count Scheme, controlling for variation in different count areas. Trends are given for those sites that have been counted prior to 1999 (long-term sites) and those sites that have been counted since 1999 (new sites)

The conservation of a once widespread species of farmed land such as the grey partridge requires bottom-up intervention, so remedial measures need to be appropriate for farmers and their advisors. Therefore, such solutions needed to be costed, ideally supported by subsidy (agri-environmental schemes), pragmatic and simple. They involve various habitat improvements and the control of generalist predators during the breeding season. The GWCT has provided experimental evidence of the effectiveness of providing brood–rearing and nesting habitats. Brood sizes nearly doubled where 'conservation headlands' were used (Sotherton 1991). In this technique, the outer 6 m of cereal crops are sprayed only with selective insecticides and herbicides to protect the insect foods of young partridge chicks and their insect host plants. Raised mid–field strips of perennial grasses ('beetle banks') provide appropriate nesting cover (Thomas et al. 1991; Sotherton 1995). Many of these habitat improvements have been included in the UK's agri–environment schemes, in which farmers are asked to deliver environmental improvements on their farm including improvements to biodiversity and the conservation of declining species of farmland wildlife in return for farm support. This has resulted in a GWCT grey partridge management package based on scientifically proven recommendations. The success of these schemes to support grey partridges depends on having in place all the options that restore the multiple habitat requirements of the partridge over its annual cycle (Ewald et al. 2010), including legal predator control during the breeding season to protect nests, sitting hens and young chicks (Tapper et al. 1996).

We now need to persuade farmers and land managers to adopt the management package nationally and reverse the decline of this species. An important factor in achieving adoption is attracting subsidy support for the management options and providing demonstration sites where visiting farmers can see how appropriate and practical management leads to successful recovery of grey partridges. In this paper we describe how, at a landowner-led demonstration site in West Sussex (Fig 18.3), southern England, targeted management has resulted in the re–establishment of a wild grey partridge population after nearly 50 years of decline.

2 Materials and Methods

The Sussex study area (32 km^2) is located on the Sussex Downs between the rivers Arun and Adur in West Sussex. The land is primarily arable farmland, with cereals and break crops interspersed with grass fields and scattered patches of traditional chalk grassland managed by grazing sheep (see Ewald et al. this volume – Chap. 11, Fig. 11.1) (Fig. 18.3). Since 1968, the Game & Wildlife Conservation Trust (GWCT) has monitored grey partridges (Potts 1986; Potts and Aebischer 1991, 1995) after harvest, usually in early September. Partridge coveys (family groups) are located by driving systematically across stubble fields in four-wheel-drive vehicles at dawn and dusk. We record the number of young in each covey, as well as the number of adult males and females. This allows us to calculate the autumn density of grey partridges. Also, spring breeding density is calculated from these autumn counts, with each male and any single female representing a spring pair. Other parameters of grey partridge demography are also calculated, namely, chick survival rate (the percentage of chicks that survive up to 6 weeks of age), young-to-

Fig. 18.3 Typical landscape of the Sussex study area. (Copyright: Peter Thompson)

old ratio (number of young produced per old bird) and brood production rate (the percentage of spring pairs that produced at least one chick, Potts 1986; Ewald et al. 2009). In this paper we use two measures of chick production, chick survival rate and young-to-old ratio (YTO). Chick survival rate can be susceptible to the effects of small sample size, whereas YTO is less so (Aebischer and Ewald 2004).

2.1 Management

In 2003, one of the 12 landowners within the study area set out to restore grey partridge numbers, eventually on an area of 1052 ha making use of all possible habitat creation and legal predation control measures. The remaining 22 km^2 of the study area forms the 'conventional area' where only limited habitat management or no predator control occurred. Habitat management included increases in nesting, insect-rich brood-rearing and overwinter covers. Nesting habitat was improved through the addition of 25 km of beetle banks and new hedgerows fringed by grassy banks on either side of the hedge. The beetle banks are linear strips of grass grown on a bank 0.5 m high and 2.0 m wide and placed across large tracts of open arable land. They are planted mainly with cock's-foot (*Dactylis glomerata*), a dense perennial tussock-forming grass, with the addition of shrubs such as hawthorn (*Crataegus monogyna*) and blackthorn (*Prunus spinosa*) and some other species such as holly (*Ilex aquifolium*) at intervals for cover. This additional nesting cover means that the managed area now has 8.2 km of nesting cover for each square kilometre.

Brood-rearing cover (suitably structured insect-rich strips of vegetation) increased to nearly 9% of the area, with a total of 97 ha of appropriate cover such as conservation headlands put in place. In addition, there is no summer insecticide used on any part of a cereal crop and at least one third of the conservation headlands are not harvested, providing some cover and seeds into the autumn. As well as the winter cover provided through the unharvested conservation headlands, 2.1 km/km^2 of wild bird cover strips were sown, incorporating various mixes that included kale (*Brassica oleracea*), chicory (*Cichorium intybus*), millet (*Panicum miliaceum*) and canary grasses (*Phalaris canariensis* and *P. arundinacea*). These provide cover through the winter months as well as some food resources throughout the year. Cropping patterns have been adjusted to ensure that there is a different crop on either side of a field boundary. Where possible, fields are sown in autumn on one side of a field boundary with spring sowing on the other side, ensuring that on at least one side of each boundary there is vegetative cover at all times of year. Feeders containing wheat were placed at intervals along the beetle banks and field boundaries, at a rate of two feeders for every pair of grey partridges or 40 per 100 ha, and are filled from October into June. Some of the new habitats were originally funded through agri–environment options of Environmental Stewardship, specifically the Higher-Level Scheme (HLS, Natural England 2010) and now through Countryside Stewardship Higher Tier (Natural England 2010). The remaining habitats and the keepering were funded privately by the landowner.

Further to these habitat improvements, two full-time and one part-time gamekeepers are deployed from February to July. Most of their time is devoted to legal predation control of common generalist predators targeted at reducing predation on grey partridge adults, eggs and chicks. This involves the culling of foxes (*Vulpes vulpes*), stoats (*Mustela erminea*), weasels (*Mustela nivalis*), rats (*Rattus norvegicus*), carrion crows (*Corvus corone*) and magpies (*Pica pica*). Methods include shooting with a rifle at night, stopped snares (with a breakaway for ease of release of non–targets) and Larsen traps (specifically for corvid control). Predator species legally protected in the UK (e.g. all raptors) are not removed. For the rest of the year the gamekeepers establish and manage the specially created habitats and provide the supplementary food.

Results from this demonstration are compared with the GWCT's own national database on partridges called the Partridge Count Scheme (PCS). The PCS is a voluntary non-random recording scheme whereby participants (farmers/gamekeepers) are asked to count the partridges on their land twice a year to enable the GWCT to monitor the number of breeding pairs and their productivity.

2.2 Statistical Analysis

We compared the density of spring pairs and the autumn density of grey partridges, two measures of grey partridge yearly production (chick survival rate and YTO) and a measure of nesting success (brood production rate), on the managed area with those on the conventional area in two time periods, the 10 years before (1993–2002) and the 10-year period after (2009–2018) grey partridge management was implemented fully across the managed area.

Generalized Linear Mixed models (GLMM) with a Poisson error and logarithmic link were used to examine the effect of management on spring pairs and the number of partridges counted in the autumn (all with an offset of the area counted, transformed using natural logarithms, so that the quantities analysed are effectively densities), while GLMMs with a normal error and identity link were used to compare chick survival rate (logit transformed, restricted to years when more than two broods of chicks were counted), and YTO (transformed using natural logarithms). The model included pre- and post-management entered as categorical variables describing the temporal element of management (0 = pre- management, 1 = post-management), year as a categorical variable (nested within the temporal element of management and entered as a random variable) and a categorical variable for the spatial element of management (managed versus conventional area, coded managed = 1, conventional =2). The effect of management was tested using the interaction between the temporal and spatial elements of management. Model parameters were estimated using the method of Schall (1991) and fixed effects were estimated by restricted maximum likelihood. Significance of model terms was tested using the Wald statistic.

All statistical analysis was carried out in Genstat Release 19.1 for Windows.

3 Results

The breeding pair density on the conventional area averaged 3.1 pairs/km^2 in the 10 years before grey partridge management began (1993–2002), while the managed area averaged 2.3 pairs/km^2 over the same time period. Breeding density on the managed area increased to an average of 15.4 pairs/km^2 from 2009 to 2018, while on the conventionally managed area pair density declined to 1.6 pairs/km^2 (Fig. 18.4). There was a significant increase on the managed area compared to the conventional area (Wald statistic = 267.02, df = 1, $P < 0.001$). Autumn densities in the 10 years before management averaged 8.3 birds/km^2 on the area that was subsequently managed, with 10.6 birds/km^2 on the area that remained conventionally managed. In the 10 years after management was established autumn density on the managed area averaged 104.4 birds/km^2, compared to 6.9 birds/km^2 on the conventionally managed area, with management significantly affecting autumn density (Wald statistic = 22.41, df = 1, $P < 0.001$) (Fig. 18.5). On the managed area the increase in nesting habitat, combined with the control of predation at the nest, has resulted in brood production rates (71% on average) that are more than double what they were before the management began (28% on average). On the conventional area, brood production rates averaged 27% from 1993 to 2002 compared to 45% from 2009 to 2018, with a significant difference in the change on the two areas (Wald statistic = 16.24, df = 1, $P < 0.001$). Comparing chick survival over the managed area to the rest of the Sussex study area from 1993 to 2002 showed that the rate on the managed area was 32% on average, while on the conventional area it was 29% before management, rising to 44% on average on the managed area in 2009 to 2018 compared to 30% on the conventional area, though there was no significant effect of

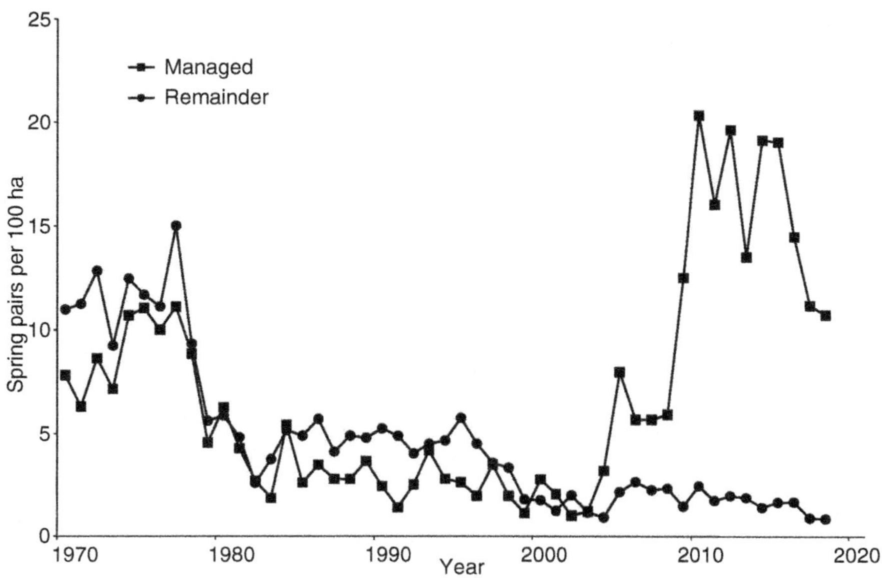

Fig. 18.4 Density of breeding pairs of grey partridge on conventionally farmed and managed areas

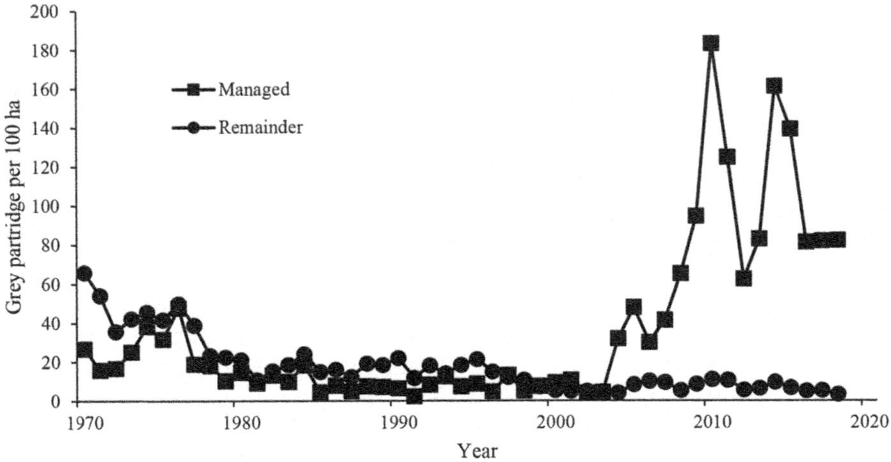

Fig. 18.5 Autumn density of grey partridge on conventionally farmed and managed areas

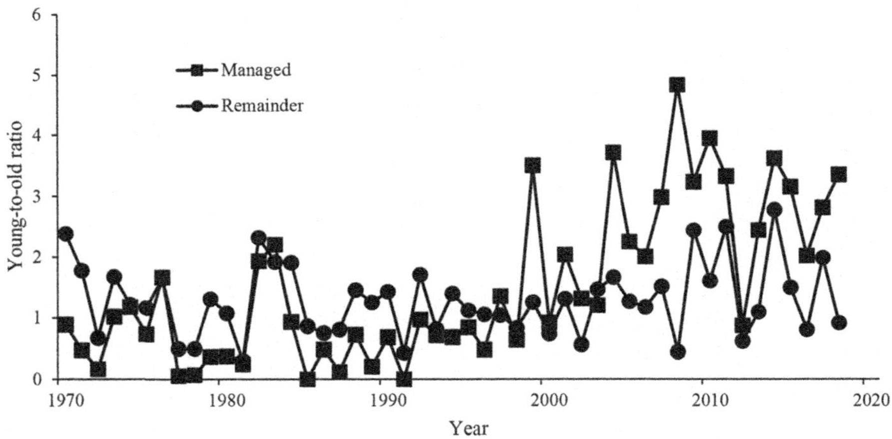

Fig. 18.6 Grey partridge young-to-old (YTO) ratio on conventionally farmed and managed areas

management (Wald statistic = 1.65, df = 1, P = 0.215). Before management, the conventionally managed area averaged 0.96 YTO, compared to 1.04 YTO on the area that went on to be managed. In the 10 years after management was established the managed area averaged 2.68 YTO (average brood size was 7.2 chicks per brood), compared to 1.44 YTO on the conventionally managed area (average brood size was 6.0), with a significant effect of management (Wald statistic = 5.89, df = 1, P = 0.026) (Fig. 18.6).

Invertebrates were sampled annually from over 100 fields on this same study area since 1970 and changes in the abundance of various taxa were tracked through time. This is discussed in Chap. 11 of this volume (Ewald et al. 2010).

4 Discussion

The recovery of grey partridges on part of the Sussex study area is a result of the hard work of the landowner and his team, with strong collaboration between the farming and keepering staff. The grey partridge went from nearly extinct on this area to densities that currently support sustainable driven shooting of a wild population. The landowner considers that the ability to shoot wild grey partridges 1 or 2 days a year justifies the substantial investment that he has made into partridge recovery. The landowner also sees this project as a means of encouraging others through demonstration of the art of the possible. Certainly, the public funds currently available for agri–environmental work would allow others to put in place the habitat management required to turn around the fortunes of the grey partridge on a much wider scale.

How do the results reported here compare to other areas where specific partridge habitat management has been undertaken? The average spring pair density on the managed area (15.4 pairs/km^2) matches the densities (an average of 14.9 pairs/km^2 from 2006 to 2010) reached on a previous GWCT demonstration project on farmland on chalk in Hertfordshire, where both habitat management and predator control were used (Aebischer and Ewald 2010). It is double the density achieved on the Salisbury Plain experimental area, where only predator control was used to boost grey partridge numbers (Tapper et al. 1996). It does, however, fall far short of the densities (80 pairs/km^2) locally reached in some hunting estates in northern France, where similar habitat management and predator control appropriate for the region was undertaken (Bourdouxhe 2002; Bro et al. 2005).

The English grey partridge model (Potts 1986) shows that predation control approximately triples the equilibrium population level expected from increases in insect–rich habitat and nesting cover alone (Aebischer 1991). At 15 pairs/km^2, the results on the managed area are in line with what the model predicts for a fully managed shot population. Without the predation control (and without shooting), the expectation would be around 5 pairs/km^2. The chick survival rates seen on the conventional portion of the Sussex study area are lower than the long–term average reported for the post–decline era (32.3%; Potts and Aebischer 1995), indicating that the situation on conventional sites may have deteriorated. Chick survival rates on the managed area are close to the level recorded before grey partridges began declining in number with the onset of herbicide use in cereals (48.6%; Potts and Aebischer 1995). They surpass those of the original experiments that verified the ability of conservation headlands to provide chick–rearing habitat, and that produced a mean grey partridge brood size of 6.4 on the original conservation headland areas (Rands 1985). The cereal area covered by conservation headlands in the Sussex managed area exceeds the recommendations first made when conservation headlands were developed (6%; Boatman and Sotherton 1988) and this possibly explains the higher chick survival rates.

How likely is it that the management described here could become widely adopted across the arable areas of the UK? Given the agri–environment options available, the habitats established in Sussex could all be incorporated into farming regimes across the country through the use of the current agricultural subsidy scheme. In the case of the GWCT's Partridge Count Scheme (PCS), increases in chick survival rates on the area managed by PCS members are associated with their use of in–field agri–environment options that are designed to provide chick–rearing and nesting habitats (Ewald et al. 2010). The area managed does not have to be of the order of 10 km^2 and could be made up of ground owned by one landowner or several working together. However, our experience is that at least one dedicated gamekeeper is needed for every 400 ha. The Salisbury Plain experiment has shown that, in the absence of habitat management, predation control alone on an area of between 4 and 5 km^2 can increase densities to allow for a shootable surplus of grey partridges (an average of 23% of autumn numbers counted; Tapper et al. 1996), but with numbers too low to provide revenue. Having a shootable surplus of partridges allows some return on a landowner's dedication towards grey partridge conservation and ensures the long–term viability of the management. It is crucial that the landowner is motivated to carry on this work, whether through conservation interest or though shooting, and this is the driving force behind the success of the Sussex project. However, we should note that although all the habitat and feeding requirements for grey partridge management can be funded through farm subsidy, the control of predators receives no publicly-funded support at this time.

The message that the GWCT's PCS tries to convey to its members across the UK is that it is possible to restore a wild grey partridge shoot on a modern, productive arable farm, provided about 9% of the total area is given over to partridge management. The agri–environment options available to help with this have never been better and all that remains is for more farmers and landowners to take up the challenge and restore grey partridges on their ground. The agri–environment programme in England is considered to be one of the most comprehensive in Europe, with its system of options allowing for greater flexibility than might be the case in other European countries. This complexity can be used to good effect for wildlife generally. A review of EU–wide agri–environment options found that just under half of all rural development programmes had options directed at management for wildlife including actions specifically aimed at providing food, nesting and breeding areas (Keenleyside et al. 2011). Only a quarter of the programmes in the newer EU countries had such options. The results here indicate how useful these options can be in restoring grey partridge numbers, emphasizing the need to include them in deliberations on the next round of reform of the Common Agricultural Policy, and in the UK's new farm subsidy system post-Brexit.

Despite the proven value of the science, the historical support of the farm subsidy system in Britain to deliver biodiversity, the body of evidence to show the positive impacts of legal seasonal predator control and the willingness of land managers to restore a much-loved quarry species, grey partridge recovery still faces an uncertain future. Grey partridge restoration is long-term, needs high levels of skill and management from the team on the ground and an agriculture with a stable future. Without a Brexit deal we face an uncertain future where the prospect of a farmer taking 10% of arable land out of production for conservation could seem foolhardy. At least the science is in place should we return to more certain times.

Acknowledgements We would like to thank all the landowners, farmers and gamekeepers within the Sussex study area. They have allowed us access to count partridges, collect invertebrates and cheerfully responded to requests for information for some 50 years. In the long run, they and land managers like them are responsible for the future of grey partridge conservation across Europe.

References

Aebischer NJ (1991) Sustainable yields: gamebirds as a harvestable resource. Gibier Faune Sauvage 8:335–351

Aebischer NJ, Ewald JA (2004) Managing the UK Grey partridge *Perdix perdix* recovery: population change, reproduction, habitat and shooting. Ibis 146(Supplement 2):181–191

Aebischer NJ, Ewald JA (2010) Grey partridge *Perdix perdix* in the UK: recovery status, set–aside and shooting. Ibis 152:530–542

Aebischer NJ, Ewald JA (2012) The grey partridge in the UK: population status, research, policy and prospects. Anim Biodivers Conserv 35(2):203–212

Aebischer NJ, Green RE, Evans AD (2000) From science to recovery: four case studies of how research has been translated into conservation action in the UK. In: Aebischer NJ, Evans AD, Grice PV, Vickery JA (eds) Ecology and conservation of lowland farmland birds. British Ornithologists' Union, Tring, pp 43–54

Anon (1995) Biodiversity: the UK steering group report, Volume 2: action plans. Her Majesty's Stationery Office, London

Boatman ND, Sotherton NW (1988) The agronomic consequences and costs of managing field margins for game and wildlife conservation. Asp Appl Biol 17:47–56

Bourdouxhe L (2002) Cent quintaux, cent perdreaux. Chasse et Nature 94:21–24

Bro E, Reitz F, Landry P (2005) Grey partridge *Perdix perdix* population status in central northern France: spatial variability in density and 1994–2004 trend. Wildl Biol 11:287–298

DEFRA (Department for Environment, Food and Rural Affairs) (2011) UK Biodiversity indicators in your pocket 2011. www.jncc.defra.gov.uk/biyp

Eaton MA, Aebischer NJ, Brown AF, Hearn RD, Lock L, Musgrove AJ, Noble DG, Stroud DA, Gregory RD (2015) Birds of conservation concern 4: the population status of birds in the UK, Channel Islands and Isle of Man. Br Birds 108:708–746

Ewald JA, Kingdon NG, Santin-Janin H (2009) The GWCT Partridge Count Scheme: a volunteer–based monitoring and conservation promotion scheme. In: Gamebird, 2006: 27–37 (SB Cederbaum, BC Faircloth, TM Terhune, JJ Thompson, JP Carroll, Eds). Warnell School of Forestry, Athens

Ewald JA, Aebischer NJ, Richardson SM, Grice PV, Cooke AI (2010) The effect of agri–environment schemes on grey partridges at the farm level in England. Agric Ecosyst Environ 138:55–63

Keenleyside C, Allen B, Hart K, Menadue H, Stefanova V, Prazan J, Herzon I, Clement T, Povellato A, Maciejczak M, Boatman N (2011) Delivering environmental benefits through entry level agri–environment schemes in the EU. Report prepared for DG environment, Project ENV.B.1/ ETU/2010/0035. Institute for European Environmental Policy, London

Massimino D, Woodward I D, Hammond MJ, Harris SJ, Leech DI, Noble DG, Walker RH, Barimore C, Dadam D, Eglington SM, Marchant JH, Sullivan MJP, Baillie SR, Robinson RA (2019) Bird trends 2019: trends in numbers, breeding success and survival for UK breeding birds. Research Report 722. British Trust for Ornithology, Thetford

Natural England (2010) Higher level stewardship, Environmental stewardship handbook, 3rd edn

PECBMS (Pan–European Common Bird Monitoring Scheme) (2017) Trends of common birds in Europe, 2017 update. European Bird Census Council, Prague

Potts GR (1986) The partridge: pesticides, predation and conservation. Collins, London

Potts GR (2012) Partridge: countryside barometer. In: New naturalist library book 121. Collins, London

Potts GR, Aebischer NJ (1991) Modelling the population dynamics of the Grey Partridge: conservation and management. In: Perrins CM, Lebreton J-D, Hirons GJM (eds) Bird population studies: relevance to conservation and management. Oxford University Press, Oxford, pp 373–390

Potts GR, Aebischer NJ (1995) Population dynamics of the Grey partridge *Perdix perdix* 1793–1993: monitoring, modelling and management. Ibis 137(Supplement 1):S29–S37

Rands MRW (1985) Pesticide use on cereals and the survival of grey partridge chicks: a field experiment. J Appl Ecol 22:49–54

Schall R (1991) Estimation in generalized linear models with random effects. Biometrika 78:719–727

Sotherton NW (1991) Conservation headlands: a practical combination of intensive cereal farming and conservation. In: Firbank LG, Carter N, Derbyshire JF, Potts GR (eds) The ecology of temperate cereal fields. Blackwell Scientific Publications, Oxford, pp 373–397

Sotherton NW (1995) Beetle banks–helping nature to control pests. Pestic Outlook 6:13–17

Tapper SC, Potts GR, Brockless MH (1996) The effect of an experimental reduction in predation pressure on the breeding success and population density of grey partridges *Perdix perdix*. J Appl Ecol 33:965–978

Thomas MB, Wratten SD, Sotherton NW (1991) Creation of island habitats in farmland to manipulate populations of beneficial arthropods – predator densities and emigration. J Appl Ecol 28:906–917

Chapter 19
The Changing Status of Steppe-Land Birds in the Lleida Plain of Catalonia

Santi Mañosa, Gerard Bota, David Giralt, and Joan Estrada

Abstract The Lleida plain is a flat and dry region in Western Catalonia (NE Spain), at the eastern edge of the Ebro Valley depression. The central part of the area is now devoted to intensively irrigated farmland, but a surrounding belt of steppe and pseudo-steppe and low intensity farming, mainly occupied by winter cereal, olive trees and almond trees, remains. The area still hosts a rich community of steppe land birds. Some are dependent on low intensity dry cereal farming and fallow, such as the little bustard (*Tetrax tetrax*), Montagu's harrier (*Circus pygargus*) and calandra lark (*Melanocorypha calandra*). Other, such as the pin-tailed sandgrouse (*Pterocles alchata*) and black-bellied sandgrouse (*Pterocles orientalis*) inhabit the few remaining large spots of fallow land, while the Dupont's lark (*Chersophilus duponti*) and lesser short-toed lark (*Alaudala rufescens*) are both associated to the remaining patches of natural steppes or halophytic vegetation. The greater short-toed lark (*Calandrella brachydactyla*) and the European thick-knee (*Burhinus oedicnemus*) favour areas of little disturbed wastelands, fallow or ploughed land. Although more generalist, the lesser kestrel (*Falco naumanni*) or the European roller (*Coracias garrulus*) also find in these low intensity farming areas one of their last strongholds in Catalonia. The single remaining Iberian population of lesser grey shrike (*Lanius minor*), on the verge of extinction, is also located within these steppes. All these species experienced severe population declines during the last decade of the twentieth century and the first decade of the twenty-first. Strong opposition to the *Segarra-Garrigues* irrigation system, launched at the start of the twenty-first century and aiming to transform most of the remaining steppes of the Lleida plain into

S. Mañosa (✉)
Departament de Biologia Evolutiva, Ecologia i Ciències Ambientals, Facultat de Biologia, Institut de Recerca de la Biodiversitat (IRBio), Universitat de Barcelona, Barcelona, Catalonia, Spain
e-mail: smanosa@ub.edu

G. Bota · D. Giralt
Centre de Ciència i Tecnologia Forestal de Catalunya (CTFC), Solsona, Catalonia, Spain

J. Estrada
Institució per a l'Estudi, Gestió i Recuperació dels Ecosistemes Lleidatans (EGRELL). Centre Cívic del Centre Històric, Lleida, Catalonia, Spain

C. Hurford et al. (eds.), *The Changing Status of Arable Habitats in Europe*,
https://doi.org/10.1007/978-3-030-59875-4_19

intensified irrigated crops, has boosted many conservation actions, which are reversing or halting the negative trends for some of these species. These actions include, among others, the declaration of eight Special Protection Areas (SPAs), the renting or acquisition and management of steppe-land and fallow land, the provision of nesting sites, the implementation of breeding programs to reintroduce or reinforce some populations, or the mitigation of power lines to reduce collision risk. Although there is room for some optimism, the persisting intensification of agriculture outside and inside the SPAs, and the industrial and urban development of the area are still of great concern. We need a change in the rural development concept, from one that values only economic services to one that values social and ecological services as well, if we want some room for steppe-land birds in the Lleida plain.

Keywords Agricultural intensification · Bird conservation and management · Mitigation and compensatory measures · Pseudo steppes · *Segarra-Garrigues* irrigation system · Steppe land birds

1 Introduction

Steppe-land birds in Europe are adapted to live in flat areas where climatic conditions (low precipitation mainly) and/or poor soil, often in combination with some kind of low intensity farming or grazing activity, prevent the development of trees and keep open landscapes with high visibility. In such conditions, natural vegetation and crops are limited to natural grass, a few scrubs or grassy crops (mainly cereal) and orchards (like almond and olive), which give refuge to a bird community specifically adapted to such kind of habitats and conditions (Valverde 1958; De Juana et al. 1988; Tellería et al. 1988; Bota et al. 2005). In these areas, called pseudo-steppes (Suárez et al. 1997), fallow fields inserted into an agricultural matrix dominated by extensive winter cereal crops, play a critical role for the conservation of the steppe bird community (Traba and Morales 2019). Therefore, changes of farming practices, such as land abandonment or farmland and rangeland intensification, produce rapid habitat modifications at the landscape and field scale, which breaks the delicate equilibrium between the needs of most of these species and the prevailing habitat conditions. We found a good example of this in the steppe-land birds in Catalonia, mainly located in the Lleida plain, which have experienced strong population changes in the last decades.

The Lleida plain, in Western Catalonia (Fig. 19.1), is a flat and relatively dry (rainfall less than 450 mm/year) region within the Ebro Valley, in the North-East of the Iberian Peninsula. Formerly covered by dry oak forest or steppes (Calvet et al. 2004), the area was historically farmed and grazed for centuries, and most forests were probably removed for wood or charcoal production. Up to the end of the nineteenth century, dry Mediterranean agriculture and sheep grazing were the only possible farmland outcomes (except near the rivers, where some irrigation developed for centuries), and the landscape was dominated by a mosaic of traditional crops

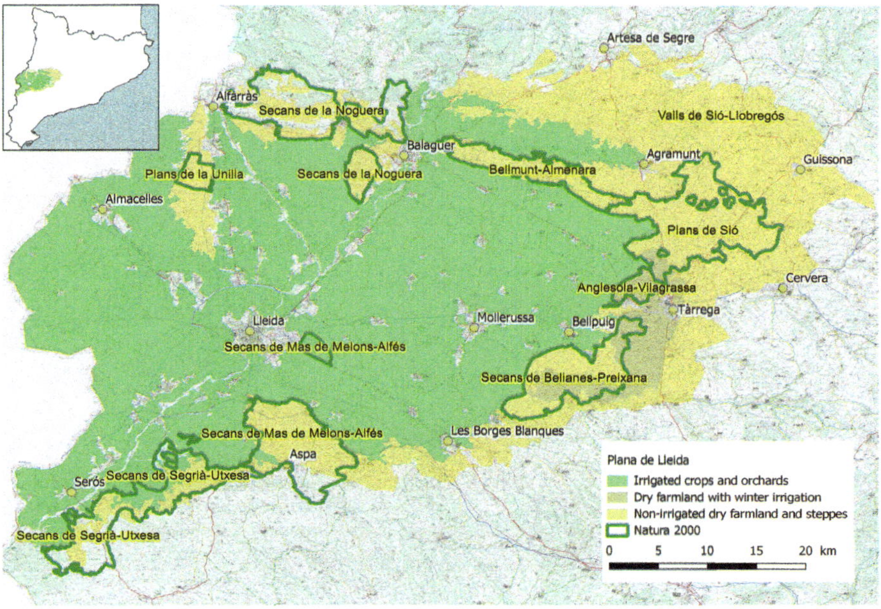

Fig. 19.1 Map of the Lleida plain in Catalonia, showing the general distribution of the irrigated areas (green) and the remaining belt of pseudo-steppes (brown). The limits of the Special Protection Areas (SPAs, Natura 2000) for steppe land birds are also shown

(wheat, barley, almond trees, olive trees, vineyards), fallow land and natural steppes, which allowed the presence of rich botanical and ornithological steppe communities (Fig. 19.2). By the end of the nineteenth century and the first half of the twentieth, however, the development of the *Canal d'Urgell* irrigation system allowed the transformation of some 70,000 hectares of the central parts of the Lleida plain into irrigated land. Subsequently, during the second half of the twentieth century, new irrigation systems (i.e. *Canal Catalunya-Aragó, Elevacions de Torres de Segre, Canal Algerri-Balaguer*) further reduced the extent of steppe and pseudo-steppe habitats, which were limited to a narrow belt of dry farmland surrounding a central patch of irrigated land dominated by alfalfa fields, corn fields and tree orchards (Fig. 19.1). By the final quarter of the twentieth century, the intensification of cereal farming progressively reduced the quality of the remaining pseudo-steppes, and most steppe-land bird species faced important population reductions, which already raised the concern of ornithologists and ornithological organisations. The final blow occurred in 2002–2010, when a new irrigation project was launched, the *Canal Segarra-Garrigues* system, aiming to irrigate a further 70,000 hectares of land, including most of the above-mentioned remaining belt of pseudo-steppes. This project faced strong opposition by environmental and ornithological organizations and, after a long process (Muñiz 2010) involving politicians, farmer associations and

Fig. 19.2 The Lleida plain steppes and cereal pseudo-steppes are a mosaic of cereal crops, fallow land and remains of semi-natural steppes which has traditionally been maintained by low intensity agriculture and sheep grazing. (Photo: Joan Estrada)

environmental NGOs, as well as the European Union, several mitigation and compensatory measures were established in order to reduce the severe impacts of the *Segarra-Garrigues* irrigation system, to preserve the pseudo-steppe and to maintain viable steppe bird populations within the Lleida plain. Our aim here is to review the current status, threats and trends of steppe-land bird populations in the Lleida plain, and to describe the several actions that have been undertaken to protect them, as well as to discuss the weaknesses, strengths and future directions of improvement.

2 Population Trends, Current Status and Threats of Steppe-Land Birds in the Lleida Plain

The land transformation and agricultural intensification (Fig. 19.3) that has taken place in the Lleida plain during the last 40 years has reduced the extent and quality of the habitat available for farmland birds. However, although reduced compared to what it used to be, the region still hosts a relatively varied steppe-land bird community. Specifically, 81% of the species considered as steppe-land birds in Spain (Traba et al. 2007) are regularly present, with variable abundance, in the Lleida plain. Most steppe-land bird populations in this area have attracted the attention of ornithologists since the 1980s, and are monitored periodically, so there is relatively good

Fig. 19.3 Irrigation leads to homogenization of the landscape, strong use of pesticides, and the loss of biodiversity. These modern irrigated tree plantation of almond trees or other fruit trees offer no opportunities for wildlife. (Photo: Joan Estrada)

information on the status and population trends of the most relevant bird species in the community. The steppe-land bird indicator (ICO 2019) shows a moderate decline of 30% in steppe-land bird populations in 2002–2018, making this the group of birds which has suffered the steepest decline in Catalonia.

The little bustard (*Tetrax tetrax*) is a ground nesting, fundamentally herbivorous bird, which, in the area, is absolutely dependent for successful breeding on the existence of traditionally managed and little disturbed cereal crops and herbaceous fallow land, used for nesting and raising the chicks. The species spends the breeding season in the steppes, and overwinters mainly in the irrigated alfalfas in the centre of the plain, as the cereal pseudo-steppes remain devoid of vegetation during autumn and winter (Ponjoan et al. 2011). In 2014, the population of breeding males was estimated at 649 breeding males, well below the 1293 males estimated in the previous count in 2002 (Bota et al. 2015). The number of females is difficult to estimate, because of their secretive behaviour, but the same authors gave estimates of 656 in 2002 and 232 in 2014. Reduction of the habitat available for breeding, as a result of land use changes (irrigation, urbanization…), is worsened by the reduced quality of the remaining patches, mainly by agricultural intensification. This results in very low breeding success (Lapiedra et al. 2011), which is probably the main problem currently faced by this population. In spite of the dramatic 50% decline between 2002 and 2014, partial standardized annual counts indicate that it has remained more or less stable afterwards (Guillem et al. 2019), probably reflecting a positive response to the management actions being carried out.

The Eurasian thick-knee (*Burhinus oedicnemus*) is a dweller of little disturbed wastelands, fallow, ploughed land and natural steppes, but may also occupy other habitats such as tree orchards and even irrigated crops, as long as they provide patches of bare ground. It feeds on any sort of invertebrates that it catches on soft soils. During the breeding season, the species occurs all over the Lleida plain. During the non-breeding season, some individuals migrate to other Iberian areas and some remain within the plain, forming non-breeding concentrations. No precise estimate of the current population breeding in the Lleida plain is available, but results of the Catalan Common Bird Monitoring Scheme (SOCC) indicates a 63% population decline in 2002–2019 (ICO 2019). Its population is probably suffering from the spreading of cereal monoculture, which replaces traditional vineyards, olive or almond tree orchards, and the ongoing decline of fallow land. Increased mechanization and frequent tillage also destroy many nests, and the use of pesticides has largely reduced the availability of soil invertebrates they need for food.

The pin-tailed sandgrouse (*Pterocles alchata*) and the black-bellied sandgrouse (*Pterocles orientalis*) are two seed eater specialists that place their nests on the ground, preferentially in dry natural steppes or old fallow land (Fig. 19.4). Besides their specific population trends, both species exhibit some inter-annual population variations, probably linked to their nomadic behaviour. The breeding population of pin-tailed sandgrouse in the Lleida plain since 2000 has been of only 60–80 birds (about 30–40 pairs), all of them within the SPAs (Giralt et al. 2019), but recent data indicate that it is now increasing, coinciding with the management actions being

Fig. 19.4 Typical sandgrouse habitat in the Lleida plain include fallow land and small shrub pastures in flat plateaus. (Photo: Joan Estrada)

carried out since 2014 (Giralt et al. 2020). The numbers in winter oscillate between 75 and 150 birds. However, GPS tagging indicates that almost all breeding birds overwinter in the neighbouring area of Aragon, so the origin and composition of the Lleida plain wintering population remains uncertain. The number of black-bellied sandgrouse breeding annually in the Lleida plain (2000–2018) oscillated between 9 and 34 birds, while the numbers present in winter range from 3 to 34 individuals (Giralt et al. 2019). However, winter and breeding populations seem to fluctuate independently, which, as reflected by a GPS tagging study conducted by Giralt et al. (2019), indicates that black-bellied sandgrouse in the Lleida plain commute from this area to the nearby region of Aragon. The abandonment of grazing practices, the reduction of fallow availability and the substitution of pseudo-steppes habitats by irrigated crops has considerably reduced the habitat available for both sandgrouse species. As a result, current black-bellied sandgrouse populations would not be viable in their own if considered in isolation from the larger nearby Ebro Valley population.

Montagu's harrier (*Circus pygargus*) is a ground nesting bird of prey which, in the area, used to nest in winter cereal crops (irrigated or not) and fodder crops (such as rye-grass *Lolium* spp. and tall fescue *Festuca arundinacea*). Nest destruction associated with the mechanization of harvest in cereal fields led the population to the edge of extinction in the late 1980s. Since then, extensive nest protection and a rescue campaign deployed by the *Generalitat de Catalunya* (regional autonomous government) has allowed the recovery of the population, which now fluctuates around 49–70 pairs, depending on the year (Varea et al. 2019).

The lesser kestrel (*Falco naumanni*) is a small colonial and migratory falcon which feeds mainly on large insects or small vertebrates. In the Lleida plain, it used to nest below tiles or in wall holes in traditional farm buildings, but became extinct as a breeding species in Catalonia in 1986. The population subsequently recovered following a reintroduction program started in 1990 by the *Generalitat de Catalunya*. Now, the population in the Lleida plain fluctuates between 40 and 100 breeding pairs, with a maximum of 101 in 2014 and a subsequent progressive decline to 58 breeding pairs in 2019 (Pont et al. 2020). The decline in the last 5 years after a long period of uninterrupted increase is consistent with similar fluctuations observed in other areas in Iberia and seems to be associated with conditions in the wintering grounds. As many nesting sites have been provided for the species in the last years (see below), these do not seem to be currently limiting the number of pairs in the area. Besides problems in the wintering range, at a local level the lack of food resulting from the elimination of fallow land and field margins, the spread of cereal monoculture and the increasing use of pesticides can compromise the future of the species in the area.

The European roller (*Coracias garrulus*) is a migratory bird that nests in natural holes on trees (Fig. 19.5) or on farmhouse walls. It mainly feeds on large insects and small vertebrates, captured on pastures, fallow or plough lands. The population in the Lleida plain in the nineties was estimated at 100–120 breeding pairs (Folch 2004) and was considered to be declining as a result of the disappearance of traditional farm buildings and felling of large olive and almond trees, which was reducing the

Fig. 19.5 In the Lleida plain pseudo-steppes, large almond trees on field margins provide nesting sites for many birds, such as the European rollers (*Coracias garrulus*), the little owls (*Athene noctua*), the Great spotted cuckoo (*Clamator glandarius)* or the hobby (*Falco subbuteo*), some taking advantage of the abandoned nests of the Iberian green woodpecker (*Picus sharpei*) or magpies (*Pica pica*). (Photo: Joan Estrada)

availability of holes for nesting. Counts in 2011 and 2017 reported 208 and 188 potential breeding territories respectively (CTFC 2018). Standardized annual counts conducted in 2010–2019 showed that the population is experiencing some increase in the Lleida plain, although subject to large inter-annual fluctuations (Giralt et al. 2020).

The lesser grey shrike (*Lanius minor*) is currently ranked as critically endangered in Spain and in Catalonia. In 2019, the total Spanish population was only five birds, involving a single pair, all of them in a unique location in the Lleida plain. This constitutes the most western location in Eurasia for the species, on the limit of its world distribution range. This passerine builds its nest on tree branches and feeds on large grasshoppers and beetles. The Lleida plain population is more than a hundred kilometres far from the nearest population in SE France and nowadays is mainly threatened by a low apparent survival and by demographic stochasticity associated to isolation and small population size (Giralt 2015). The long-term decline of the lesser grey shrike in the area has been linked to the disappearance of large trees for nesting and farmland intensification, though other possible causes such as climate change and droughts may have also contributed (Krištín 2008; Giralt and Valera 2007).

Dupont's lark (*Chersophilus duponti*) is a near-endemic passerine bird of the Iberian Peninsula. It is a strict dweller of flat and natural dry shrub steppes, which has been almost completely transformed to cropland in the Lleida plain. A single 100 ha

Fig. 19.6 The flat thyme steppe in the *Timomeda d'Alfès* gives refuge to the last Dupont's lark population in Catalonia. It was traditionally maintained by sheep grazing. (Photo: Joan Estrada)

patch of shrub steppe dominated by *Thymus vulgaris* and other shrubs, known as *Timoneda d'Alfés*, used to give refuge to up to 40 pairs when first reported in the eighties (Gómez-Catasús et al. 2018) (Fig. 19.6). The preservation of this single patch of thyme steppe was probably the result of a combination of factors, including the presence of a local military aerodrome, which prevented the area of being cultivated. This population, already prone to extinction because of small population size and isolation, gradually declined as a result of grazing abandonment and increasing leisure activities in the area, until the last Dupont's lark was reported in 2005 (Gómez-Catasús et al. 2018). However, the species was seen again in spring 2015, and subsequent censuses suggest a constant population of 7–8 males.

The calandra lark (*Melanocorypha calandra*) is a grassland passerine which favours cereal crops and herbaceous fallow land. According to the Catalan Common Bird Monitoring Scheme, SOCC (ICO 2019), in 2002–2019 the species has experienced a 31% population increase in Catalonia, a trend which is very apparent in the Lleida plain, where it can be found almost everywhere. In many cases, this species is the last steppe-land bird to disappear as cereal monoculture intensifies.

The greater short-toed lark (*Calandrella brachydactyla*) is a small summer visitor farmland passerine which used to be common in farmland habitats in Catalonia, but which is now restricted to a few localities, with more than 50% of the Catalan population hosted by the Lleida plain. It favours fallow land, undisturbed vineyards and natural steppes with high proportion of bare ground. According to the Catalan Common Bird Monitoring Scheme SOCC (ICO 2019), this species has experienced a 69% decline between 2002 and 2018. The population in the Lleida plain is less than 180 individuals and about 82% is within the Special Protection Areas (Sazatornil et al. 2019). Intensive monitoring in the study area indicates that, in spite of the overall bad situation of the species, it remains stable or is slightly increasing in the last few years (Sazatornil et al. 2019), partly as result of the conservation actions that are being put in place.

The lesser short-toed lark (*Alaudala rufescens*) is also a scarce bird in the Lleida plain, tied to the remaining patches of halophytic vegetation, as well as the few remaining patches of thyme or rosemary steppes. Its population in the Lleida plain is probably fewer than a few hundred individuals, but it has received much less attention that its cousin, the greater short-toed lark, because in Catalonia the bulk of the species is found in the Ebro Delta, where it is considered secure. This species has experienced a sharp decline in the Lleida plain during the last few decades, as a result of land reclamation for agriculture, which destroyed many of the few remaining patches of halophytic vegetation. However, in the last 10 years it seems to have experienced local increases, as a result of conservation actions, particularly from the creation and management of long-term fallows.

3 Conservation and Management Initiatives: Opportunities and Constraints

Most of the steppe bird species present in Lleida plain are included in Annex I of the Birds Directive (Council Directive 2009/147/EC), so EU member states (and, in Spain, regional governments as well), are required to declare Special Protection Areas (SPAs) to guarantee the conservation of these species. The progressive intensification of cereal farming in the Lleida plain took place mainly during the eighties and the nineties of the last century. This involved an increased mechanisation of farming practices, combined with an increased use of fertilizers and herbicides, and the loss of rotational fallow and extensive sheep grazing. This induced the reduction of many steppe-land bird populations so that, by the end of the century, both sandgrouse, the lesser kestrel, the Montagu's harrier and the Dupont's lark were extinct, or on the verge of extinction. In that context, the development of the *Segarra-Garrigues* irrigation system at the beginning of the twenty-first century represented the definitive threat to the steppe-land habitats and steppe-land bird populations in the Lleida plain and, by extension, in Catalonia (Brotons et al. 2004). After decades of neglect and inattention by the local authorities, the strong opposition faced by this project from environmental NGOs and the scientific community lead to conviction of the Catalan and the Spanish authorities for the lack of commitment to the EU environmental rules. This finally forced the local authorities to address the issue and to establish some legal and administrative instruments to preserve the steppe-land areas in the Lleida plain, opening the door to many opportunities for management and conservation of steppe-land birds in the area. Now, some legal and management actions are being implemented, which are beginning to deliver results in terms of the conservation of bird populations. According to their objectives, these actions can be grouped as:

1. Actions aimed to establish and manage a network of protected areas.
2. Actions aimed to improve the quality of farmland habitats for birds.
3. Actions aimed to increase key limiting resources for some species.
4. Actions aimed to reintroduce or reinforce populations.
5. Actions aimed to reduce man-induced mortality.

Although the current situation of many species is far from being optimal and conservation actions need much improvement, the current scenario is better than it would have been in the absence of any legal or active measure (Brotons et al. 2004). We describe the several actions that are being taken, and provide a brief critical review of their successes and shortcomings.

4 Actions Aimed to Establish and Manage a Network of Protected Areas

The delimitation of a network of protected areas is often a necessary, although not always a sufficient step, to preserve biodiversity in a given area. In the Lleida plain, this network was established by means of the delimitation of Special Protection Areas (SPAs) for birds (action 4.1) where land use and human activities can be regulated in order to prevent habitat transformations. This has been combined with the acquisition and management of some emblematic, highly valued and critically endangered estates in terms of steppe-land habitat conservation (action 4.2 and 4.3), as a way to give full preservation of the singularities of these sites.

4.1 Designation of Special Protection Areas (SPAs) for Birds and the Special Plan of Protection and Management of the Natural Areas of the Lleida Plain

After many years of conflict and negotiation, as well as two EU convictions (one for insufficient designation of protected areas and another for the lack of mitigation measures for steppe-land birds within the *Segarra-Garrigues* irrigation system), eight SPAs were finally delimited in 2009, to protect the main steppe-land habitats in the Lleida plain (Fig. 19.1). The delimitation of these areas was largely based on the Important Bird Areas (IBAs) proposal of SEO/BirdLife (Viada 1998). This involved a total of 47,360 ha, on eight separate SPAs, each one with its own habitat and landscape peculiarities, and hosting a different array of bird species (Table 19.1).

Table 19.1 Composition of the ornithological communities on every steppe land SPA in the Lleida plain. The number of crosses indicates the relative density in every area where each species is reported as a breeder (x: rare; xx: common; xxx: abundant). (Based on several sources and own data)

	Anglesola-Vilagrassa (ES0000321)	*Secans de Belianes – Preixana* (ES0000479)	*Bellmunt-Almenara* (ES0000477)	*Secans de Mas de Melons-Alfés* (ES0000021)	*Secans de la Noguera* (ES5230021)	*Secans del Segrià-Utxesa* (ES0000480)	*Plans de Sió* (ES0000478)	*Plans de la Unilla* (ES5130035)
Area (ha)	**1122**	**6521**	**4039**	**7619**	**8960**	**7727**	**10,382**	**989**
Tetrax tetrax	xxx	xxx	xx	xx	x	x	xx	xx
Burhinus oedicnemus	xx	xx	xx	xx	x	xx	xx	x
Pterocles alchata				xx		x		
Pterocles orientalis					x	x		x
Coracias garrulus	xx	xx	xxx	xx	xx	xx	xx	x
Falco naumanni		x	x	x	xxx			x
Circus pygargus	xx	xx	x	x	x		x	
Lanius minor				x				
Alaudala rufescens				xx		x		
Calandrella brachydactyla		x		xx	x	xx		x
Chersophilus duponti				x				
Melanocorypha calandra	xx	xxx	xx	xxx	xxx	xx	xxx	xxx

Within these areas, the transformation into high intensity irrigation farming, which would have resulted in the complete disappearance of pseudo-steppes and steppe-land birds (Brotons et al. 2004), was prohibited. Measures to preserve steppe-land birds have also been promoted and any activity contrary to their conservation is forbidden or regulated according to a Special Plan of Protection and Management of the Natural Areas of the Lleida Plains (Limonium 2010) approved by the Catalan Government (Generalitat de Catalunya 2010). Although this was a fundamental step towards the conservation of steppe-land habitats in Catalonia, the functionality and effectivity of this network of protected areas is impaired by the fragmentation and isolation of the different SPAs from one another, which reduces connectivity and increases the border effects. Moreover, partial transformation (winter irrigation of cereal or orchard irrigation with minimum water supply) is allowed in some areas within the SPAs, particularly where historical irrigation rights already existed. Although theoretically less impacting, this partial transformation promotes plot concentration, monoculture and an increased use of inputs. By the time of declaration, some of the SPAs (i.e. *Belianes-Preixana*, *Anglesola-Vilagrassa*, *Plans de Sió*) had already been severely affected by the landscape transformations involved in the *Segarra-Garrigues* irrigation system.

An additional problem is that part of the breeding ranges of some steppe-land bird species (i.e. black-bellied sandgrouse, calandra lark, European roller, Eurasian thick-knee, greater short-toed lark) is not included within these protected areas. Although this may not pose a severe problem for the commonest species, it is particularly worrying for the rare greater short-toed lark or the black-bellied sandgrouse. Another paradigmatic example of this situation is provided by the Montagu's harrier. The breeding colonies of this hawk were initially included within the SPAs limits but, as a result of changes in land use and phenological conditions, most nest are now outside the protected areas (Varea et al. 2019). Even for those nests that are within the limits of a SPA, the birds concentrate their foraging activity on alfalfa fields on irrigated land outside them (Guixé and Arroyo 2011).

The SPAs network was devised considering only the distribution of the species during the breeding season. For that reason, the whole or part of the wintering range is not included within the SPAs for those species that perform local seasonal movements. Some of them move to nearby irrigated areas in the centre of the plain (green in Fig. 19.1), where they face increased risks of anthropogenic death (Marcelino et al. 2018). This is the case of the little bustard, the European Eurasian thick-knee or the black-bellied sandgrouse (not to mention the migratory species) which spent most of the non-breeding season outside the SPAs. Including the local wintering areas within the SPA network would be extremely difficult and would face strong opposition, because they encompass irrigated areas. However, these should not be an impediment for active management actions and regulations being implemented to reduce or eliminate the risk faced by the species on these sites. In spite of these constraints, for the most critically endangered species, such as the Dupont's lark, the lesser kestrel, pin-tailed sand-grouse or the little bustard, the areas designated as SPAs give refuge to the bulk or the totality of the breeding stocks. For these species, the current protected areas network offers the minimum necessary framework to undertake conservation initiatives for the breeding populations of most steppe-land birds in Catalonia.

4.2 Acquisition of the Mas de Melons *Estate*

Mas de Melons is an iconic 1340 hectares estate of steppe and pseudo-steppes in Catalonia, valued for being the main stronghold for pin-tailed sandgrouse in the Lleida plain, but also hosting good populations of larks, thick-knees and European rollers. The estate is also famous as being one of the best dispersal areas for juvenile Bonellis's or golden eagles (*Aquila fasciata* and *Aquila chrysaetos*, respectively). The estate was acquired by the *Generalitat de Catalunya* in 2004, with the objective to devote the entire reserve to steppe-land conservation. This was important because it ensured the preservation of one of the largest pieces of semi-natural steppes in the Lleida plain. The area is currently managed to promote fallow land and semi natural steppes to recover, which is giving particular benefit to the pin-tailed sandgrouse.

4.3 Acquisition and Management of the Timoneda d'Alfès

The *Timoneda d'Alfès* is a flat 100 ha estate hosting the last thyme steppe in the Lleida plain and the last Dupont's lark population in Catalonia (Fig. 19.6). The area was formerly occupied by a small military airport, which in part explained why it was not cropped and the steppe grazed and maintained. Things changed as plans to build the Lleida airport in the area were put forward on the nineties. After many years of popular fight, these plans were abandoned, the regional government acquired the estate and the aerodrome activity was stopped in 2015. Efforts to preserve the habitat included the restoration of grazing, as well as selective cuttings, to prevent vegetation encroachment, combined with the regulation of leisure activities at sensitive times of the year and areas. Future actions also include the restoration of thyme steppe in current cereal plots within and beside the estate, to increase the amount of habitat available for larks and pin-tailed sandgrouses.

5 Actions Aimed to Maintain and Improve the Quality of Farmland Habitats

The persistence of pseudo-steppe and steppe-land habitats in the Lleida plain is intimately tied to the persistence of low intensity farming and grazing practices. Delimiting protected areas would be useless if farming was abandoned or further intensified. In dry farmland, for instance, the maintenance of fallow fields is essential for the preservation of steppe-land bird populations (Giralt et al. 2018; Traba and Morales 2019). For instance, fallows are a key habitat for the survival of little bustard broods (Lapiedra et al. 2011) and a priority nesting and feeding substrate for the Eurasian thick-knee, both sandgrouse, and the greater short-toed lark, among others. However, the progressive intensification of dry cereal farming and the elimination within the CAP of the incentives to set-aside land, produced a dramatic reduction of this essential habitat, which was correlated with the decline of farmland birds in Spain (Traba and Morales 2019). In the Lleida plain, the maintenance of these essential bird-friendly farming practices at large is being promoted by means of an agri-environmental scheme (action 5.1), while a more specifically devised program to increase the carrying capacity for steppe-land birds is achieved by the leasing and management of fallow land (action 5.2).

5.1 *Development of Agri-Environmental Schemes (AES)*

An agri-environmental scheme of voluntary practices was developed within the SPAs to promote conservation-friendly farming practices. The farmers entering the program commit not to undertake nocturnal farming practices and to keep natural elements such as field margins, ponds and other elements of interest for birds and wildlife. In cereal fields, harvesting should be delayed some days (until 1st June–22th June, depending on the SPA), and stubble needs to be left undisturbed until September. Moreover, farmers on the scheme can choose to leave fallow fields for wild life, which should not be treated with herbicides or pesticides, and no tillage is allowed on these fallow fields until September. Alternatively, they can choose to leave unsown 5 m field boundary strips in cereal crops, on up to 10% of the farm. These strips should not be treated or disturbed during the main breeding period of birds (from 15th April to 1st September). These measures are subsidized by the European Agricultural Fund for Rural Development (EAFRD) and, in 2018, amounting to a total investment of about € 600,000 and benefiting more than 4100 hectares and 141 farmers. The specific environmental benefits obtained from this scheme have not yet been fully evaluated, but increased involvement and compliance with the measures are probably necessary to achieve maximum benefits for wildlife.

5.2 *Land Lending and Management*

Because of their very voluntary nature and their general scope, agri-environmental schemes are not always enough to guarantee the needs of more demanding and specialist bird species (Kleijn et al. 2006). As a compensatory measure of the

Segarra-Garrigues irrigation system, but also as a compensatory measure of the gravel-pit industry, and with the objective to guarantee the presence of fallow fields in optimal areas and to adapt their management to the specific requirements of the target species, in 2014 a program of lending and managing plots of land in the SPAs was initiated. This programme has achieved its objective of managing plots of land for bird conservation. By 2019, there were c.3400 ha within the SPA (Fig. 19.7), with an additional 200 ha already being managed by the gravel-pit industry. These plots are leased by the *Generalitat de Catalunya* Government and managed as fallow fields (Fig. 19.8) or as uncut alfalfa fields for wildlife (Fig. 19.9), for periods of 3–5 years in order to produce green cover for steppe-land birds. Fallow and uncut alfalfa fields are a key element for the conservations of steppe birds (Giralt et al. 2018; Mañosa et al. 1996) and orthopterans (Estrada et al. 2017) in the area. This fallow land, however, needs careful management by means of tillage, shredding or grazing, so as to increase habitat suitability for the target bird species (Sanz-Pérez et al. 2019). In the eastern areas (*Belianes-Preixana*, *Bellmunt*, *Anglesola Vilagrassa* and *Plans de Sió*), the target species is mainly the little bustard. There, fallow management aims to provide good cover for displaying males in April and May as well as good nesting cover for females and good brooding cover in late May, June and July. This demands a heterogeneous management programme, in order to obtain enough plots to satisfy these variable needs. In the western areas (*Alfés-Mas de Melons*, *Secans de la Noguera*, *Plans d'Unilla*, *Secans de Segrià Utxesa*), the sandgrouse and the

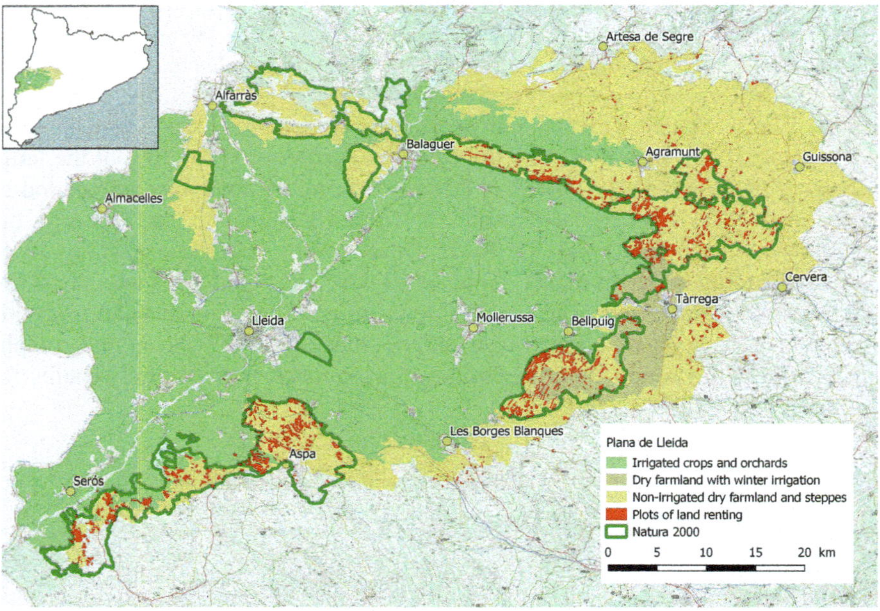

Fig. 19.7 Distribution of field plots that are being rented to promote steppe-land bird habitats, fundamentally fallow land and uncropped alfalfa

Fig. 19.8 Fallow field managed as habitat for steppe-land birds. In this case, this sort of green cover facilitates the presence of incubating female little bustard. (Photo: Joan Estrada)

Fig. 19.9 Fallow plot sown with alfalfa for the benefit of many steppe-land birds in the cereal steppe. This habitat is being much favoured by little bustard groups in summer and winter, when cereal crops are devoid of vegetation in the steppes (Photo: Santi Mañosa)

greater short-toed lark are the main target, and fallow fields are managed in order to obtain very low height and vegetation cover, and a high percentage of bare ground. This is done by means of well-planed cutting and tillage at the right time of the year, combined with grazing. Displaying little bustard males and sand-grouse show a clear preference for these managed plots (Guillem et al. 2019; Giralt et al. 2019; Sanz-Pérez et al. 2019). The number of little bustard families recorded in the areas where these managed plots have been implemented shows a remarkable upward trend and broods clearly select these plots. Another interesting result is that little bustard wintering flocks are again seen within the SPAs during the winter, after many years of absence, mainly using the managed alfalfa plots. This will help to keep little bustard within protected areas all year round, preventing them from needing to travel to unsafe areas during the non-breeding season. Also pin-tailed sandgrouse females show a clear preference to nest or to spend the breeding season in these managed plots. These results are encouraging, and future directions of improvement include our ability to select the best plots to manage and a better understanding of how to obtain the desired vegetation structure of each plot, which is very dependent on past history and weather conditions.

6 Actions Aimed to Provide Specific Limiting Resources for some Species

In traditional pseudo-steppes, large trees or farm buildings provided nest sites for lesser kestrels or European rollers, while sheep ponds provided water for sandgrouse which, as strictly granivorous birds, need to have access to a water supply for their chicks. Agricultural intensification and grazing abandonment lead to the loss of these elements, subsequently reducing the availability of nesting sites and water, and limiting the carrying capacity for these species. Providing artificial nest sites for kestrels and rollers (action 6.1) or water ponds for sandgrouse (action 6.2) aimed to restore the availability of these essential resources.

6.1 Building Nesting Towers and Boxes for the Lesser Kestrel and the European Roller

Agricultural modernization and land consolidation caused the traditional farm buildings to become abandoned and derelict. Old trees or almond-tree lines also disappeared as a consequence of the removal of field margins or the substitution of traditional tree orchards. As a result, there is scarcity of adequate nesting cavities available for the lesser kestrel or the European roller in the Lleida plains. Repairing these old buildings is often difficult or economically inefficient, so building secure nesting towers was seen as a more cost effective way to increase availability of

Fig. 19.10 Nesting tower specifically built to host lesser kestrels and European rollers. (Photo: Joan Estrada)

breeding sites. A total of 32 nesting towers have been built in the SPA and about 130 nest boxes put on trees or poles, to provide nesting cavities for these species. The nesting towers (Fig. 19.10) are 3 × 3 m square shaped buildings some 8 m of height which have nesting boxes on the inside of the walls, specifically adapted for the European rollers and lesser kestrel. In 2018, of a sample of 20 towers inspected, 5 were occupied by lesser kestrel, 15 by European rollers, 9 by barn owl (*Tyto alba*) and 19 by the Eurasian jackdaw (*Corvus monedula*) which was, by far, the main beneficiary of these structures. As a sedentary bird, it occupies the boxes before kestrels or rollers arrive from their wintering grounds in Africa. The entrances of the boxes are now being reduced in diameter and managed to prevent early occupancy by jackdaws. While these towers are now the main nesting substrate for the lesser kestrels (72% of the pairs in the Lleida plain nest on them), most European rollers use alternative substrates, natural or artificial boxes set on trees or poles (Pont et al. 2020; CTFC 2018).

6.2 Arrangement and Construction of Water Ponds

Water ponds were traditionally scattered across the Lleida plain pseudo-steppes, as a water supply for sheep flocks. These ponds were usually small, but numerous across the landscape. They were strategically placed to collect rainwater, and they were kept and cleaned so that water was permanently available. They were an excellent habitat for amphibians and crustaceans (such as *Triops* sp.) in an otherwise dry land, as well as an essential source of water for steppe-land birds during the summer. Contrary to other birds, which can fulfil their water needs from their insect or green vegetable diet, strictly granivorous birds such as sandgrouse have to drink fresh water to satisfy their needs. These birds have particular adaptations to obtain and transport water to chicks, but there have to be water bodies in the proximity of their nesting area. Consequently, the disappearance of the sheep herds affected these species twice, through the disappearance of grazing, that maintained a low vegetation cover, and also by the disappearance of water ponds. Up to 2019, six water ponds (Fig. 19.11) have been repaired or rebuilt in the western SPAs. Camera-trapping as well as GPS-tagging studies show that they are regularly visited by sandgrouses and other species in the area.

Fig. 19.11 Traditional sheep water holes restored as drinking points for sandgrouse and other wildlife. (Photo: Joan Estrada)

7 Actions Aimed to Reintroduce or Reinforce Populations

Reintroducing or reinforcing populations can help to restore an already lost population or delay the extinction of a very small and declining one. This was exactly what was done in the case of the lesser kestrel (action 7.1) or the lesser grey shrike (action 7.2) in the Lleida plain, with variable results.

7.1 Captive Breeding and Release Programs of Lesser Kestrel

A captive breeding and release program of lesser kestrel was started in Catalonia in 1989, aiming to restore the already extinct population. Chicks, born in captivity by cross-fostering, were released by hacking (leave the chicks on an artificial nest and provide them with food until they naturally fledge) in the Lleida plain. Initially the sites for hacking were traditional farm buildings, but predation and weather condition frequently compromised the reproduction of the few birds that returned to breed, so hacking is now conducted using the brand new breeding towers build for that purpose (action 6.1). More than one thousand chicks have been released up to 2019 and the current population amounts to 50–100 breeding pairs, depending on the year. However, they are still distributed in less than 15 colonies (depending on years) and annual fluctuations of the number of breeding pairs occur (Pont et al. 2020).

7.2 Captive Breeding and Release Programs of Lesser Grey Shrike

A lesser grey shrike captive breeding and release program was started in 2008, aiming to reinforce the small population of only a few remaining birds. The first captive-bred chicks were released by hacking in 2009. Since then, 606 chicks have been released and the single pair present in the Lleida plain in 2019 was formed by two captive hatch birds, which raised three chicks (www.trenca.org). Contrary to the lesser kestrel situation, in spite of all this efforts the lesser grey shrike population in the Lleida plain continues on the verge of extinction, probably because it is located at the very western edge of its palearctic breeding distribution and completely isolated, which probably hinders its recovery (Giralt 2015).

8 Actions Aimed to Reduce Man Induced Mortality

Agricultural intensification is often associated to mechanization of farming practices or the proliferation of man-made structures, which may cause accidental death of birds and, eventually, seriously compromise the viability of some populations. This was the case of the Montagu's harrier, whose populations were almost extinct as a result of nest destruction by harvesting machines (action 8.1) and of little bustards, which are prone to fall victim of collision with overhead power lines (action 8.2).

8.1 Location and Protection of Montagus's Harrier Nests

The main conservation problem of the Montagu's harrier in the Lleida plain is that, in the absence of active protection, harvesting or mowing destroys more than 95% of the nests. This has become even worse in recent years when, as a result of some drought episodes, many breeding territories have moved from the pseudo-steppes, where they used to nest on barley crops, to irrigated intensive crops, where more than half the nests are found in fodder crops, which are subject to even earlier mowing, increasing the risk of nests being destroyed. In order to protect the species, between the end of April and mid-June, all the Lleida plain is prospected in search of the pairs that breed throughout, both in the steppe-land and in the irrigated areas. When a nest is found, the land owner is contacted and is asked to delay mowing or harvesting of a quarter of a hectare around the nest until the chicks have fledged, receiving as compensation an amount of money equivalent to the market price of the unharvested crop. In addition, the nest is secured with a metal fence 1.5 m high to protect it from predation by land mammals. Up to 46 nests out of 65 were protected in this way in 2019, both within and outside the protected areas, 40 of which were successful (Varea et al. 2019). However, this population is absolutely dependent on this expensive and difficult to sustain nest rescue and protection programme (Cardador et al. 2015). Unless alternative, less expensive or self-sustaining harrier-friendly farming systems are devised, the long term persistence of the population is not guaranteed in the absence of this intensive conservation programme.

8.2 Installing Bird Diverters on Power Lines

Collision with power lines is a relevant cause of mortality for little bustard in Iberia, including Catalonia (Marcelino et al. 2018), and may reduce habitat quality and have a significant demographic impact on little bustard populations (Silva et al. 2010, 2014). Unfortunately, new power lines deployed on the SPAs or their vicinity to fulfil the *Segarra-Garrigues* irrigation system's power needs were not buried. To reduce the risk involved by these lines, 24 sections, totalling 39.5 km of power lines within the SPA (those that were identified as potentially more risky for birds), were provided with catadioptric bird diverters. However, the area is prone to fog (particularly in winter) and the Little bustard or the Eurasian thick-knee are partially nocturnal or crepuscular birds. In these situations, bird diverters might be ineffective in preventing these sorts of casualties, so we need to find a more effective and practical solution.

9 Conclusions and Future Directions

In recent years, many conservation actions have been implemented in the Lleida plain to halt and to reverse the decline of steppe bird populations, which has been vertiginous as a result of agricultural intensification and land use changes. The results of these actions leave room for some optimism, as the declining trends have been stopped in many cases. Paradoxically, only a few of them were deployed before the *Segarra-Garrigues* irrigation system was launched (action 5.1, 7.1, 8.1), while most (actions 4.1, 4.2, 4.3, 5.2, 6.1, 6.2, 7.2, 8.2) were set only thanks to the financial support associated to the mitigation measures of a project that, otherwise, plans to eliminate half of the steppe-land habitat remaining in the Lleida plain. Steppe-land bird populations in the Lleida plain are currently far from what they were 40 years ago and there is little hope that they can recover to these previous levels, as the suitable habitat area extent has been seriously reduced. Most steppe-land bird species can however still maintain viable populations in the region if adequate management is conducted. However, the persisting intensification of the agricultural matrix outside and inside the SPAs as well as the continuous industrial and urban development of the area are still of great concern. Although agri-environmental measures can play an important role in recovering the functionality of the agricultural landscape matrix, both within and between the SPAs, more specifically devised measures are needed. Only a change in the rural development concept, from one that values only economic services to one that values social and ecological services as well, is needed if we want some room for steppe-land birds to be left in the Lleida plain.

Acknowledgements Many people and organizations have contributed to the study and conservation of the Catalan steppes during the last 30 years which has been summarized here, and it is not possible to list all of them here. However, we want make a specific mention to Jaume Bonfil, Manel Pomarol and Francesc Sardà-Palomera for their engagement in the study and monitoring of steppe bird populations in the Lleida plain. We also specially thank Teresa Fainé (*Infraestructures.cat*) and Carme Bernat (*Aigües Segarra Garrigues, S.A.*) for their deep implication to the implementation of the *Segarra-Garrigues* mitigation and compensation measures.

References

Bota G, Morales M, Mañosa S, Camprodon J (2005) Ecology and conservation of steppe-land birds. Lynx edicions & Centre Tecnològic Forestal de Catalunya, Barcelona

Bota G, Sardà F, Cardador L (2015) Cens poblacional de sisó (*Tetrax tetrax*) en les zones de secà de la Plana de Lleida: seguiment quinquenal de la població de sisó CTFC-Infraestructures.cat, Solsona

Brotons L, Estrada J, Mañosa S (2004) Modelling the effects of irrigation schemes on the distribution of steppe birds in Mediterranean farmland. Biodivers Conserv 13:1039–1058

Calvet J, Estrada J, Mañosa S, Moncasí F, Solans J (eds) (2004) Els ocells de la Plana de Lleida. Pagès editors, Lleida

Cardador L, Brotons L, Mougeot F, Giralt D, Bota G, Pomarol M, Arroyo B (2015) Conservation traps and long-term species persistence in human-dominated systems. Conserv Lett 8:456–462

CTFC (2018) Cens de gaig blau a la Plana de Lleida. CTFC-Infraestructures.cat, Solsona

De Juana E, Santos T, Suárez F, Tellería JL (1988) Status and conservation of steppe birds and their habitats in Spain. In: Goriup D (ed) Ecology and conservation of grassland birds. ICBP Technical Publication 7, pp 113–123

Estrada J, Mañosa S, Bota G (2017) Importància dels guarets ambientals com a font d'insectes per a la conservació del sisó (*Tetrax tetrax*) i altres ocells esteparis en els secans cerealistes. But. de la Institució Catalana d'Historia Natural 81:23–29

Folch A (2004) Gaig Blau *Coracias garrulus*. In: Calvet J, Estrada J, Mañosa S, Moncasí F, Solans J (eds) Els Ocells de la Plana de Lleida. Pagès Editors, Lleida, pp 284–286

Generalitat de Catalunya (2010) Acord GOV/185/2010, d'11 d'octubre, pel qual s'aproven definitivament el Pla especial de protecció del medi natural i del paisatge dels espais naturals protegits de la plana de Lleida i el Pla de gestió d'aquests espais. Diari Oficial de la Generalitat de Catalunya 5775(15/11/2010):83489–83571

Giralt D (2015) Decline of the lesser Grey shrike (*Lanius minor*) at the western limit of the distribution area: causes, mechanisms and conservation proposals (doctoral dissertation). University of Barcelona, Barcelona, Spain

Giralt D, Valera F (2007) Population trends and spatial synchrony in peripheral populations of the endangered lesser Grey shrike in response to environmental change. Biodivers Conserv 16:841–856

Giralt D, Robleño I, Estrada J, Mañosa S, Morales MB, Sardà-Palomera F, Traba J, Bota G (2018) Manual de gestión de Barbechos para la conservación de aves esteparias. Fundación Biodiversidad. Centre de Ciència i Tecnologia Forestal de Catalunya

Giralt D, Sardà-Palomera F, Santiesteban C, Bota G (2019) Estudi i seguiment de la població catalana de Ganga (*Pterocles alchata*) i Xurra (*Pterocles orientalis*) a la Plana de Lleida. Any 2018. CTFC-Infraestructures.cat, Solsona

Giralt D, Pou N, Pla M, Sardà-Palomera F, Bota G (2020) Xarxa de seguiment d'ocells i hàbitats en l'àmbit Segarra-Garrigues (FARMDINDIS). Any 2019. CTFC-Infraestructures.cat. Solsona

Gómez-Catasús J, Pérez-Granados C, Barrero A, Bota G, Giralt D, López-Iborra GM, Serrano D, Traba J (2018) European population trends and current conservation status of an endangered steppe-bird species: the Dupont's lark *Chersophilus duponti*. PeerJ 6:e5627. https://doi.org/10.7717/peerj.5627

Guillem R, Rocaspana R, Mañosa S (2019) Estudi i seguiment. Cens de mascles de sisó en Àrees Pilot dels secans Orientals. Mesures correctores i compensatòries del regadiu del sistema Segarra-Garrigues. Campanya 2018–2019. Infraestructures.cat i GESNA Estudis Ambientals, Barcelona

Guixé D, Arroyo B (2011) Appropriateness of special protection areas for wide ranging species: the importance of scale and protecting foraging, not just nesting habitats. Anim Conserv 14:391–399

ICO (2019) Disetè informe del Programa de Seguiment d'Ocells Comuns a Catalunya (SOCC). Institut Català d'Ornitologia, Barcelona

Kleijn D, Baquero R A, Clough Y, Díaz M, De Esteban J, Fernández F, Gabriel D, Herzog F, Holzschuh A, Jöhl R, Knop E, Kruess A, Marshall E J P, Steffan-Dewenter I, Tscharntke T, Verhulst J, West T M, Yela J L (2006) Mixed biodiversity benefits of agri-environment schemes in five European countries. Ecol Lett 9:243–254

Krištín A (2008) Lesser Grey Shrike (*Lanius minor*). In: Del Hoyo, Elliott, Christie (eds) Handbook of the birds of the world, vol 13. Barcelona, p 783

Lapiedra O, Ponjoan A, Gamero A, Bota G, Mañosa S (2011) Brood ranging behaviour and breeding success of the threatened little bustard in an intensified cereal farmland area. Biol Conserv 144:2882–2890

Limonium (2010). Plan de gestión y Plan especial de protección del medio natural y del paisaje de los ENP de la Plana de Lleida. Generalitat de Catalunya. Barcelona, Spain

Mañosa S, Estrada J, Folch A, Orta J, Gonzàlez-Prat F, Bonfil J (1996) Bird-habitat relationships in the Catalan Steppes. In: Fernández Gutiérrez J, Sanz-Zuasty (eds) Conservación de las aves esteparias y sus hábitats. Junta de Castilla y León, pp 153–161

Marcelino J, Moreira F, Mañosa S, Cuscó F, Morales MB, García de la Morena E, Bota G, Palmeirim JM, Silva JP (2018) Tracking data of the little bustard *Tetrax tetrax* in Iberia shows high anthropogenic mortality. Bird Conserv Int 28:509–520

Muñiz S (2010) Agricultura i natura? Anatomia d'un conflicte. Claus i perspectives de futur del canal Segarra-Garrigues. Societat Catalana d'Ordenació del Territori, Barcelona

Ponjoan A, Bota G, Mañosa S (2011) Sisó *Tetrax tetrax*. In: Herrando S, Brotons L, Estrada J, Guallar S, Anton M (eds) Atels dels ocells de Catalunya a l'hivern 2006–2009. Institut Català d'Ornitologia & Lynx edicions, Barcelona

Pont F, Guillem R, Rocaspana R (2020) Estudi i seguiment de la població de Xoriguer Petit de la Plana de Lleida, anàlisi de l'ocupació de torres i caixes niu. Mesures correctores i compensatòries del regadiu del sistema Segarra-Garrigues. Campanya 2019-2020. Informe inèdit. Infraestructures.cat i GESNA Estudis Ambientals, Barcelona

Sanz-Perez A, Giralt D, Robleño I, Bota G, Milleret C, Mañosa S, Sardà F (2019) Fallow management increases habitat suitability for endangered steppe bird species through changes in vegetation structure. J Appl Ecol 56:2166–2175

Sazatornil V, Giralt D, Sardà-Palomera F, Bota G (2019) Cens de terrerola vulgar (Calandrella brachydactyla) a la Plana de Lleida. CTFC-Infrastructures.cat, Solsona

Silva JP, Santiago-Santos MG, Queirós L, Cabral JA (2010) Estimating the influence of overhead transmission power lines and landscape context on the density of little bustard *Tetrax tetrax* breeding populations. Ecol Model 221:1954–1963

Silva J-P, Palmeirim JM, Alcazar R, Correia R, Delgado A, Moreira F (2014) A spatially explicit approach to assess the collision risk between birds and overhead power lines: a case study with the little bustard. Biol Conserv 170:256–263

Suárez F, Naveso MA, De Juana E (1997) Farming in the drylands of Spain: birds of the pseudosteppes. In: Pain DJ, Pienkowsky MW (eds) Farming and Birds in Europe: the common agricultural policy and its implications for bird conservation. Academic Press, London, pp 297–300

Tellería JL, Suárez F, Santos T (1988) Bird communities of the Iberian shrubsteppes. Holarct Ecol 11:171–177

Traba J, Morales MB (2019) The decline of farmland birds in Spain is strongly associated to the loss of fallowland. Sci Rep 9:9473

Traba J, García de la Morena EL, Morales MB, Suárez F (2007) Determining high value areas for steppe birds in Spain: hot spots, complementarity and the efficiency of protected areas. Biodivers Conserv 16:3255–3275

Valverde JA (1958) Las aves esteparias de la Península Ibérica. Instituto Biología Aplicada 27:41–48

Varea A, Estrada J, Bonfil J, Rodríguez J, Pont F (2019) Estudi i seguiment de la població d'arpella cendrosa a la Plana de Lleida, localització de nius. Mesures correctores i compensatòries del regadiu del sistema Segarra-Garrigues. Campanya 2019-2020. EV-E7-MA-1990. Infraestructures de la Generalitat de Catalunya & Departament de Territori i Sostenibilitat, Barcelona
Viada C (1998) Áreas Importantes para las Aves en España. Seo/Birdlife, Madrid

Part V
Applications, Considerations and Recommendations for the Future

Chapter 20
Monitoring Arable Landscapes Using Free Satellite Imagery

Alan Brown and Iain Cameron

Abstract The first part of the chapter tells the reader new to satellite remote sensing how optical imagery is captured and processed, with practical exercises looking at freely available imagery from the Sentinel 2 satellite. Elements of sensor design and satellite orbits are used to show how satellite imagery differs from digital photographs, with enhanced abilities to see in unfamiliar wavelengths using multiple channels at different spatial resolutions. The exercises start with visual interpretation of visible wavelengths, and then look at how the arable landscape can be displayed in false colour images, including infra-red channels. The accumulation of satellite imagery means we cannot simply rely on visual interpretation, but need strategies for automated analysis.

The second part of the chapter introduces the Copernicus and Landsat programmes, and outlines three common approaches to remote sensing of arable landscapes, including the classification of land parcels, followed by a brief review of the potential role of Synthetic Aperture Radars. The chapter finishes with a summary of the advantages of remote sensing, with a survey of future developments, not only of satellite systems but the wider range of sensors including satellite swarms and drones. A final section suggests practical next steps for the interested reader.

Keywords Remote sensing of arable landscapes · Arable landscape classification · Free satellite imagery · Infra-red channels · NDVI (normalised difference vegetation index) · Landsat · Role of synthetic aperture radars · Sentinel 2

A. Brown (✉)
Upper Thorn, Evenjobb, Presteigne, Powys, UK
e-mail: alb115@aber.ac.uk

I. Cameron
Environment Systems, Aberystwyth, Ceredigion, UK
e-mail: iain.cameron@envsys.co.uk

C. Hurford et al. (eds.), *The Changing Status of Arable Habitats in Europe*,
https://doi.org/10.1007/978-3-030-59875-4_20

1 Introduction

Cinema sometimes uses satellite remote sensing in exposition. Hat-wearing military types look on as the chosen satellite is instantly moved into place, zooming in to fill the screen with recognisable, labelled objects showing people moving under tree canopies and inside buildings, all in total darkness. The control desk has lots of serious-looking buttons, all with bad consequences. Almost all of this is wrong. The most useful satellites for monitoring arable landscapes follow fixed orbits passing near the north and south poles, orbiting once roughly every 90 min. Instruments do not zoom, though some can be pointed to look either side of the track; most rely on reflected sunlight and pass over the target once every few days, close to midday, when there are least shadows. Most cannot see through clouds or tree canopy and most might show a house or larger tree, but not an individual person. Only one image is taken on each overpass, not a live video stream, and a complex chain of processing steps and decisions is needed to extract the observations.

Nevertheless, what we have now is still revolutionary. What if we could just look at a familiar landscape and remember precise colour measurements, giving us the ability to compare every detail of what we see with the same view last week? Or last month, last year or 40 years ago? What if, by opening our eyes wider, we could look using new parts of the electromagnetic spectrum, including infra-red frequencies which show contrasts in the photosynthetic activity of plants; thermal bands showing warm and cool surfaces; or longer radar wavelengths sensitive to plant structures and soil moisture which can see through clouds and tree canopy? All this and more is possible with *satellite remote sensing* (RS).

2 Remote Sensing Instruments and Sensors

Remote sensing is any means of measuring or detecting objects at a distance, without touching them, using reflected electromagnetic radiation or pressure waves. Remote sensing is done with instruments or *sensors* carried on *platforms*. *Optical* sensors have a set of lenses which focus light on a *detector* consisting of an array of thousands of precisely calibrated *elements* or *pixels*.

Filters are used to restrict each pixel to part of the optical spectrum, e.g. green or blue light. The pixel measures the intensity of electromagnetic radiation. Processing in the sensor converts the intensity to numbers which are transmitted to a receiving station, and later processing converts it to the proportion of light reflected from the ground. Each two-dimensional set of measurements is stored as a single *band* (or *channel)* which can be displayed on a computer screen as a monochrome image; or two or three bands can be combined into a colour image.

As well as collecting visible light, sensors can target other parts of the optical spectrum to extend the range of what we can see with our eyes, for example measuring the reflection of light by plants in the infra-red, or measuring heat radiated at night by warm surfaces. These are commonly broad bands, spanning a range of frequencies. *Hyperspectral* sensors split the visible and infra-red range up into hundreds of narrow bands to give a *spectral* profile (or *spectral signature*) which tells us about the chemical composition of surfaces, including leaves, stems and soils.

While these *passive* sensors work in basically the same way, using lenses to collect and focus reflected sunlight or radiated heat, there are also *active* radar instruments which emit a pulse of radiation towards the earth. This pulse is scattered by objects on the ground, and an image is created from the radiation that is reflected back towards the satellite (*backscattered*) as it moves through space. These *synthetic aperture radars* (SAR) are especially important in mapping crop types using seasonal variation in vegetation structure because, unlike passive optical sensors, they can see through clouds.

2.1 Platforms: Scale, Mobility, Revisits

A tripod can be used to fix the frame of a landscape photograph. Most remote sensing instruments are carried on some sort of equivalent mobile *platform* which moves along the *track* of the orbit, and look at a narrow or wide *swathe* of the ground (as this suggests, like mowing a hay field), depending on the design and purpose. For example, the Sentinel 2 satellite — which we will focus on as the single most important source of free optical imagery — passes over the poles, travelling north to south 700 km above Europe several times each day, acquiring an image roughly 290 km wide and hundreds of kilometres long in a matter of a few minutes. Each track drifts slightly to the west so that the next swathe overlaps only a little (more towards the poles: imagine wrapping a basketball in paper tape), allowing coverage of large areas as the satellite follows the sun with the earth turning underneath. This continuous stream of data is transmitted to a ground station as the satellite passes over, where it is *pre-processed,* re-projected into a standard grid or *reference frame* and cut up into *scenes* or *data granules* for on-line distribution. Re-projection means of course that the pixels we see on screen are not those captured by the sensor, but a new grid made by resampling and combining them.

While satellites and aircraft cover very large areas in a few hours, drones and moving vehicles (such as tractors) cover smaller areas closer to the ground, enabling their sensors to capture more spatial detail.

Box 20.1
Exercise 1. We will start with Sentinel Playground (https://www.sentinel-hub.com/explore/sentinel-playground). Load the site and click on the big green button to start playing. Put in the name of your local village or town in the search box 'Go to Place' and press return. By default this brings up the latest Sentinel 2 image, at processing level L1C (this means it has been calibrated and re-projected to fit on a map), showing natural colour: RGB = red, green, blue = bands (4,3,2). To see older images you can click on the calendar (on the top bar) and select any of the dates with a circle. Light circles have less cloud than a set limit (which you can adjust by clicking on the cloud symbol) the darker circles more cloud cover, and hovering the cursor shows the actual cloud cover. Notice that because the images are so big, the place you are interested in may still be cloud free even if most of the image is cloudy. Scrolling back through the dates and years in the calendar shows just how much imagery is available, and the revisit intervals. If you zoom out, you can also see the edges of the image swathes as apparent diagonal joins running to the west of due north to south.

2.2 Spatial Resolution

With the same digital camera we can photograph a landscape or take a portrait of a few plants at our feet by changing the lens which focusses light onto the sensor. A landscape photograph creates a wider field of view than a close range portrait by increasing the area of ground seen by each pixel. While the landscape photograph still measures the light reflected by every plant, these cannot be picked out as separate objects: *spatial resolution* is the size of the smallest object which can be seen individually. The contrast in colour and intensity between the object and its background also influences this, as even a very small bright object can be visible against a dark background.

Unlike a digital camera, most optical satellite instruments typically have one-dimensional lines of detectors, one for each band, which sweep across the track as the satellite platform moves forward, creating a two-dimensional image by adding one data row at a time. In this *pushbroom* design, many bands can be placed next to one another to look at the ground in turn, making multiple image *layers*. The design allows the number and size of sensors (elements) in each row to be varied, giving layers in different bands which cover the same area but with different spatial resolutions. For example, the Sentinel 2 MSI (MultiSpectral Instrument) sensor has bands at 10 m, 20 m and 60 m (Fig. 20.1).

The detector size is designed to fit the expected amount of reflected light, which depends in turn on the total ground area seen by the lenses on the instrument and the width of the filter band. In those parts of the spectrum where more sunlight is available, the detector elements (pixels) can be smaller, allowing more spatial detail

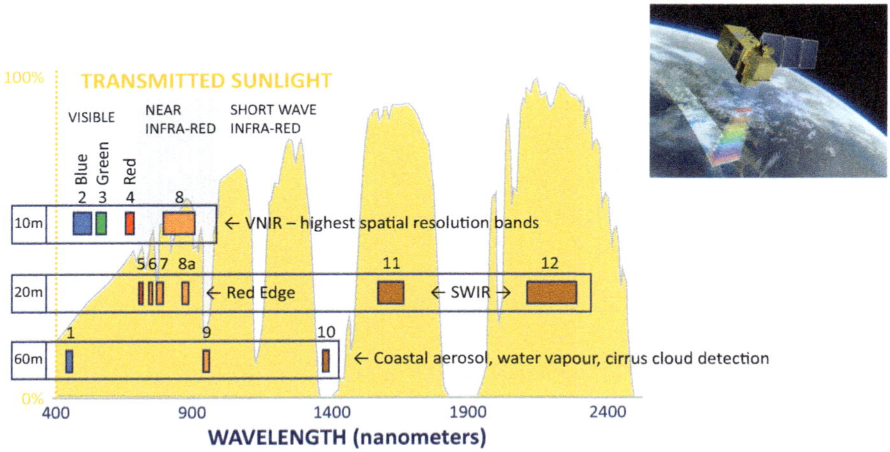

Fig. 20.1 Sentinel 2 MSI bands. Three groups correspond with 'atmospheric windows' where sunlight penetrates the atmosphere: high (10 m) spatial resolution visible and infra-red; lower (20 m) spatial resolution narrow 'red edge' bands between red and near infra-red (NIR); and two 60 m short-wave infra-red (SWIR) bands. The width of the band in the figure is proportional to the interval of wavelengths (the 'bandwidth') in this channel. The three bands for mapping coastal aerosols (band 1), water vapour (9) and cirrus cloud (10) correspond with wavelengths absorbed in the atmosphere. The VNIR bands are used for 'natural colour' imagery, false colour infra-red imagery and producing vegetation indexes. The SWIR bands map snow and ice, soils, living and dead vegetation and burn scars. The red edge bands are designed to map vegetation properties such as leaf area index (LAI) and chlorophyll content. Even though there is high transmittance of SWIR, in sunlight the amount of energy in these wavelengths is much less than in the visible and NIR. (Redrawn from Ferran et al. 2017; Image of Sentinel 2 satellite: ESA/Copernicus)

while maintaining an acceptable signal-to-noise ratio. The detector design can also filter a wider or narrower range of wavelengths, in some cases focusing on a very small part of the spectrum where interesting features of – for example – light absorption by plants can be detected. MSI has the so-called 'red edge' channels (5 to 7 and 8a) sensitive to crops and natural vegetation cover shown in Figs. 20.1 and 20.3b.

Other broader channels need broader filters or larger pixels to collect enough photons from parts of the spectrum with less light available, especially in the short-wave infra-red (SWIR) region. Remote sensing instruments are designed to make precise measurements, not just create pictures, so precision, bandwidth and spatial resolution are traded off against each other to collect the most informative set of image layers.

Satellite sensors which look across a wide track (or *swathe*) from high above the earth (Landsat, Sentinel, IRS) tend to have a lower spatial resolution, but even from a high orbit some satellites (Quickbird, Rapideye) can resolve objects as small as a few tens of centimetres, with a pixel size of half a metre or less. To do this they need to collect more light from a narrow swathe (typically 15 km) targetted at the ground by pointing ('tasking') *off-nadir* — and this means the imagery is commercial and often expensive. More recently, higher resolutions have been achieved by using swarms of hundreds of very small satellites in low earth orbits known as 'CubeSats' (https://www.cubesat.org).

Box 20.2
Exercise 2. We can also look at individual bands as monochrome images by dragging the same band to all three colour channels: e.g., for 'blue', RGB = band (2,2,2); for near infra-red, RGB = band (8,8,8). Go to the 'custom' option at the top of the menu to drag one band onto all three RGB circles. Zoom in as far as you can. While the images look a bit blurry as the software smoothes out neighbouring pixels, if you press 'Generate' you get a downloadable copy which shows the actual pixels on the L1c level image, the closest we can get to the original data.

What scale of linear features can you see, how big must trees be to appear in a field (check against air photos). In winter scenes, can you see the shadows of trees and hedges?

Try comparing what you see with the air photography of the same area from Google maps or Google earth. Notice the difference in pixel size/spatial resolution between the RGB and Near Infra-Red (NIR) band 8 (10 m × 10 m) and the short-wave infra-red bands 11 & 12 (20 m × 20 m), and how the blue band 2 is less sharp than NIR band 8.

2.3 *Atmosphere and Topography*

The higher the platform above the earth, the more the sensor mixes the light reflected from the ground surface with light scattered from small particles (*aerosols*) and water droplets in the atmosphere. These *atmospheric effects,* which also modify the *downwelling* sunlight, can vary from day to day and need to be minimised in each image in order to make calibrated measurements and reveal the clearest view of the ground surface. When atmospheric water vapour condenses, it forms mist and clouds which absorb light in visible and infra-red wavelengths, blocking the view of the ground. Clouds also cast shadows which both reduce the brightness of the ground reflectance and minimise local shadowing in the vegetation – which can be informative about habitat structure.

In the Northern Hemisphere, light intensity varies between bright south-facing slopes and shaded northern slopes, most obviously in the winter when the sun has a low elevation. While the extremes are less important in the case of arable landscapes which typically occupy flatter ground, these *topographic effects* can be minimised by modelling slope and aspect with a *digital elevation model* (DEM) or by looking at ratios of bands (such as NDVI, see below).

2.4 Geometry and Repeatability

Cameras are 'framing' instruments which expose all the pixels in a two-dimensional image at the same time. This gives photographs a pattern of radial displacement, where objects closer to the camera are shown larger and pushed towards the edge of the frame. Air photographs show this especially well for tall chimneys: it is what allows us to create stereo images. The geometry of pushbroom instruments differs as the sensor always tracks overhead down the centre line of the image, and tall objects and relief are displaced towards the edges in a symmetrical fashion. The great advantage of this geometry – for a satellite travelling north to south – is that the sun angle remains the same for all objects in the image strip, which makes calibration and the comparison of images much more straightforward, and avoids risking the solar disk reflecting directly into the sensor. Some recent swarms of miniature satellites have an alternative design, similar to a digital camera, each contributing a small part of a mosaic of overlapping frames. These tend to be limited to 3 or 4 colours since both dimensions of the detector are taken up with the two-dimensional image, rather than having the parallel rows of detectors for different channels seen in the pushbroom sensor.

3 Overwhelmed by Data Choices?

With the rapid growth of free satellite imagery, it is easy to suppose that if we have enough data the observations we want will fall out by themselves. In practice, the data captured in hundreds of thousands of images can lead to something known as the 'curse of dimensionality': in other words, we cannot look at the images without making choices – and cannot see all of the information content at once.

Supposing we have a single Sentinel 2 image. The MSI optical sensor has 12 bands, stored as 12 data layers. Using software, we select three of these and our computer screen combines them to form an image using *additive colour mixing* of red, green and blue (RGB) channels. (To see how this mixing is done, look at your LCD screen with your botanical hand lens!).

We can mix bands to show the 'natural' colours we imagine seeing from high above the ground, by assigning the RGB colour channels to bands 4 (=collecting red reflected light), 3 (=green) and 2 (=blue). Since we can assign any of the 12 bands to any of the three channels, we can also produce a *false colour* image, where the colour and colour combinations show the *intensity* of reflected light in each band, relative to other bands. For example, a yellow tone shows high values of the bands mapped onto the green and blue channels, but a low value for the third band.

Even though the MSI has 12 bands, we are limited to seeing only 3 at a time. This gives a total of 1320 (= 12!/(12–3)!) possible three band combinations. On screen, the intensity in each channel depends on a value in a colour look-up-table (LUT) between 0 (dark) and 255 (bright): this is 8 bits per channel, i.e. 256 = 2^8. To compare intensities from three bands, we also need to balance them to ensure the 2^8 display range is not saturated by what is, on average, a very bright band.

Older satellite sensors also converted the intensity of radiation into the same range of digital numbers, giving 8 bits per channel, just as digital cameras do: the range of the instrument and the screen is the same. Newer instruments, including the MSI, convert to a range of $2^{12} = 4096$ values, stored as 2 bytes or 16 bits.

Sunlight also varies in intensity, with most around the visible and NIR wavelengths, so the satellite instrument collects more reflected light from some parts of the spectrum. This is partly taken into account by sensors with different spatial resolutions (coarser where there is less light) and wider filter bandwidths (wider bands to let through more light), and setting the amplification or gain to optimise exposure as the photons are converted to digital numbers (DNs). Having a 12 to 16-bit range of values makes it easier to cope with contrasts between dark forests and highly reflective urban surfaces. If we put this whole 12 bit range on screen, the smallest difference in the LUT shows a jump of 4 values in the image as we squeeze 2^{12} into 2^8. This is already a loss of detail and does nothing to balance the colours from different bands.

3.1 Selecting and Enhancing Part of the Data

We need to show just part of the DN range to highlight the landscape differences we are interested in. So, our task is to work out which bands to map to each colour, squeezing 12 into 3; and which part of the 12 bit range to use. Fortunately, software gives us a help with common band combinations and clever highlighting of the generally best intervals by *clipping* out an interval and *stretching* it across 8 bits. The best interval is the one which allows the screen to see the best contrasts in colour, showing how intensities of the chosen three bands change together.

Bands can also be combined algorithmically in different ways to make *indexes*. The most widely used is NDVI (normalised difference vegetation index) which is a ratio of (NIR-red)/(NIR + red): for MSI (band 8 - band 4)/(band 8 + band 4). As photosynthesis increases both reflection of NIR radiation by the spongy mesophyll tissues, and absorption of red radiation by chlorophyll, NIR-red and so NDVI increases as the amount of active photosynthetic plant tissue increases. The denominator normalises this by cancelling out topographic differences in illumination.

Box 20.3

Exercise 3. Back in the 'rendering' menu, select 'Colour Infrared (vegetation)', where RGB = band (8,4,3). These bands are the same as traditional colour infra-red photographs, where the red channel shown infra-red reflectance, the remaining green and blue channels show the red and green bands!

(continued)

Box 20.3 (continued)
Take a moment to understand this. Photosynthetic vegetation absorbs red light and very strongly reflected near infra-red, so this band combination will probably be saturated by red and we need to rebalance the channels. Go to the 'custom' option at the top of the menu and recreate the same bands by dragging 08, 04 and 03 onto the RGB circles. This should look the same as before. Now select the small hand symbol and you will get a box with a string of code:

```
return [B08*2.5,B04*2.5,B03*2.5].
```

Edit this to reduce the weighting on B08 (infra-red) to 1, so the string now reads:

```
return [B08*1,B04*2.5,B03*2.5]
```

And press 'Refresh'. You can see the dominance by the red channel is softened and you can see more detail.

Figures 20.2 and 20.3 show two arable landscapes from Wales in natural colour and false colour, highlighting how the additional NIR and SWIR bands add spectral (colour) differences which allow us to separate classes of land cover and land use, once we can interpret the landscape in more familiar natural colours. NDVI is contrasted with true colour and radar images in Fig. 20.6.

Box 20.4
Exercise 4. In the 'rendering' menu, select Vegetation Index, which gives a colour-coded screen where low values (e.g. unvegetated surfaces such as ploughed fields) are red, higher values blue and green. Since clouds absorb NIR but reflect red light, clouds also have low values and appear red. Water absorbs both equally and appears black. NDVI is a good way of using satellite imagery to show the most productive arable systems in the spring, contrasted with ploughed ground. The rise of NDVI/productivity in the spring nicely contrasts less productive extensive grasslands, candidate areas for higher biodiversity, with intensive, agriculturally improved grasslands. To map low productivity systems, we also need to check that these fields were not ploughed and seeded in the winter, showing how we can build a set of rules out of the seasonal changes in the appearance of objects, interpreted here as patterns of growth and phenology.

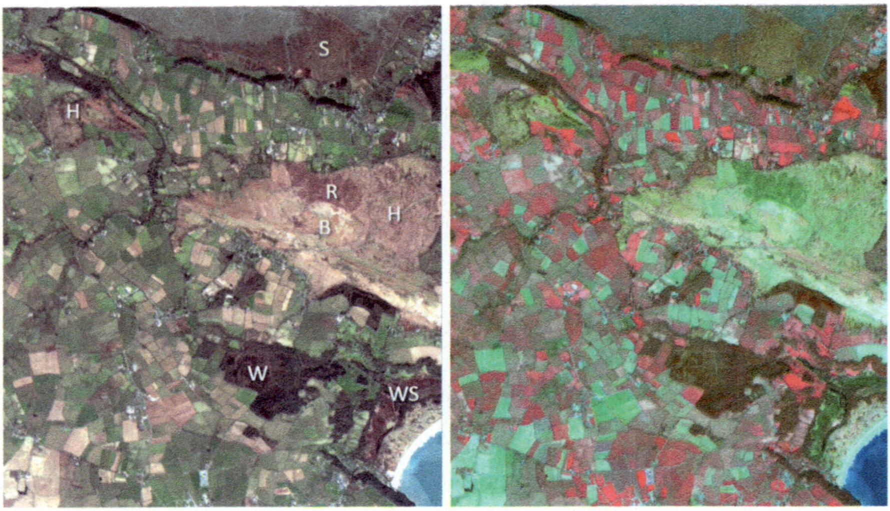

Fig. 20.2 An old agricultural landscape on part of the Gower Peninsula, South Wales, close to the sea with diverse rocks and soils including limestones. The 'natural' colour Sentinel 2 MSI image from early March, left, shows obvious contrast between the low-intensity hay meadows which have not yet started growing and the brighter green agriculturally improved fields, interspersed with browner bare soils or aftermath from the summer hay crop. Imagery from later in the year will show which fields have now been drilled for cereal crops. There are also deciduous woodlands (W) not yet in leaf, wetland and scrub (WS), heathland with patches of bracken (B) and dense rushes (R), part of several 'common lands' with low grazing pressure but often burned. To the north, the fields run down to intertidal saltmarsh (S) and mud flats. The false colour image, right, shows near infra-red (NIR) in the red channel, the brighter colours suggesting actively growing vegetation as NIR is strongly reflected by plant tissues. Notice the contrast between actively growing fields (showing bright red) and the deciduous woodlands and heathlands, indicating a landscape where semi-natural elements are mixed with more intensive arable management, even in clusters of small, irregular fields of similar size, bounded by hedges and stone walls. The connectivity of patches of less intensive land-use and penetration of linear features suggests the potential for high biodiversity. (Satellite image: Copernicus)

4 The Satellite Remote Sensing Menu

Only a decade ago, the menu of NASA and commercial satellite imagery was strictly à la carte, and correspondingly expensive. While expensive sensors, bespoke platforms and specialist software are still essential for many applications, for example precision farming (Higgins, Schellberg & Bailey (2019), Charania & Li (2020)), we can now also choose the remote sensing equivalent of popular fast food from the cheap end of the menu. This is because satellite imagery has become much more available and — thanks to the European Space Agency (ESA) — cheap and simple to download and use. There are still limits on their utility, but these are now mainly from cloud cover, fixed acquisition times (e.g. limiting for intertidal habitats) and the lack of understanding and expertise in the wider user community.

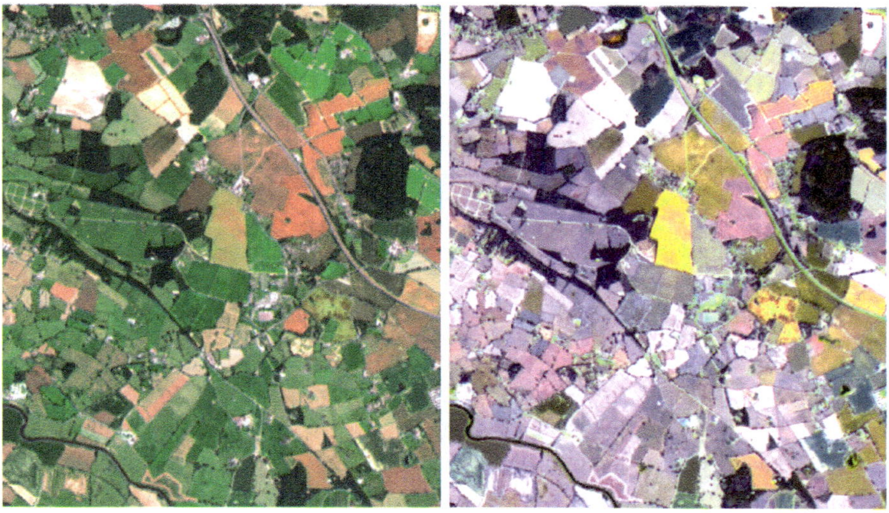

Fig. 20.3 A mixture of old and new agricultural landscapes, east Wales. The 'natural' colour Sentinel 2 MSI image on the left is from mid-July, and shows large fields either side of the major road which runs along the valley floor on rich alluvial soils. The top half of the image includes recently cut cereal fields (wheat and barley) and very large maize (corn) fields with obvious striping. Some fields — for example at the centre of the image — have a light green colour and have been drilled for the winter crop, with a mottled appearance as the new green crop starts to grow patchily against the background of soils. By contrast, the south-west part of the image has a mixture of small cereal fields and grasslands, with hedges and standing trees. Notice how the change in scale of enclosures and loss of hedges and trees at the boundaries shows more intensive agricultural management. Woodland is still present, but less connected. The bright green fields at the top right are root crops, with obvious bright patches showing bare soil and erosion, often around gateways. One of the problems in the more intensive part of this landscape is winter run-off of soil and artificial fertilizer into the river. The false colour image on the right, from the same date, uses the MSI 'red edge' bands, hinting at how much additional spectral information Sentinel 2 has for developing rules to classify crop types, once we learn to interpret the images. (Satellite image: Copernicus)

The growth in the availability and utility of satellite imagery is not, of course a coincidence: satellites are designed to be useful. By far the most important sources of remote sensing fast food are the European Copernicus programme and — if we want to go back in time — NASA's Landsat series of satellites. ASTER is another highly useful source of archival imagery.

Other highly successful commercial satellites include the French SPOT and Indian IRS, but these have to be paid for, as do a whole series of very high resolution satellite instruments such as Quickbird and Rapideye (for details of many more see Appendix A1 in Vihervaara et al. (2017) and on-line directories such as https://directory.eoportal.org).

Satellite data has not only become more available in recent years due to the proliferation of options, it has also become easier to use by non-remote sensing experts due to the introduction of Analysis Ready Data (ARD) standards (CEOS,

2019). ARD provides a minimum set of processing standards that need to be met in order for a non-specialist user to be able to easily pick up and use any remote sensing image. While users can access the raw satellite data directly from the data provider (e.g. scihub.copernicus.eu), and use the ARD standards as guidelines for processing, there are an increasing number of commercial services that provide ARD image products ready for download and use by non-specialists (e.g. data.envsy.co.uk). This standards- driven approach supports users as they navigate through an increasingly complex remote sensing landscape.

4.1 Copernicus and Sentinel Satellites

Copernicus, originally known as GMES, is part of a programme to make remote sensing data freely available to anyone wanting to develop commercial services around *land cover*, environmental monitoring, disaster response and security. The purpose and design of Copernicus is explained vividly in ESA's online videos (https://esa.int).

The seven Sentinel satellites — several of each design — collect a range of complementary observations of the atmosphere and ground surface over the entire earth on a near-daily basis. The imagery is centrally processed and made available as an on-line archive, from which individual scenes can be downloaded. Copernicus includes not only Sentinel satellites and instruments but a whole set of matching free software, on-line resources and training programmes. The same SNAP software will work with imagery from other satellites, including Landsat, and can process SAR radar imagery.

4.2 Landsat

NASA's Landsat played a key role at the end of the 'cold war' by monitoring agricultural crop production in the former USSR. The Landsat archive is now freely available online, going back to the early 1980s and giving us a limited but important ability to travel backwards in time over a period covering the intensification of agriculture in Europe. Both legacy Landsat imagery and new acquisitions can be freely downloaded from sites including the US Geological Survey's earth explorer (earthexplorer.usgs.gov).

Box 20.5
Exercise 5. Sentinel Playground allows you to look at Landsat 8 imagery, noting that this has different band numbers which you seen if you hover over the choices in the 'Custom' menu. For example, the equivalent of false colour is RGB = band (5,4,3). You will have to reduce the weight on the NIR band 5, as before. It's worth looking at the thermal and panchromatic layers, but note that downloaded Landsat 8 imagery is much better and sharper than the version which appears in Sentinel Playground.

5 Just Add Expertise: From Remote Sensing Observations to Information

Remote sensing images are pictures made up of carefully calibrated sets of measurements which can be repeated to show the state and change of surfaces, objects, habitats and land cover. Understanding and trusting these measurements starts with visualisation and leads on to working out how to use these observations to monitor the target variables that interested us. For example, in arable habitats, how do we convert routine visible and near infra-red (VNIR) images to a repeatable measure of cropping regime and vegetation productivity?

In fact, few of our key variables are measured directly. We are given reflected light but want to know the performance of different crops; real-time yield forecasting; agricultural land use monitoring (crop classes, weeds, improvement and reversion etc) and ecosystem services such as soil health, surface water, carbon sequestration and biodiversity. Weiss, Jacob & Duveiller (2020) summarise three approaches for using remote sensing observations to infer variables of interest, all of which use local knowledge and ground observations:

1. We can make purely *statistical* links using empirical methods and indexes, combining the remote sensing observations with observations of our target variables on the ground. For example, if we have known fields falling into a class of (agriculturally) unimproved grasslands according to ground observations, we can use these to define common values of a vegetation index such as NDVI for this class to find a threshold which separates unimproved from semi-improved grasslands. More typically, we will have a whole set of labelled classes, perhaps including habitats, crop types and intervals of variables such as tree canopy cover, and use these to parameterise a set of variables measured on a stack of remote sensing images. See the Living Wales project (https://wales.livingearth.online) and Australia's TERN project (https://www.tern.org.au).
2. A second approach uses *physical models*, known as 'radiative transfer models' which simulate how particular surfaces of leaf cover and structure reflect radiation. The model is then inverted to predict the structure from the observed reflectance. These models aim to predict important economic variables such as

future crop yield. Rather than making only statistical links, variables such as leaf area index are modelled directly. These models need a great deal information about, for example, vegetation structure which in turn need a great deal of information measured on the ground. Jones and Vaughan (2010) is a good introduction; Thenkabail et al. (2019) more challenging.

3. The third approach tries to capture our existing expert knowledge about the arable landscape, using *contextual* information about ground surfaces to classify imagery. This starts with finding familiar 'objects' on the image, such as fields, marginal strips, boundaries and hedges, pools and patches of woodland and wetland. In many cases, we already have *vector* maps of fields, but, if not, *segmentation* software can be used to find relatively homogeneous candidate *objects* automatically, grouping together adjacent pixels which seem to belong to the same class of object such as a field filled evenly with a uniform crop. Having objects allows us to build sets of expert rules which can use average values of the object, the shape and context. Figures 20.4 and 20.5 show a simple example of how classification rules can be devised, and the conversion of an image into raster objects.

Fig. 20.4 Building classification rules from time-series imagery. The 'natural' colour images show changes in the appearance of a small field, from bare soil in mid-May to a crop greening up over the summer and autumn. By September, this is indistinguishable from the surrounding grasslands. However, in the winter the green vegetation is gradually reduced to a narrowing strip surrounded by bare soils, with a straight line along the sharp contrast in colour. On the ground, we see the straight line is an electric fence which is moved every few days to allow a flock of sheep to feed on a root crop of brassicas. This locally characteristic pattern allows us to say something about the crop, the land-use and stock on the farm (we know cattle are not fed in this way) and the size of the flock. No single image tells us this, but we can easily make a set of classification rules by stacking and joining up images, allowing us to generalise from our local knowledge and local observations. Time-series imagery compensates very well for the relatively poor information content of just a few spectral bands. (Satellite image: Copernicus)

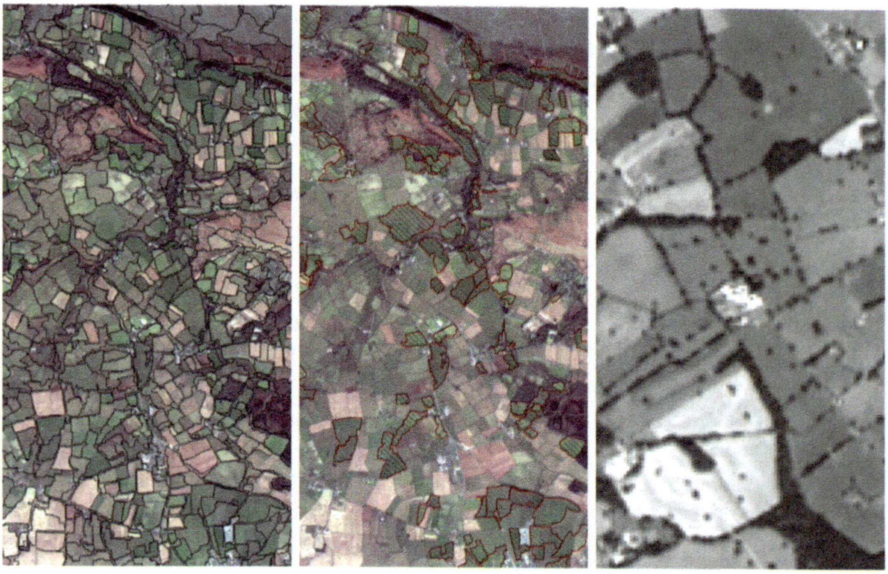

Fig. 20.5 Segmentation and objects. Image processing software includes segmentation algorithms which divide the image up into areas of relative homogeneity. A simple segmentation of part of the Gower Peninsula, left, shows how often — but not always — these correspond to management units such as fields, especially in an arable landscape. While objects are groups of pixels in a raster image, they can just as easily be converted to vector objects and handled in GIS as vector layers. We can also import an existing set of polygons from a vector map and convert these to raster objects in order to extract band values from the image. However we get them, objects can be used to summarise and average pixel values in different bands and produce indexes such as NDVI. The middle image shows NDVI thresholded to highlight those objects that appear to be most productive — typically the improved reseeded (*Lolium perenne*) grasslands. In the analysis, converting millions of pixels to a few thousand objects allows the analysis to be carried out using tables, separating the band values from the spatial data. This makes the analysis of large stacks of imagery possible, including the use of machine learning methods. The right-hand red band monochrome image shows how Sentinel MSI images are also capable of mapping individual trees and narrow linear features, important components of the landscape which can be lost in the segmentation process as their pixels are incorporated into neighbouring objects. For these small features, monitoring and change detection will also need to compare pixels as well as objects. (Satellite imagery: Copernicus)

These methods can also be combined, for example using samples from an expert classification of part of a landscape to train a statistical analysis of more recent imagery for the same set of objects. While current trends include using meta-machine learning methods (such a random forest algorithms and support vector machines) to classify and monitor land cover without having to devise explicit expert rules, these methods also rely on ground observations and expert training data. Managed landscapes can include extensive areas of semi-natural habitat, for example in the marginal uplands; areas of low intensity agriculture in a mosaic with structural elements (such as field margins, road verges, hedgerows, patches of scrub and wetland, small rivers corridors); and the same linear elements and patches of semi-natural vegetation persisting where arable crops are intensively managed.

These landscape elements can be mapped and monitored from satellite imagery, both by combining remote sensing and sample-based ground observations of landscape function and biodiversity, and by extension using coarser satellite imagery to produce maps based on parallel analysis of, for example, key attributes of landscape heterogeneity (Maskell et al. 2019). Intensification of grasslands systems can be mapped from seasonal changes in productivity indexes such as NDVI (Franke, Keuck & Siegert 2012; Gómez Giménez et al. 2017), and a wide range of remote sensing can help map intensification in crops (See Table 1 in Hunt, Blackburn & Rowland 2019). Combined satellite and ground observations can help map essential biological variables (see Table 2 in Vihervaara et al. 2017) and ecosystem function (Pettorelli et al. 2017). A nicely illustrated review of how different types of remote sensing can be combined in land cover mapping is Minasny et al. (2019).

Box 20.6

Exercise 6. Looking at natural colour optical imagery for a familiar place, get used to the scale and appearance of objects. Which objects and boundaries can you see? Scrolling through the dates, (small arrows on calendar) see how the appearance changes seasonally and through the crop cycle. What different types of vegetation, habitat, land cover and land use can you distinguish? Can you use similar distinctions to classify next season's imagery? Which areas look similar in nearby parts of the landscape, with similar colours and seasonal patterns (bright yellow oil seed rape fields are a good start) that you might expect, and can you start to write down some rules to summarise this? Now you are starting to think of an expert system or classification tree.

5.1 Accuracy

All remote sensing measurements have uncertainty. This uncertainty can come from multiple sources such as noise in the sensor itself, atmospheric interference, and errors in the statistical, physical or classification models used to convert remote sensing observations into actionable data. Whenever we are using remote sensing to understand earth surface processes, we should use ground observations to help understand the accuracy of our remote sensing outputs.

A good example of this approach is widely used during automatic classification of satellite imagery. Classifications produce a thematic map of the earth's surface, separating the image into different groups such as fields, buildings, streets etc. Classes are entirely semantic definitions, and examples of each class need to be presented to a computer algorithm in order to teach it how to separate the image into each class. These examples are provided from ground observations, and it is standard practice to keep back a reasonable percentage of these observations (typically

25–35%) to evaluate the predictive capability of the classification. As well as understanding the overall accuracy of the classification, it is just as important to know how these errors are distributed across the map and across the classes we are most interested in.

6 The Special Role of Synthetic Aperture Radars

Synthetic Aperture Radar (SAR) is a type of sensor that uses pulses of microwave radiation to construct a two-dimensional image of the earth's surface. All radars are active sensors, meaning that they generate their own electro-magnetic radiation rather than relying on external source, such as the sun, and can operate day or night. The radar antenna emits a pulse of energy that is scattered by features on the ground, some of which returns back to the antenna where it is detected. SAR systems are side looking, meaning that they make observations orthogonal to the platform flight path, and an image of the ground is built up using the history of pulses recorded as the sensor travels along its flight path. They can be carried on an aircraft or satellite platform.

The strength of the returned signal, or backscatter, primarily depends upon the size and orientation of features on the ground. Generally speaking, a rougher surface (e.g., dense vegetation or a city) will generate higher backscatter than a smoother one (e.g., the surface of a lake). However, backscatter is also modified by the electrical conductivity of a feature, with higher conductivity leading to increased backscatter. This latter point can be particularly important when comparing SAR observations over time as changes in surface moisture can have an appreciable effect on backscatter.

Microwave radiation operates at wavelengths ranging from ~1 mm to ~1 m, and is split into different bands to allow for ease of reference as Table 20.1 shows. However, the military origins of radar have led to the band names having no relationship to the wavelength making this at times a confusing convention.

The wavelength determines the scale of feature that can be observed, with features of a similar or larger size relative to the wavelength being detectable. As air molecules are substantially smaller than microwave wavelengths, this leads to a widely quoted advantage of radar, i.e. that it "sees through clouds". Different SAR sensors use wavelengths selected to best reflect the desired use case. As shorter wavelengths interact with smaller objects, the exact wavelength used by each SAR system is tuned to its desired use case. In a vegetation context, L and P band penetrates through leaves, twigs and small branches but are reflected by tree large branches/trunks. This means that L and P band backscatter measurements can be used to infer forest biomass, and is the primary motivation for the planned ESA BIOMASS mission. C-band is routinely used in marine applications as the interaction between C-band and both sea-surface capillary waves and sea ice is well understood. Since the launch of Sentinel-1 there has been a proliferation in the use

Table 20.1 Radar wave bands

Band name	Wavelength (cm)	Satellite sensor
P	30–100	BIOMASS (planned)
L	15–30	ALOS PALSAR 1 & 2
S	7.5–15	NovSAR-1
C	3.75–7.5	Sentinel-1, RADARSAT-2
X	0.25–3.75	TerraSAR-X, Cosmo SkyMed

of C-band for vegetation monitoring, particularly for crop and grassland monitoring as C-band is well suited for measuring leaf and stem biomass.

In addition to the strength of backscatter, the orientation, or polarisation of the returned wave can also be measured. Through comparing the polarisation of the transmitted and received waves, we can gain additional information regarding the structure of target features. Most current SAR systems use linearly polarised waves which oscillate in vertical (V) or horizontal (H) planes. If the wave is scattered by a flat surface, or series of flat surfaces at ~90° angles (such as a building or ship), then the orientation of the wave will not change and it will be received in the same polarisation as transmitted. However, if the wave is scattered by a randomly oriented collection of features, such as a woodland, the wave will be scattered in all orientations. The degree to which it has been re-orientated (or de-polarised) can be quantified by measuring the return using the opposite polarisation to the one transmitted. The convention for referring to image polarisation is to denote the transmit and receive polarisation for each data layer (or channel) in the final image, e.g. VV denotes transmitted V and received V. Polarisation can be further separated into co-polarised, where the same polarisation is transmitted and received (i.e. VV or HH), versus cross-polarised where the transmitted and received polarisation are different (i.e. VH or HV). For vegetation monitoring the cross-polarised channels are typically of most interest as they are more sensitive to variations in vegetation structure (Fig. 20.6).

From an operation context, the SAR data provided by suppliers typically requires specialist processing to generate an analysis-ready product, and requires powerful workstations to do this. While the rise of analysis-ready data providers is helping make SAR data more readily accessible by non-specialists, it can still be costly to generate at large scales. SAR data also typically appears noisier than a similar resolution optical image due to the presence of speckle. Speckle is a noise-like phenomenon that creates the characteristic "salt and pepper" appearance of SAR images. It can be suppressed during processing by applying adaptive filters, or by using multi-looking to average neighboring pixels together. However, both of these approaches sacrifice spatial information to reduce the influence of speckle.

SAR satellites have been deployed for over 40 years since Seasat in 1978, but they have largely been either science missions or commercial which has restricted the scope for routine use. With the launch of Sentinel-1 in 2014, global monitoring with SAR at a six- or twelve-day tempo has become viable. When combined with SAR's all-weather capabilities, and high spatial resolution (nominally 10 m), this

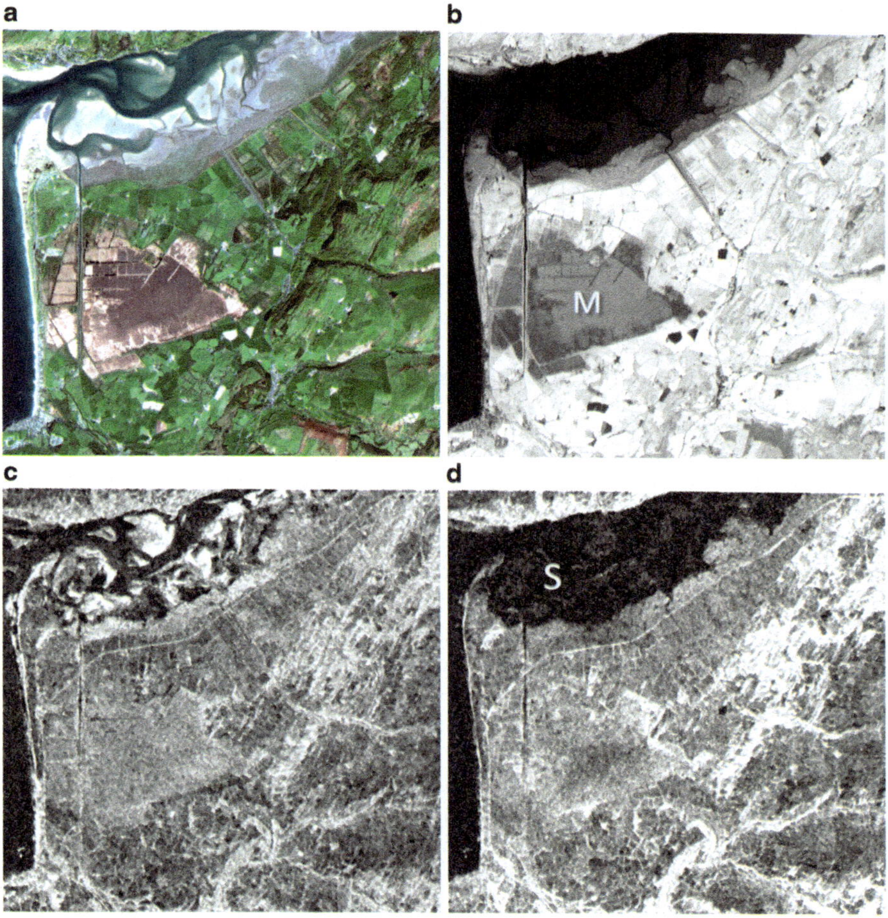

Fig. 20.6 Comparison between Sentinel-1 and Sentinel-2 images captured on 14th April 2020 over the Dyfi Estuary in Wales. Sentinel-2 true colour (**a**) and NDVI (**b**) clearly show the difference between the low chlorophyll content of brown marshland (M) in the centre of the colour images versus the surrounding higher productivity green pasture, with dark (low values) and light (high values) of NDVI. This difference is not apparent in the Sentinel-1 VV (**c**) or VH (**d**) as both the marsh and pasture contain plentiful biomass, and hence have strong radar responses. Comparing the Sentinel-1 VV (**c**) and VH (**d**), you can also see that wet sand in the estuary (S) has a similar response to vegetation in the VV, but is very low in the VH. This is because sand is a flat surface and does not change the orientation of the radar wave, hence it has a low VH response compared to vegetation. (Imagery: Environment Systems Data Services © 2020, CC BY-SA 4.0. Contains modified Copernicus Sentinel data 2020)

provides unparalleled opportunities for reliable vegetation monitoring throughout the season. A number of studies have demonstrated the potential for using time series of SAR observations to monitor crop development or identify crop type (Veloso et al. 2017), and have shown particular success when combining SAR time series with optical imagery. Moreira et al. (2013) is an accessible tutorial on SAR basics illustrating the features of radar imagery.

Box 20.7
Exercise 7. In Sentinel Playground, select 'Sentinel-1' to look at the VV and VH polarisations. Notice how these images are more difficult to interpret visually because of the poorer spatial resolution and 'speckle', even though they are extremely informative when a stack of images is combined in analysis.

7 The Advantages of Remote Sensing

Remote sensing systems are designed to be useful, and to allow planned applications to be supplemented by new uses opening up as imagery, ground observations and algorithms accumulate and are shared between practitioners. This is greatly facilitated by free imagery and free, open-source software. Universal coverage means that solutions developed in one country can freely be adopted in another. The Copernicus programme is designed to spread understanding of both remote sensing and the environment; and of how remote sensing can monitor the environment, creating an open market for new 'downstream' service-providers across Europe.

As we have seen, satellite remote sensing has clever technology. Unlike conventional air photography, which is optimised for stereo visual interpretation, pushbroom scanners have discrete colour channels with no overlapping frequencies — so there is no mixing of, say, green and blue light — and include the near and short-wave infra-red, supporting spectral indexes which greatly extend our use of the electromagnetic spectrum. NDVI is just one example of this. We hope this short introduction has shown that satellite imagery is accessible, and that, starting with optical imagery and 'natural' colours, and looking at familiar places, we can extend our ecological interpretation using the less obvious infra-red and thermal channels. This gives us insights into the same processes of energy capture that plants use to grow, including seasonal effects, phenology, and the contrast in plant growth between sunny and shaded slopes, low and high ground. By understanding these patterns, we can create rules that allow us to interpret satellite images of unfamiliar landscapes and detect landscape change across a stack of images. Synthetic aperture radar further extends our ability to look at plant structure and properties such as soil moisture (using different wavelengths), and see through the clouds that cover the wet fringes of western Europe and high mountains of continental Europe. New software technologies go hand in hand with data mining, automated change detection and machine learning finding patterns in the growing archive of millions of remote sensing observations.

The recent high frequency of data capture and rapid online delivery of imagery makes it possible to collect and compare simultaneous ground observations, facilitating near real-time monitoring, disaster response (e.g. flood mapping) and economically important predictions such as crop yield. As imagery accumulates we can also go backwards in time, for example to see when a meadow was mown or fertilised as candidate explanations of the distribution of scarce plants recorded in the field. Similar remote observations can be used to stratify and prioritise a campaign of fieldwork, and provide *a posteriori* statistical weightings of survey data. By sharing the same access to imagery and software, different observers can share similar metrics, compare monitoring data and build up hypotheses to explain large-scale geographical trends in biodiversity. This ability to scale up and share both observations and predictions will be critical for conserving all aspects of biodiversity and habitat condition in the face of agricultural improvement, pollution and climate change.

While not the focus of this chapter, hyperspectral instruments, normally flown in aircraft, give very sophisticated spectral indexes for identifying and mapping surfaces, including cover-forming grasses, tree canopies and crops. More capable and affordable drones will soon be available carrying these sensors, and others including LiDAR (ranging instruments which show structure) and radars, giving access to highly detailed spatial, spectral and structural characterisation of small areas.

7.1 The Future of Arable Remote Sensing

While at the time of writing (2020) fewer studies have made practical use of Sentinel 2 imagery, we can expect significant improvements in modelling from the better temporal and spatial resolution over Landsat. We can also expect better and cheaper sensors will be available for small drones, including hyperspectral and LiDAR (ranging systems which measure structure), and multiple narrow-band cameras which give accurate vegetation indexes. Inexpensive near real-time observation will be carried out by swarms of small optical satellites in stationary orbits, each one continuously observing a small part of the ground at high spatial resolution; and by long endurance drones flying at a high altitude, both able to carry the latest sensors — unlike conventional satellites which have decade-long design and build times. Cheap small sensors will also be available to measure ground conditions such as temperature and soil moisture, complementing more sophisticated instruments carried on agricultural machinery. Interpretation will benefit from machine learning software and cloud-based data cubes able to stack and analyse multiple images at different scales. Most importantly of all, expert users including ecologists, land managers and farmers will start to become familiar with satellite imagery and make these analyses their own.

8 Next Steps

In the practical exercises, we have been working with imagery processed remotely, which is instant but has limited options. For do-it-yourself remote sensing, the next step is to download Sentinel 2 images (data granules) and analyse these on your computer. Good places to start are GIS packages such as Quantum GIS (https://www.qgis.org), working in the raster menu and ESA's own dedicated Sentinel Toolbox software (SNAP/STEP, available on http://step.esa.int).

Box 20.8
Exercise 8. Go to the Sentinel Open Access Hub (https://scihub.copernicus.eu), register an account and download images that you can load into QGIS or SNAP. There are plenty of online tutorials showing how, while other excellent commercial services give easy access to the same imagery.

A useful first exercise is to load more than one image (e.g. in QGIS), look at the differences between them by simple visualisation, then try constructing a false colour image from the same band on different dates, in which cases changes will show up as colour differences. It is also worth getting to know the shared library of raster tools, GDAL (https://gdal.org) which can be used independently of packages to subset, resample and re-projecting imagery. This is quick and can be scripted in Windows. Apart from these free packages there are excellent — but quite expensive — commercial image analysis packages such as IDL ENVI (https://www.harrisgeospatial.com) and ERDAS Imagine (https://www.hexagongeospatial.com). These are recommended for automating serious analysis, as is cloud-based processing using Google Earth Engine (https://www.earthengine.google.com).

An alternative approach is to look for projects and collaborations with partners that are already experts in remote sensing, including government agencies and universities, who might be looking for study sites. Europe also has excellent specialist remote sensing consultancies and service-providers, including suppliers of pre-processed remote sensing optical and radar imagery in the form of analysis-ready data (ARD).

Finally, there are many useful texts in remote sensing, though some of them are very technical. A good modern introduction to Remote Sensing and GIS is Liu & Mason (2016).

References

CEOS (2019) CEOS analysis ready data for land (CARD4L) description document, http://ceos.org/document_management/Meetings/Plenary/30/Documents/5.5_CEOS-CARD4L-Description_v.22.docx Last accessed 18 Mar 2020

Charania I, Li X (2020) Smart farming: Agriculture's shift from a labor intensive to technology native industry. Internet of Things 9

Ferran G, Bouzinac C, Thépaut O, Jung M, Francesconi B, Louis J, Lonjou V, Lafrance B, Massera S, Gaudel-Vacaresse A, Languille F, Alhammoud B, Viallefont F, Pflug B, Soullier V (2017) Copernicus sentinel-2A calibration and products validation status. Remote Sens 9 (6):1–81

Franke J, Keuck V, Siegert F (2012) Assessment of grassland use intensity by remote sensing to support conservation schemes. J Nat Conserv 20(3):125–134

Gómez Giménez M, De Jong R, Della Peruta R, Keller A, Schaepman ME (2017) Determination of grassland use intensity based on multi-temporal remote sensing data and ecological indicators. Remote Sens Environ 198:126–139

Higgins S, Schellberg J, Bailey JS (2019) Improving productivity and increasing the efficiency of soil nutrient management on grassland farms in the UK and Ireland using precision agriculture technology. Eur J Agron 106:67–74

Hunt ML, Blackburn GA, Rowland CS (2019) Monitoring the sustainable intensification of arable agriculture: the potential role of earth observation. Int J Appl Earth Obs Geoinf 81:125–136

Jones HG, Vaughan RA (2010) Remote sensing of vegetation: principles, techniques, and applications. Oxford University Press, Oxford/New York

Liu J-G, Mason J (2016) Image processing and GIS for remote sensing: techniques and applications, 2nd edn

Maskell LC, Botham M, Henrys P, Jarvis S, Maxwell D, Robinson DA, Rowland CS, Siriwarden G, Smart S, Skates J, Tebbs EJ, Tordoff GM, Emmett BA (2019) Exploring relationships between land use intensity, habitat heterogeneity and biodiversity to identify and monitor areas of high nature value farming. Biol Conserv 231:30–38

Minasny B, Berglund Ö, Connolly J, Hedley C, De Vries F, Gimona A, Kempen B, Kidd D, Lilja H, Malone B, McBratney A, Roudier P, O'Rourke S, Rudiyanto Padarian J, Poggio L, Caten Ten A (2019) Digital mapping of Peatlands – a critical review. Earth Sci Rev 196:102870

Moreira A, Prats-Iraola P, Younis M, Krieger G, Hajnsek I, Papathanassiou KP (2013) A tutorial on synthetic aperture radar. IEEE Geosci Remote Sens Mag 1(1):6–43

Pettorelli N, Schulte to Buhne H, Tulloch A, Dubois G, Macinnis-Ng C, Queirós AM, Keith DA, Wegmann M, Schrodt F, Stellmes M, Sonnenschein R, Geller GN, Roy S, Somers B, Murray N, Bland L, Geijzendorffer I, Kerr JT, Broszeit S, Leitão PJ, Duncan C, El Serafy G, He KS, Blanchard JL, Lucas R, Mairota P, Webb TJ, Nicholson E (2017) Satellite remote sensing of ecosystem functions: opportunities, challenges and way forward. Remote Sens Ecol Conserv 4 (2):71–93

Thenkabail, Srinivasa P, Lyon JG, Huete A (2019) Hyperspectral remote sensing of vegetation. Volume I, Fundamentals, sensor systems, spectral libraries, and data mining for vegetation (2nd edn) CRC press

Veloso A, Mermoz S, Bouvet A, Le Toan T, Planells M, Dejoux J-F, Ceschia E (2017) Understanding the temporal behavior of crops using Sentinel-1 and Sentinel-2-like data for agricultural applications. Remote Sens Environ 199:415–426

Vihervaara, Petteri, Auvinen A-P, Mononen L, Törmä M, Ahlroth P, Anttila S, Böttcher K, Forsius M, Heino J, Heliölä J, Koskelainen M, Kuussaari M, Meissner K, Ojala O, Tuominen O (2017) How essential biodiversity variables and remote sensing can help national biodiversity monitoring. Glob Ecol Conserv 10.C:43–59

Weiss M, Jacob F, Duveiller G (2020) Remote sensing for agricultural applications: a meta-review. Remote Sens Environ 236

Chapter 21
The Art of Agri-environment in the UK, an Advisor's Perspective

Emily Swan

Abstract Championing a group of species that economics, science, technology and society have comprehensively tried to control for the best part of half a century is no mean feat. This was not conspiracy, at the time it was considered progress. The plight of many of these plants is also compounded by the physiology of the species themselves. Many are small and easily overlooked. Conversely, some such as the cornflower and corn cockle are regularly included in flower mixtures such that many people probably don't realise these species once thrived in the wild. Others such as poppies, are held by many with fondness, thanks to their long association with remembrance and respect, since the fields of Flanders.

Since the 1990s Agri-environment schemes (government-led initiatives and paid incentives for introducing conservation into agriculture) started to feature in the agricultural story. It provides a mechanism for some of these species to persist and potentially regain a foothold as part of a diverse farmland landscape. Whilst helpful, to a certain extent the financial incentive is of lesser importance. The key to Agri-environment rests with those who advise on it and how they influence hearts and minds to steer progress in a nature friendly way.

Keywords Arable plants · Agri-environment schemes · Practical conservation · Farming economics · Farmer motivation · Management advice

1 "What do you do?" the Question I Dread

I have been working in farmland conservation for over 16 years and yet still I find myself without a pithy answer to this question. In part this is down to the inevitable dynamic of working in the public sector and frequent re-organisation that comes with it. Aside from that, the sector within which I work is one of the most dynamic, changeable and possibly contradictory ones in the UK. Traditional practices, passed

E. Swan (✉)
Philip Wilson, Pennyhayes Farm, Shute, Axminster, Devon, UK

C. Hurford et al. (eds.), *The Changing Status of Arable Habitats in Europe*,
https://doi.org/10.1007/978-3-030-59875-4_21

down through the generations persist to this day and yet at the other end of the spectrum it is one of the most cutting edge industries, with driverless tractors and genetic crop modification. Agriculture, and by extension farmland conservation is endlessly influenced through time and by an almost imperceptibly broad range of factors. Such is the dynamism, it is often unquantifiable and in turn that can make it exceptionally difficult to decide on a suitable conservation destination, let alone pick a path to get there.

There are a great many reports which detail why we need to conserve. Indeed the State of Nature 2019 (Hayhow et al. 2019) provides a statistical look at the reality facing the Nature of the UK. There are also a wide range of studies and statistics which analyse the successes and failures of many arable conservation initiatives. Indeed, scientific evidence to inform decision-making at all levels is critical to ensure that the best possible decisions are made. However, an understanding of the softer aspects of working in and delivering arable conservation is just as critical if agri-environment schemes and the conservation of arable habitats are to be successful. When working in farmland conservation for years you start to get a feel for some of these.

In this chapter I hope to combine some of the anecdotal stories, scientific evidence and the successes and failures that I have experienced as a passionate conservationist of arable land and in particular arable plants.

2 External Pressures

In 2019 there are a multitude of pressures and influences which can affect the willingness and ability of farmers and land managers to engage in farmland conservation to a high standard. In no particular order and not exhaustively, these include:

- Increasing demand for land – other projects such as housing, infrastructure projects and energy generation all need land
- Commodity prices, food production and supply chains
- Weather and climate change with extreme weather events and changing weather patterns
- Public access
- Changing political situations
- Competing conservation projects and initiatives – such as woodland creation or 'rewilding'.

I have seen the direct impact of commodity prices on the uptake of conservation measures on arable land. When feed wheat prices went up to nearly £200/tonne in 2010, as a result of a changeable global wheat market due to a wet Australian harvest and a drought in South America there was a distinct slowdown in the uptake of arable conservation options in England. People were telling me they simply wanted to be able to grow. Fast forward a few years and wheat was back down to £60/tonne and uptake levels of schemes increased. The schemes it should be noted at the time were paid at a price that was set for 10 years and therefore unable to rise/fall in accordance with the commodity prices.

Perhaps unsurprisingly weather and climate can affect the delivery of arable conservation options in much the same way that they affect crop production and agriculture in general. The prolonged summer drought of 2018 in the UK resulted in numerous crop failures of seed mixtures sown to provide winter bird food. Despite being established within the prescribed deadlines, the lack of rain meant that many did not germinate well and were overtaken by pests or more prolific weed species that were better able to adapt to the drought conditions. A changing climate and more extreme weather events are likely to present a number of future challenges. However the deployment of farmland conservation measures themselves may well have a critical role to play in helping mitigate some of these impacts.

Despite the well evidenced environmental benefits and potential future opportunities for climate change adaptation, incentive schemes, (those which provide grants for environmental measures in an arable setting) will often not hold their own against other economic forces. Land prices which developers are willing to pay for prime sites for development and infrastructure projects can be phenomenal. Social and economic pressures upon individuals and their families have to be taken into account and any incentive schemes or funding measures need to be appropriately financed and targeted to ensure people are encouraged to adopt the most appropriate measures. Coupled with this there must be a suitable level of support available in the form of good, impartial technical advice as well as feedback. All too often in the past as an adviser I have not been able to do the follow up work that I would like after the initial scheme set up and so the ongoing benefits of encouragement and at times troubleshooting are not always available.

Conservation initiatives can often compete with themselves. Not only in terms of resources such as the scenario above but also spatially. With an increasing focus on Carbon capture, tree planting, rewilding and natural flood management farmland is often the first place we look at to provide these benefits. In 2018 I began working on a project in Norfolk to create a large contiguous stretch of semi natural habitat through a process of natural regeneration, a version of 'rewilding' if you will. Much of this was to be created on arable land which also supported a good population of grey partridge (*Perdix perdix*) and also a nationally significant suite of arable plants. In this instance such was the richness and abundance of the arable plant population we were able to deploy a significant package of targeted measures towards maintaining and increasing these plants across the rest of the arable estate acreage. This created not only a large scale habitat creation project but also left enough space to ensure that the wildlife rich farmland of west Norfolk was enhanced as well. In this instance we were able to strike a strong balance, but it was not an easy thing to do and in other situations it may be more complicated to reach an acceptable conclusion.

3 Motivation

When it comes to conservation and especially farmland conservation, people are driven to conserve for a variety of reasons. Whether it be market forces, land prices, incentive schemes, environmental passion or competitiveness to name but a few. It may be assumed that those who are driven by environmental motives deliver the best conservation, and in many instances this may be true. However one study into social and ecological drivers of success (McCracken et al. 2015) found that some of the most efficient, production orientated farmers delivered the best quality wild bird seed mixes to feed our declining farmland birds. Similarly those who begin their farmland conservation journey for financial reasons I have seen grow into some of the best environmental advocates I know. Peer to peer encouragement between farmers can also be an extremely powerful positive environmental influence. Understanding and working with these individual motivations is critical to delivering successful conservation measures. Therefore it is vitally important to understand what drives people, quite aside from the conservation requirements themselves and ensure that the measures which are available to them are both effective but deliverable. This is something which can be missed with an approach that focuses solely on the nature conservation requirements being imposed.

Knowing your audience is critical. My personal experiences of working extensively on the conservation of arable plants (formerly known as weeds!) demonstrates nicely how important this is. I arrive at a farm armed with a suite of available options to offer farmers to conserve farmland wildlife. Hedgerow planting, wildflower margins and winter bird food are staples for many. They enhance the farm's landscape, provide shelter and food for a range of species and can often make farming easier. All are 'quick wins' with multiple benefits. At this point I have to decide whether or not to brace the question 'have you considered a cultivated margin or two'. When I describe what this actually entails (cultivating land on an annual basis and leaving nature to take its course) many folk look at me aghast. I have just asked them to 'grow weeds' and encourage what they have spent 40 years trying to eradicate!

After my first few unsuccessful attempts I refined my approach and am ready now with the responses to people's concerns, dispelling any fear that surrounds the option and making it a practical and viable activity for farmers to deliver. Taking farmers back out to see the rare species that have recovered from the seed bank and explaining the importance of them really fires people's enthusiasm and opens their eyes beyond the thistles. Continuing to foster the motivations and enthusiasm over time is such an important job in arable conservation. Other priorities and pressures can soon over take a small weedy corner or a billowing hedgerow thick with berries if people don't hold them in a higher regard.

Maintaining the motivations of those engaged in arable, or indeed any conservation measures is essential. The vast majority of people want to do the best they can across the piece. Conservation is no different. However for many people it is not necessarily clear cut what success may look like. A good crop of wheat or barley or a prize cow is something many farmers can relate to and recognise easily and the endless challenge is to make sure that conservation objectives are clearly understood and valued.

4 Language and Narrative

Anyone who has worked in farmland conservation will be intimately familiar with the vast range of interchangeable terms which are used to describe the various measures that are deployed for conservation around the edges of fields. Headlands, margins, buffer strips, all of these can mean the same thing, the six-ish metres or so around the edge of the field where conservation is done. There have been attempts, only partially successful, to resolve this terminology (Greaves and Marshall 1987). They also mean very different things depending on the scheme context against which they sit. When so many schemes run concurrently at any one time it can become overwhelming and can cause confusion in delivery. I am not sure what the solution is going forwards, or whether it can ever be simplified, but an awareness of the vast, yet often specialist glossary that farmers and conservationists alike are exposed to will be essential if future initiatives are going to be effectively communicated and readily adopted.

A positive example of simple and effective communication is the Plantlife Important Arable Plant Area (IAPA) (Byfield and Wilson 2005) classification work. Based on the relative scarcity of declining arable plants (reviewed in 2015) this system ranks species on a scale of 1–9. There are formal definitions for these based on the species red list status (Cheffings and Farrell 2005). However the ones I use when working in the field and discussing with land managers is 1 being a 'declining but still fairly common' species and 9 being 'rare as hens teeth' (Fig. 21.1). The sense of achievement that people clearly display when you tell them they have a species that is a 7 or above is both obvious and heartening in most instances. They don't need to know the biological ins and outs necessarily, there are specialists that can do that but they do relate to the very human feelings of pride and accomplishment when they realise they have got and maintain something special.

In much the same way I have spent over a decade trying to encourage farmers to think of arable annuals as rare arable plants (as opposed to weeds) I think many conservationists must also amend the way they view the farmed environment. If we are to truly reverse the declines in farmland birds, increase pollinator numbers, restore our soil health and diversify rare arable plant populations we must move preservationist thinking on a bit. Much focus has been placed on important and designated sites in the past and these are without doubt some of the 'jewels in the crown'. However in placing these sites on such a pedestal we run the risk of devaluing the sumptuous velvet of the wider countryside on which these jewels sit. With over 70% of the UK's land area as farmland we overlook its intrinsic value at our peril. Farmland has its own rich and varied biodiversity with a range of species which rely on agriculture to persist. Connectivity between special sites, across

Fig. 21.1 Rare species found in cultivated field margins established under Agri-environment agreements in Eastern England. *Silene gallica* (IAPA score 8), *Legousia hybrida* (IAPA score 3)

farmland is of course vital but so too is the creation of habitat hotspots and species recovery within the farmed environment.

5 Evolution

It has struck me as inherently strange that conservation thinking and planning does not always embrace evolution. Many of the goals we set ourselves as conservationists are the desirable end state of a habitat or population. These decisions are nearly always based on good scientific data but they remain an end goal. We want to see an increase in farmland birds, a more diverse suite of plants and so forth. We set ourselves timeframes and targets for this but these are fixed in space and time. When we do not achieve them it is viewed as failure. Personally I think that this is rather de-motivating and in any case nature does not realise it has been put on notice in this way.

On the first ever agri-environment scheme I was involved with I noticed 2 corn marigold plants in the edge of a barley field. I didn't at the time even know what they were. However, I went away, found out and put in place the aforementioned cultivated margin. Fast forward 15 years real time (not the 5 or 10 year chunks we

Fig. 21.2 Cultivated field margins in Eastern England established for rare arable plants under Agri-environment agreements

measure Agri-environment in) and those margins contain 5 vulnerable 1 endangered and one species considered extinct in Norfolk for over 50 years (Fig. 21.2). It has been an exciting and unpredictable journey, with hurdles along the way. However approaching it with good ecological principles, a good dose of agronomic reality and an open mind has yielded phenomenal results all round. The farmer now advocates to other farmers about how to deliver the option successfully and from a practical perspective. He is very proud of what they have achieved.

Evolution of schemes is just as important as the evolution of habitats and species populations they seek to conserve. Agri-environment schemes emerged in the mid-late 1980s but could never have been pitched where they are today. It was a new venture then, and completely contrary to the incentives and policy drivers that had gone before them. Indeed there was a period of time when some initiatives overlapped, for example UK government grants. Payments to increase agricultural production ran from the mid 1960s until the late 1980s (Pretty 1999), while the Broads Grazing Marsh scheme, the forerunner of modern agri-environment came in 1985, followed by the first Environmentally Sensitive Areas in 1987 (Natural England 2012). The shifting focus had therefore to be incremental and achievable. Schemes were simple, with a small range of options revolving around a geographically distinct narrative. The Broads Environmentally Sensitive Area was the first of these that I encountered when I started my career and it is a lifetime away from the

Schemes I work with now in terms of aspiration, complexity and process. Some elements of these schemes have changed for the better with a more flexible range of objectives. However the corresponding impact has been to develop a more complicated and some would say onerous process which has had a knock on impact on uptake.

6 Conclusion

True harmonisation of all of the factors that can influence farmland conservation is nigh on impossible. However, in my experience success comes through careful alignment and balance between these factors. This could be at varying scales, from increasing a population of Corn Buttercup from 10s to 100s of plants in one margin up to pollinator superhighways that stretch across a whole county.

With an eye to the future and an ear to the past we must learn from our previous experiences, to understand what has caused the well reported declines in our farmland wildlife not only ecologically but the sociological and societal factors as well. People did not plough up chalk grasslands on the South Downs in the 1940s to eradicate chalk grassland, they did it to feed the country during the war years. Societal pressures and economics continued to drive the policy makers and land managers through the years and gradually people started to value land differently. This is illustrated nicely by the South Downs becoming a fully functioning National Park on 1st April 2011.

Respecting the evolution of the conservation journey is another key element of how we must develop. Nobody runs a marathon without training first and the analogy applies to conservation. For many people the creation of that 'weed strip' is a big step away from what they know and where they feel comfortable. They may take time to get there, but if given the stepping stones to do so, more often than not they do get there and then go beyond. Policy, schemes and incentives must facilitate this incremental approach. We must respect the fact that conservation is a journey and not a destination and that everyone starts that journey from a different place and in a different vehicle.

We have travelled through the eras of over production and agricultural mechanisation, into herbicide resistance and extreme climatic events. In turn these affect our willingness and ability to conserve the habitats and species that thrive in the farmed environment. We can use technology to help us in conservation as well as production whilst at the same time respecting the tried and tested methods of species and habitat management.

Ultimately the complexity of striking a balance between how we use land in the UK, the need to conserve our natural and cultural heritage, provide space for people, grow food and future-proof ourselves against the effects of a changing climate is probably more complex now than it ever has been. Therefore adaptability and flexibility of approaches will be critical as is an understanding into the communities and locations in which they will deployed.

One of my colleagues summed the nature of this up wonderfully in his retirement speech in 2007. He had worked for the government's various agricultural departments his entire career.

> I started my career in 1976 as a land drainage engineer with the Ministry of Agriculture, draining and 'improving' the grazing marshes of Broadland. I retire today as a Conservation adviser having spent the last 10 years providing advice and schemes for how to 're-wet' the broads to conserve breeding wading birds. Therefore the net contribution of my career on agriculture and the environment is zero! (C Hitchman pers. comm).

It is my greatest wish that when I retire my words do not offer the reverse sentiment to this tongue in cheek, yet thought provoking anecdote from one of my greatest mentors.

References

Byfield AJ, Wilson PJ (2005) Important arable plant areas: identifying priority sites for arable plant conservation in the United Kingdom. Plantlife International, Salisbury, UK. http://www.plantlife.org.uk/publications/important_arable_plant_areas/

Cheffings CM, Farrell L (eds) (2005) Dines TD, Jones RA, Leach SJ, McKean DR, Pearman DA, Preston CD, Rumsey FJ, Taylor I (2005) The vascular plant red data list for Great Britain. Species Status 7:1–116. Joint Nature Conservation Committee, Peterborough

Greaves MP, Marshall EJP (1987) Field margins: definitions and statistics. In: Way JM, Grieg-Smith PW (eds) Field margins, BCPC monograph, vol 35, pp 85–94

Hayhow DB, Eaton MA, Stanbury AJ, Burns F, Kirby WB, Bailey N, Beckmann B, Bedford J, Boersch-Supan PH, Coomber F, Dennis EB, Dolman SJ, Dunn E, Hall J, Harrower C, Hatfield JH, Hawley J, Haysom K, Hughes J, Johns DG, Mathews F, McQuatters-Gollop A, Noble DG, Outhwaite CL, Pearce-Higgins JW, Pescott OL, Powney GD, Symes N (2019) The State of Nature 2019. The State of Nature Partnership

McCracken ME, Woodcock BA, Lobley M, Pywell RF, Saratsi E, Swetnam RD, Mortimer SR, Harris SJ, Winter M, Hinsley S, Bullock JM (2015) Social and ecological drivers of success in agri-environment schemes: the roles of farmers and environmental context. J Appl Ecol 52:696–705

Natural England (2012) Evolution of Agri-environment schemes in England (NE373) http://publications.naturalengland.org.uk/publication/3567470

Pretty J (1999) The living land: agriculture, food and community regeneration in rural Europe. Earthscan Publications, London & Sterling VA

Chapter 22
The Future of Europe's Arable Wildlife

Phil Wilson, Clive Hurford, Marian Reed, Chris Stoate, and Jonathan Storkey

Abstract The chapters of this volume describe the transformation of Europe's arable biodiversity during the twentieth century and the early years of the twenty-first. A large number of species have declined and disappeared in response to the huge changes in farming methods especially since the 1940s. Many of these species are totally reliant on the continuation of low-intensity farming practices, and if we are not to suffer further deterioration of our farmland biota, changes in farming practice are essential.

There has been much recent work on the plants and birds of arable farmland although rather less on invertebrates and other soil organisms, and we have much of the scientific knowledge required to stop their declines. Much of this work has involved the farming community, and the involvement of land-owners and managers is essential for effective conservation.

Payments to farmers through agri-environment schemes to extensify management for the benefits to wildlife are now available throughout the EU and in other European countries. It should be emphasised that these are not subsidies, but payments for ecosystem services which are of benefit to the whole population.

Despite improved scientific knowledge, the enthusiastic engagement of some sectors of the farming community and the availability of payment for environmental goods, the declines of some groups of farmland organisms are continuing. It is clear that there are some gaps to be filled in our understanding of the biology of many

P. Wilson (✉) · M. Reed
Pennyhayes Farm, Shute, Axminster, Devon, UK
e-mail: cstoate@gwct.org.uk

C. Hurford
Serapias (Ecological Monitoring Consultancy), Ty Llaeth, Y Fraich, Fishguard, UK
e-mail: clive.hurford@serapias.net

C. Stoate
Game and Wildlife Conservation Trust Allerton Project, Loddington, Leicestershire, UK

J. Storkey
Sustainable Agriculture Sciences, Rothamsted Research, Harpenden, Hertfordshire, UK
e-mail: jonathan.storkey@rothamsted.ac.uk

C. Hurford et al. (eds.), *The Changing Status of Arable Habitats in Europe*,
https://doi.org/10.1007/978-3-030-59875-4_22

species, that more work is required to spread awareness of the importance of farmland biodiversity and that refinements to agri-environment schemes may be needed. Pressures to concentrate conservation effort on "land-sparing" should be resisted. While some "rewilded" land in the European lowlands would be very welcome, it should not distract from the importance of existing valuable habitats that have developed under human intervention.

A deeper appreciation of farmland wildlife and its integration into farming systems is needed to ensure its future security in a world where value is increasingly expressed in terms of material profit. Contributors to this volume have shown that with appropriate advisory and financial assistance, agriculturally productive farmland can support a rich and flourishing biodiversity.

Keywords Agricultural intensification · Agri-environment schemes · Species loss · Habitat heterogeneity · Science-led conservation · Land-use change

1 Declines and Disappearances

It is clear from the results of work described in this volume, that the biodiversity of arable land throughout Europe has undergone major changes particularly during the second half of the twentieth century, and that these are continuing into the twenty-first century. The great majority of these changes have involved the loss of species, many of which are much loved and charismatic (e.g. grey partridge *Perdix perdix,* little bustard *Tetrax tetrax* and cornflower *Centaurea cyanus*), the impoverishment of species assemblages and reduced abundance of species that remain relatively common. This transformation has been the result of developments in arable farming including the adoption of broad-spectrum herbicides for weed control, the use of chemical insecticides and fungicides, massive increases in the application of nitrogen and phosphorus-containing fertilisers, the breeding of highly nutrient-responsive crop varieties, the simplification of crop rotations, loss of field boundary features, enlargement of agricultural machinery and many others. A limited number of plant species for instance brome grasses *Bromus spp,* black-grass *Alopecurus myosuroides* and cleavers *Galium aparine* have benefited from these changes and have now become the pests of contemporary arable farming.

Much work is therefore required if we are to help restore our arable habitats to anything resembling their historic importance for wildlife without sacrificing gains in productivity. This will require both changes to the political processes which drive agricultural policy and the development of techniques for de-intensifying land-management while maintaining levels of crop production.

One barrier to effective conservation of arable ecosystems has been the indifference of many ecologists and conservationists. It is hoped that this volume will help to remedy this. Forty years ago in the early 1980s, the suggestion that arable fields and the wildlife associated with them was of any conservation value and worthy of scientific study would have been met with bewilderment and sometimes laughter.

Arable land was generally regarded as the enemy, and ploughing for crop production as the destroyer of habitats of real value. There was some justice to this attitude. In the UK during the war years (1939–1945) and following, the drive for increased national food production and the development of mechanical and chemical technology had provided the incentive for the destruction of very large areas of semi-natural habitat and conversion of land rich in biodiversity to intensively managed arable and pasture farmland. Considerable national and regional investment was provided to subsidise this process. The detrimental effects of these processes of intensification resulted in arable land being considered by ecologists as devoid of interest, irrespective of the wildlife that was actually present, the intensity of the management systems or the philosophy of the farmer.

The declines of farmland and steppe birds in Europe became apparent from the nineteenth century onwards. While some of the early losses may have been the result of over-hunting (e.g., great bustard in the UK), the large-scale declines throughout the continent in the latter half of the twentieth century and particularly the early twenty-first century can be attributed to agricultural intensification and loss of habitat heterogeneity (Traba and Morales 2019; Gaget et al. 2018; Mañosa et al. this volume – Chap. 19; Sotherton et al. this volume – Chap. 18). Invertebrates also depend on habitat heterogeneity, although over a smaller scale than the larger birds and mammals. Some of the recent losses of invertebrate diversity can be attributed to the loss of interstitial habitats such as hedges, grassland patches, ditches and ponds, although they have also been profoundly affected by the direct and indirectly toxic effects of agrochemicals (Storkey et al. this volume – Chap. 13; Ewald et al. this volume – Chap. 11; Goulson 2013).

Arable plants are not as dependent on habitat heterogeneity, nevertheless individual species and species diversity have declined throughout Europe since the end of the nineteenth century, and recent work suggests that local habitat diversity may have a significant effect on arable plant assemblages (Gaba and Bretagnolle this volume – Chap. 14; Bourgeois et al. 2020). Improved seed-cleaning methods prevented the resowing of non-crop plants and caused the loss of several highly specialised species (Kornas 1961), but from the mid-twentieth century onwards as agricultural inputs increased, competitive crop varieties were introduced, rotations were simplified and fields were enlarged the decline accelerated until over much of Europe arable vegetation has now become extremely species-poor and uniform (this volume: Pinke – Chap. 3, Storkey – Chap. 2, Meyer – Chap. 6, Meyer & Bergmeier – Chap. 8, Recasens – Chap. 5, Fried – Chap. 4, Hyvonen – Chap. 7). There is some evidence that many segetal plant species are highly site-faithful under current management conditions, and that long-established arable land is richer in uncommon species than that which has entered into cultivation in recent years (Wilson 1990).

The loss of plants from arable ecosystems is not only important for botanical reasons. Plants provide directly or indirectly a food source for all other organisms, and a loss of floral productivity and diversity from arable fields has had a proven effect on a wide range of animals. The positive function of non-crop plants in agro-ecosystems is being increasingly recognised and whereas, until recently, the discipline of "weed science" was exclusively concerned with mitigating the negative impact of weeds on crops, scientists are now also working actively to also promote botanical diversity (Bretagnolle and Gaba 2015; Storkey and Neve 2018; Nichols et al. 2019). Phytophagous insects and their invertebrate predators have experienced an overall decline, although as results from the Game and Wildlife Conservation Trust's (GWCT) Sussex study show, trends differ between groups (Ewald et al. this volume – Chap. 11). The loss of particular suites of invertebrates has been the principal factor in the decline of the grey partridge and corn bunting throughout northern Europe (Sotherton et al. this volume – Chap. 18; Salek et al. this volume – Chap. 15; Brickle et al. 2000). The greatly reduced quantity of weed seed available in arable fields has been a direct cause of the decline of the turtle dove in Europe (Browne and Aebischer 2005).

2 Conservation and Restoration – How Are We Doing?

Awareness of the plight of farmland biodiversity has risen since the early 1980s, and in the years since then, there have been numerous initiatives from governmental and non-governmental organisations and concerned individuals throughout Europe.

Germany pioneered the conservation of endangered arable plants in field margins. The first attempts were carried out by Wolfgang Schumacher of the University of Bonn in the Eifel Mountains in the west of the country in 1978 (Schumacher 1980). This was initially a voluntary scheme which depended on the enthusiasm of Professor Schumacher and relationships between him and his colleagues with individual farmers. It proved very successful at maintaining and enhancing populations of endangered arable plants and was popular with farmers. Field margin management for arable plants was readily adopted as an agri-environment scheme option throughout West Germany when the first European Union-funded schemes started in 1987 and which later became compulsory for all member states in 1992 (EU Regulation 2078/92) (Batary et al. 2015). In the UK, the first arable agri-environment scheme options were introduced in 1998, and the plant conservation NGO Plantlife has been championing the cause of arable plants since 2000. In addition to devising the first method for the objective evaluation of arable plant assemblages (Byfield and Wilson 2005), they work with landowners to raise the awareness and knowledge of these plants, most recently through the Colour in the Margins project (https://naturebftb.co.uk/the-projects/colour-in-the-margins/). Some rare arable plants such as *Centaurea cyanus* and *Scandix pecten-veneris* have started to recover in the UK in recent years, but it is uncertain whether these recoveries have been the result of conservation measures, better recording or changes in farming systems. The overall picture throughout Europe however is still one of decline and impoverishment.

The relationships between the intensity of farmland management and the specialised birds of farmland habitats have received much study since the 1980s. The catastrophic declines of many species of farmland and steppe-land birds in most European countries resulting from increases in farming intensity are now well known (O'Connor and Shrubb 1986; de Juana et al. 1988; Onrubia and Andres

2005), and there have been several detailed studies demonstrating the close link between habitat heterogeneity and condition, food availability and the reproductive success and population size of single species (Potts 1986; Lapiedra et al. 2011). The mechanisms underlying these declines are known, and management strategies for their reversal are proven in practice (Sotherton et al. this volume – Chap. 18; Peach et al. 2001; Aebischer et al. 2000). Schemes have been in place throughout the EU since 1992 under the agri-Environment programme, with options directed specifically at farmland birds (de la Concha 2005; Oñate 2005) with the aims of restoring farmland habits and reversing population declines. The Farmland Bird Index (FBI) is monitored annually throughout the EU (+Norway and Switzerland) as an indicator of ecological sustainability, but despite the ability to reverse farmland bird declines and the provision of a funded mechanism to implement change in farmland management, the FBI has declined steadily since 1990, with a negative change of 35% between 1990 and 2016 (https://www.eea.europa.eu/data-and-maps/indicators/abundance-and-distribution-of-selected-species-8/assessment).

While other taxonomic groups have not been studied in such detail as arable plants and farmland birds, there appear to have been have been similar changes. Evidence suggests that there have been massive declines in insect abundance in Europe (Hallmann et al. 2017), and Ewald et al. (this volume – Chap. 11) describes a complex series of changes in invertebrate populations in one area of the UK which have resulted in a net decrease in species of importance as food for birds. Farmland mammals such as field hamster *Cricetus cricetus* (https://www.iucnredlist.org/species/5529/115073669) and brown hare *Lepus europaeus* (https://www.gwct.org.uk/research/species/mammals/brown-hare/trends-in-brown-hare-numbers/) have also declined, although there is evidence for recovery of the latter species in the UK.

It is clear therefore that pressures from intensive agriculture on farmland biodiversity are still present and constantly developing. A case study of the rapid impact of sudden, recent changes in management on a vulnerable assemblage of farmland birds is given in this volume by Mañosa et al. (Chap. 19). The overall and locally catastrophic deterioration of the biodiversity of Europe's arable land is continuing. This is of particular importance in Europe as arable agriculture has had a long history in the continent, in 2016 occupied approximately 25% of the land area of the EU, and until recently had a rich, varied and flourishing biota. Much of Europe's wildlife has co-evolved with the spread of agriculture, and it is likely that many species owe their current European range to farming. In the absence of appropriately managed farmland, Europe's biodiversity would be far poorer.

For many species we have sufficient knowledge to understand the reasons for their decline. This is described repeatedly in this volume and can be classified broadly as the loss of habitat heterogeneity and intensification of management. In many cases, restoring populations can be achieved through a reversal or partial reversal of the processes which have led to their attrition. This has been widely demonstrated on a local level. Increase in grey partridge chick survival and consequent population increases with reduction of agro-chemical use in field margins have been described since the 1980s (Potts 1986; Sotherton et al. this volume – Chap. 18), The conservation of the remaining areas of steppic grassland and arable and their associated avifauna in Catalunya are described by Mañosa et al. (this volume – Chap. 19). Other examples from the UK have shown how the promotion of nesting and feeding areas for lapwing, stone curlew and skylarks (Evans and Green 2007), and cirl bunting (Ovenden et al. 1998) by the RSPB using agri-environment scheme options have been very successful in increasing chick survival and number of breeding pairs. The local restoration of arable plant assemblages through the reduction of agrochemical and nutrient inputs and crop competition has been successful throughout western Europe (Wilson 2010; Meyer this volume – Chap. 6; Fried this volume – Chap. 4). Field margin strips designed primarily for arable plants and strips sown with nectar-rich plants have also been shown to be of benefit to invertebrates (McHugh et al. 2020).

Agri-environment schemes are the basic tool available to the majority of landowners and conservationists for the management of farmland wildlife, and in relation to farmland outside these schemes, they have proven very effective, although in some cases expensive (Batary et al. 2015). When designing these interventions however, it is essential to have a clear idea of the intended benefits and to be aware of the existing wildlife that is known to be or could reasonably be expected to be present. Appropriate design and location are of fundamental importance. There is also evidence that their efficacy is greatly improved if there is strong advisory input, and where land managers feel that their work is appreciated. There is clearly a considerable sociological element involved in the success of agri-environment agreements (Swan this volume – Chap. 21; Jarratt 2013).

A range of organisations have provided and continue to provide additional input into agri-environment schemes throughout Europe. Examples are given in this volume. In the UK the Game and Wildlife Conservation Trust introduced the concept of farmer clusters in which groups of farmers within a region liaise to share experiences, commissioning specialists and advisors (https://www.farmerclusters.com/), and such groups have considerable potential to maintain enthusiasm among landowners and to facilitate communication with experts. These groups are now also administered by local government and NGOs including County Wildlife Trusts, Areas of Outstanding Natural Beauty (AONBs), Farming and Wildlife Advisory Groups (FWAGs), National Park authorities and as farmer-led groups.

It should be noted however that the success of agri-environment schemes for these species depends on the maintenance not only of single habitat features, but of their maintenance within a varied farmland landscape that provides the range of different resources required by individual species and the community of which they are part. An ideal agri-environment agreement for a farm or region should aim to place options appropriately for maximum benefit to wildlife. This assumes that there is a good local knowledge of plant and animal distribution, which is present in some European countries, particularly in the north-west, but which in other areas may not be. Even in the UK, where biological recording is normally thought to be very comprehensive, agri-environment options are frequently inappropriately targeted. The sowing of perennial grass strips for instance on field margins with populations of rare plants is not uncommon, and is usually the result of poor communication between biological recording organisations and the statutory conservation organisations.

3 Conservation and Restoration – What Do We Need to Do?

It is probable that the introduction of agri-environment schemes in Europe from the mid-1980s onwards has been crucial to the survival of many organisms associated with arable fields and to the overall biodiversity of these habitats. It is clear however that further modifications to land management are needed if ongoing negative trends across multiple taxa are to be reversed, with a subsequent extension of successful interventions to the wider countryside, and a guarantee from funding bodies that they be continued into the future. It is essential that any new proposals have a sound scientific basis.

The ecological management needs of the majority of arable plants and birds are relatively well understood. Some refinements may be overcome by technical research. For example, one obstacle to the adoption of field margin options for arable plants in the UK and elsewhere has been the increase in quantity of invasive perennial species such as *Sonchus arvensis, Cirsium arvense* and *Eryngium campestre* and grass weeds such as *Bromus spp, Avena spp* and *Apera spica-venti* (Schumacher 1992; Critchley et al. 2006; Meyer and Leuschner 2015; McHugh et al. 2020). These species can form a dense, competitive cover to the exclusion of less competitive species and deter farmers from adopting these options in favour of sown perennial margins. Recent work has highlighted the importance of considering the landscape context of these margins in terms of the likelihood of them becoming colonised by these competitive species further emphasising the need to implement agri-environment Schemes at a farm or landscape scale as opposed to individual fields (Metcalfe et al. 2019). Some land managers in the UK have used applications of glyphosate and selective graminicides to control these species with considerable success. Without reducing the residual soil fertility on formerly intensively-managed land this may however prove to be a long struggle. Recent research by the Centre for Ecology and Hydrology has investigated methods for selective weed control, but this is unpublished at the time of writing. Improved knowledge of the seed biology (including dormancy, germination periodicity and longevity) of many species would be of considerable help when designing management regimes for individual rare species. This would also contribute greatly to our understanding of site-faithfulness and gene-flow between populations, and would be valuable information when considering attempts to resurrect populations of rare species from buried seed-banks.

Other modifications to agri-environment schemes may be beneficial. For example, field margin strips and plots can be sown with species mixtures to attract pollinators. While the recommended legume-rich mixtures are attractive to *Bombus spp* and favour a limited range of species, recent work has shown that mixtures including Asteraceae, Apiaceae and Geraniaceae are attractive to a much wider spectrum of wild pollinators and predatory invertebrates, and that field margins rich in naturally occurring wild plants can be equally beneficial (Nichols et al. 2019; McHugh et al. 2020). It is likely that other agri-environment scheme options would also benefit from refinement, and all options should be capable of modification following scientific research and recommendations.

4 Into the Future

Throughout much of Europe there has been a trend towards the increasing specialisation of land use in response to economic pressures on farming enterprises where profit margins are small. This has resulted in the loss of grazed land from predominantly arable areas and the loss of arable land from predominantly pastoral areas, in addition to the intensification of management on the remaining land. In the UK the most serious impacts of this on farmland biodiversity have been the loss of predominantly mixed farming from much of central and central-southern England, and the loss of much long-established arable land of marginal profitability especially near the coasts of western Britain. Some of these coastal fields are among the earliest known systems in the UK, are very botanically rich and locally important for birds including corn buntings and cirl buntings. One example is illustrated by Hurford in this volume (Chap. 18), where a substantial area of botanically and ornithologically-rich arable land has been converted to species-poor grassland with consequent loss of feeding habitat for a large number of wintering birds. In southern and central Europe much marginal arable land especially in mountain regions has been abandoned or converted to grassland (Bergmeier and Meyer 2018; Meyer this volume – Chap. 6) with the loss of its characteristic flora and fauna.

The expansion of urban areas in Europe also threatens valuable arable and steppe habitat. In the UK for example, additional housing for two major cities in central southern England is likely to damage and partly destroy internationally important arable plant sites with populations of nationally endangered species including *Centaurea cyanus, Scandix pecten-veneris* and *Valerianella rimosa* and associated farmland avifauna. In these cases however, it is remarkable that their value for arable flora is being considered as part of the planning process as a result of these species being listed as conservation priorities by the UK government under Section 41 of the Natural Environment and Rural Communities Act (2006), and it is likely that at least part of these urban developments will be managed for biodiversity.

Urbanisation is also affecting farmland wildlife elsewhere in Europe. Urbanisation and infrastructure development has been identified as a major threat to farmland birds, especially to already fragmented and isolated populations throughout Europe (Onrubia and Andrés 2005). In Spain for example the internationally important population of great bustards in the Madrid area is becoming increasingly threatened by urban encroachment and range fragmentation (Alonso et al. 2003).

Stimulated by the current concerns about biodiversity loss in the context of global climate change the concept of "rewilding" large areas has gained considerable popular and political support in the UK and other northern European countries (Monbiot 2013; Tree 2018; Goeren and Koenders 2020). This enthusiasm for returning large areas of farmland to their pre-agricultural state builds on the work of Vera (2000) and example sites have been established in the lowlands of northern Europe (e.g. Ooostvardersplassen in the Netherlands and Knepp in the UK). Some of the principles of rewilding are now part of UK government policy (Lawton et al. 2010). It is undoubtedly a very positive move to establish some areas in the most intensively managed parts of Europe where natural processes including naturalistic grazing, vegetation dynamics and predation by large carnivores can take place. It must however be remembered that the biodiversity of lowland Europe is that of a landscape that has been farmed for many millennia and that a high proportion of its wildlife are organisms of habitats managed by humans. When considering the conservation of the flora and fauna discussed in this book, rewilding is therefore an idea which should be approached with some circumspection. There has been much recent debate surrounding the subject of land sparing against land sharing (Fischer et al. 2013), whether biodiversity and food production are compatible in a world where demand for food is increasing, and whether biodiversity is best conserved within farmed landscapes or in discrete areas. The work described in this volume shows that many high-profile species and many more which are less appreciated are now restricted to arable land and other habitats which make up the farmland mosaic. These would suffer further severe declines should the area of low-intensity farmland decrease, with management on the remaining arable land being intensified. Increased political enthusiasm for land-sparing or rewilding may also run the serious risk of diverting resources in terms of agri-environment payments and scientific research and monitoring effort away from conservation on farmland.

The relative values to wildlife of an extensively managed habitat mosaic typical of European farmland in the years before the rise of modern farming methods when compared to a rewilded landscape are very difficult to determine with current knowledge. A research project in which comparable areas of extensively farmed land and rewilded land could be monitored would be of great value. It is likely that both would show unique and complementary benefits, but that given the current extent of productive land, relatively little input would be required to achieve disproportionately large benefits to wildlife by encouraging extensified farmland management while simultaneously maintaining the supply of high-quality food, maintaining rural employment and the links between people and nature.

It is possible that "organic" farming practices (Lampkin 1990) could represent a sustainable way forward for some landowners. These methods are well-established in Europe, although occupying a small area of the available arable land. While yields from these systems are lower than those from comparable conventionally-farmed systems, inputs are far smaller and external costs are greatly reduced. Many organic farming practices for pest control and maintaining soil fertility can be readily incorporated into conventional farming systems. While some organic systems especially for dairy production can be very intensive and offer little additional benefit to wildlife, there is considerable evidence for increased biodiversity, including soil organisms, on organically-managed arable and mixed farms (Hole et al. 2005; Rahmann 2011). The effect of organically-managed farms on biodiversity on a landscape scale is also starting to be understood (Bourgeois et al. 2020).

Ultimately in our current neo-liberal economic system, the use of arable land and the mosaic of farmland habitats of which it forms a central part will be determined by its monetary value. This can be primarily measured in terms of gross margin – the value of crops produced after the costs of their production have been accounted for. The costs of agricultural systems however also include all of the associated externalities such as water quality, atmospheric pollution and damage to the rural infrastructure. At the same time, the value of agricultural produce also includes revenue generated for example from tourism, carbon storage, flood prevention and biodiversity. The equation is therefore complicated, but includes public goods such as flower-rich field margins, hedgerows and stone curlews which cannot be accorded an immediate financial value. So much of Europe's wildlife is farmland wildlife. If we are to retain these elements in our countryside and for it to maintain its full multifunctionality, then farmers deserve to be paid for them, while in return managing their land in a way that retains and enhances its broadest diversity. The work described in this volume demonstrates that this is achievable while maintaining a sustainable supply of food.

References

Aebischer NJ, Green RE, Evans AD (2000) From science to recovery: four case studies of how research has been translated into conservation action in the UK. In: Aebischer NJ, Evans AD, Grice PV, Vickery JA (eds) Ecology and conservation of lowland farmland birds. British Ornithologists' Union, Tring, pp 43–54

Alonso JC, Martín CA, Palacín C, Magaña M, Martín B (2003) Distribution, size and recent trends of the great bustard *Otis tarda* population in Madrid region, Spain. Ardeola 50(1):21–29

Batary P, Dicks LV, Kleijn D, Sutherland WJ (2015) The role of agri-environment schemes in conservation and environmental management. Conserv Biol. https://doi.org/10.1111/cobi.12536

Bergmeier E, Meyer S (2018) Segetal plants of winter-annual crop fields in the Aegean islands – viewed in the contexts of landscape and traditional agricultural practice. Berichte der Reinhold-Tüxen-Gesellschaft 30:73–84

Bourgeois B, Gaba S, Plumejeaud C, Bretagnolle V (2020, in press) Weed diversity is driven by complex interplay between multi-scale dispersal local filtering

Bretagnolle V, Gaba S (2015) Weeds for bees? A review. Agron Sustain Dev 35:891–909

Brickle NW, Harper DGC, Aebischer NJ, Cockayne SH (2000) Effects of agricultural intensification on the breeding success of corn buntings Miliaria calandra. J Appl Ecol 37:742–755

Browne SJ, Aebischer NJ (2005) Studies of West Palearctic birds: Turtle Dove. British Birds 98:58–72

Byfield AJ, Wilson PJ (2005) Important Arable Plant Areas: identifying priority sites for arable plant conservation in the United Kingdom. Plantlife International, Salisbury

Critchley CNR, Fowbert JA, Sherwood AJ (2006) The effects of annual cultivation on plant community composition of uncropped arable field boundary strips. Agric Ecosyst Environ 113:196–205

De Juana E, Santos T, Suárez F, Tellería JL (1988) Status and conservation of steppe birds and their habitats in Spain. In: Goriup D (ed) Ecology and conservation of grassland birds. ICBP technical publication, 7:113–123

De la Concha I (2005) The common agricultural policy and the role of the rural development programmes in the conservation of steppe birds. In: Bota G, Morales MB, Mañosa S, Camprodon J (eds) Ecology and conservation of steppe-land birds. Lynx Edicions & Centre Tecnològic Forestal de Catalunya, Barcelona

Evans AD, Green RE (2007) An example of a two-tiered agri-environment scheme designed to deliver effectively the ecological requirements of both localised and widespread bird species in England. J Ornithol 148:279–286

Fischer J, Abson DJ, Butsic V, Chappell MJ, Ekroos J, Hanspach J, Kuemmerle T, Smith HG, von Wehrden H (2013) Land sparing versus land sharing: moving forward. Conserv Lett. https://doi.org/10.1111/conl.12084

Gaget E, Fay R, Augiron S, Villers A, Bretagnolle V (2018) Long-term decline despite conservation efforts questions Eurasian stone-curlew population viability in intensive farmlands. Ibis 161:359–371

Goeren M, Koenders H (2020) Green heritage – Dutch design. Staatsbosbeheer, www.staatsbosbeheer

Goulson D (2013) An overview of the environmental risks posed by neonicotinoid insecticides. J Appl Ecol 50:977–987

Hallmann CA, Sorg M, Jongejans E, Siepel H, Hofland N, Schwan H, Stenman W, Müller A, Sumser H, Hörren T, Gourson D, de Kroon H (2017) More than 75 percent decline over 27 years in total flying insect biomass in protected areas. PLoS One 12(10):e0185809. https://doi.org/10.1371/journal.pone.0185809

Hole DG, Perkins AJ, Wilson JD, Alexander IH, Grice PV, Evans AD (2005) Does organic farming benefit biodiversity? Biol Conserv 122:113–130

Jarratt S (2013) Linking the environmentally friendly farming careers of farmers to their effective delivery of wildlife habitats within the East of England. Unpublished Ph.D. thesis. University of Nottingham, Nottingham

Kornas J (1961) The extinction of the association Spergulo-Lolietum remote in flax cultures in the Gorce. Bulletin de l'Academie Polonaise des Sciences 9:37–40

Lampkin N (1990) Organic farming. Farming Press, Ipswich

Lapiedra O, Ponjoan A, Gamero A, Bota G, Mañosa S (2011) Brood ranging behaviour and breeding success of the threatened little bustard in an intensified cereal farmland area. Biol Conserv 144:2882–2890

Lawton JH, Brotherton PNM, Brown VK, Elphick C, Fitter AH, Forshaw J, Haddow RW, Hilborne S, Leafe RN, Mace GM, Southgate MP, Sutherland WJ, Tew TE, Varley J, Wynne GR (2010) Making space for nature: a review of England's wildlife sites and ecological network. Report to DEFRA

McHugh NM, McVeigh A, Bown B, Wilson PJ, Holland J (2020) Evaluation of cultivated margin option effectiveness and exploration of their natural capital. Defra Research report on Project ITT_3829

Metcalfe H, Hassall KL, Boinot S, Storkey J (2019) The contribution of spatial mass effects to plant diversity in arable fields. J Appl Ecol 56:1560–1574

Meyer S, Leuschner C (eds) (2015) 100 Äcker für die Vielfalt - Initiativen zur Förderung der Ackerwildkrautflora in Deutschland. Universitätsverlag Göttingen

Monbiot G (2013) Feral. Penguin Books, London

Natural Environment and Rural Communities Act (2006) https://www.legislation.gov.uk/ukpga/2006/16/contents

Nichols RN, Goulson D, Holland JM (2019) The best wildflowers for wild bees. J Insect Conserv 23(5–6):819–830

O'Connor RJ, Shrubb M (1986) Farming and birds. Cambridge University Press, Cambridge

Oñate JJ (2005) A reformed CAP? Opportunities and threats for the conservation of steppe-birds and the agri-environment. In: Bota G, Morales MB, Mañosa S, Camprodon J (eds) Ecology and conservation of steppe-land birds. Lynx Edicions & Centre Tecnològic Forestal de Catalunya, Barcelona

Onrubia A, Andrés T (2005) Impact of human activities on steppic-land birds: a review in the context of the Western Palearctic. In: Bota G, Morales MB, Mañosa S, Camprodon J (eds) Ecology and conservation of steppe-land birds. Lynx Edicions & Centre Tecnològic Forestal de Catalunya, Barcelona

Ovenden G, Swash A, Smallshire D (1998) Agri-environment schemes and their contribution to the conservation of biodiversity in England. J Appl Ecol 35:955–960

Peach WJ, Lovett LJ, Wotton SR, Jeffs C (2001) Countryside stewardship delivers cirl buntings (*Emberiza cirlus)* in Devon UK. Biol Conserv 101:361–373

Potts GR (1986) The partridge, pesticides, predation and conservation. Collins, London

Rahmann G (2011) Biodiversity and organic farming, what do we know? Agric For 3:189–208

Schumacher W (1980) Schutz und ehrhaltung gefährdeter ackerwildkräuter durch integration von landwirtschaftlicher nutzung und naturschutz. Natur und Landschaft 55:447–453

Schumacher W (1992) Zur effizienz der schutzmaßnahmen für ackerbiozönosen – mit empfehlungen zur vereinheitlichung der bisherigen schutzprogramme. Flora und Fauna der Acker und Weinberge, In, pp 201–217

Storkey J, Neve P (2018) What good is weed diversity? Weed Res 58:239–243

Traba J, Morales MB (2019) The decline of farmland birds in Spain is strongly associated to the loss of fallowland. Scientific reports 9, Article number: 9473

Tree I (2018) Wilding: the return of nature to an English farm. Picador

Vera F (2000) Grazing ecology and Forest history. CABI Publishing, Cambridge

Wilson PJ (1990) The ecology and conservation of rare arable weed species and communities. PhD Thesis, University of Southampton, UK

Wilson PJ (2010) Environmental stewardship monitoring project: uncropped cultivated margins (EF11/HF20). Report to Plantlife International, Salisbury

The manufacturer's authorised representative in the EU is Springer Nature Customer Service Centre GmbH, Europaplatz 3, 69115 Heidelberg, Germany. If you have any concerns regarding our products, please contact ProductSafety@springernature.com

Printed and bound by CPI Group (UK) Ltd, Croydon, CR0 4YY
28/09/2026
02224092-0006